AF676135

Novel Technologies to Enhance Energy Performance and Indoor Environmental Quality of Buildings

Novel Technologies to Enhance Energy Performance and Indoor Environmental Quality of Buildings

Editors

Alessandro Cannavale
Francesco Martellotta
Francesco Fiorito

MDPI • Basel • Beijing • Wuhan • Barcelona • Belgrade • Manchester • Tokyo • Cluj • Tianjin

Editors

Alessandro Cannavale
Polytechnic University of Bari
Italy

Francesco Martellotta
Polytechnic University of Bari
Italy

Francesco Fiorito
Polytechnic University of Bari
Italy

Editorial Office
MDPI
St. Alban-Anlage 66
4052 Basel, Switzerland

This is a reprint of articles from the Special Issue published online in the open access journal *Buildings* (ISSN 2075-5309) (available at: https://www.mdpi.com/journal/buildings/special_issues/Energy_Indoor_Environmental).

For citation purposes, cite each article independently as indicated on the article page online and as indicated below:

LastName, A.A.; LastName, B.B.; LastName, C.C. Article Title. *Journal Name* **Year**, *Volume Number*, Page Range.

ISBN 978-3-0365-2339-2 (Hbk)
ISBN 978-3-0365-2340-8 (PDF)

© 2021 by the authors. Articles in this book are Open Access and distributed under the Creative Commons Attribution (CC BY) license, which allows users to download, copy and build upon published articles, as long as the author and publisher are properly credited, which ensures maximum dissemination and a wider impact of our publications.
The book as a whole is distributed by MDPI under the terms and conditions of the Creative Commons license CC BY-NC-ND.

Contents

About the Editors

Alessandro Cannavale Assistant Professor at the Polytechnic University of Bari from 2016 to 2021; currently, he is an adjunct Researcher at CNR—Nanotec and Senior Postdoctoral Researcher at Technical University of Bari. He earned a Qualification as an Associate Professor in 2018. His research activity deals with applied thermodynamics and devices for energy efficiency. He is Associate Editor of *Frontiers in Built Environment*, and a board member of *Energies, Buildings and Clean Technologies*.

Francesco Martellotta Full Professor of Building Physics and Building Energy Systems at the Department of Civil Engineering and Architecture, Politecnico di Bari. His scientific interests include: architectural acoustics; sustainable energy use in buildings; characterization of sustainable building materials; thermal comfort; assessment of indoor environment quality; noise in offices. He contributed to securing and managing several fundings within public competitive financing programs and from private companies. He is author/co-author of more than 100 scientific papers, nearly half of them published in peer-reviewed journals. In 2007 he was Keynote lecturer at the International Symposium on Room Acoustics. He serves as an Associate Technical Editor of the *Journal of Audio Engineering Society* and as an Associate Editor of the *Journal of Acoustical Society of America*. He is currently serving as guest editor for several MDPI journals. He is full member of AIA, AES, and ASA.

Francesco Fiorito Associate Professor in Architectural Engineering at the Polytechnic University of Bari (Italy) and Senior Visiting Fellow at the University of New South Wales (Sydney, Australia). Previously he has held appointments at the University of New South Wales (Sydney, Australia) as an Associate professor of High Performance Architecture and at the University of Sydney (Sydney, Australia) as a senior lecturer in Architectural Technology. His main research expertise is in facade engineering and design. In recent years, his research has focused on the development of technologies and systems for adaptive facades and for Urban Heat Island mitigation. Prof. Fiorito is Associate Editor of the international journal Architectural Science Review and member of editorial board of several international journals, among which are *Energy* and *Buildings* (Elsevier), and *Climate* and *Energies* (MDPI).

buildings

MDPI

Editorial

Novel Technologies to Enhance Energy Performance and Indoor Environmental Quality of Buildings

Alessandro Cannavale [1,2,*], Francesco Martellotta [1] and Francesco Fiorito [3]

1 Department of Civil Engineering Sciences and Architecture, Polytechnic University of Bari, 70126 Bari, Italy; francesco.martellotta@poliba.it
2 National Research Council, Institute of Nanotechnology (CNR-NANOTEC), Via Monteroni, 73100 Lecce, Italy
3 Department of Civil, Environmental, Land, Building Engineering and Chemistry, Polytechnic University of Bari, 70126 Bari, Italy; francesco.fiorito@poliba.it
* Correspondence: alessandro.cannavale@poliba.it; Tel.: +39-080-596-3718

check for updates

Citation: Cannavale, A.; Martellotta, F.; Fiorito, F. Novel Technologies to Enhance Energy Performance and Indoor Environmental Quality of Buildings. *Buildings* **2021**, *11*, 303. https://doi.org/10.3390/buildings11070303

Received: 8 July 2021
Accepted: 9 July 2021
Published: 11 July 2021

Publisher's Note: MDPI stays neutral with regard to jurisdictional claims in published maps and institutional affiliations.

Copyright: © 2021 by the authors. Licensee MDPI, Basel, Switzerland. This article is an open access article distributed under the terms and conditions of the Creative Commons Attribution (CC BY) license (https://creativecommons.org/licenses/by/4.0/).

Here, we overview the Buildings journal Special Issue dedicated to the following topic: "Novel Technologies to Enhance Energy Performance and Indoor Environmental Quality of Buildings" (https://www.mdpi.com/journal/buildings/special_issues/Energy_Indoor_Environmental, last access on 7 July 2021) and the scientific papers it hosts. The aim of this Special Issue was to report current trends in the investigations dealing with emerging materials and devices, aiming at an increase in energy performance of buildings and indoor comfort. We are currently facing an epochal transition, involving a compelling change in the way we produce and use energy: in this roadmap, low impact buildings might play a relevant role. The enhancement of energy efficiency, in the construction sector, can be obtained by the development of new materials, with improved properties, spanning from chromogenics [1–3] to semitransparent photovoltaics [4–7], superinsulating materials [8,9], and phase change materials [10,11]. Novel technologies may also increase comfort, indoor environmental quality and safety as well. Ten original research studies have been published, with the contributions of international research groups, from Italy, Poland, Australia, United Kingdom, Croatia, Spain, and Taiwan. All these contributions address the main topics of the Special Issue, with an effective and targeted effort.

Cannavale et al. [12] proposed an innovative aerogel-based "thermal break" for window frames, to effectively reduce the frame conductance of regulation compliant reference windows. The thermal performance of this new window was assessed by finite element method. Furthermore, numerical simulations were carried out to assess energy savings for heating and cooling, for several international locations, confirming the potential of this novel building component, embodying super-insulating materials, which revealed its suitability for extremely rigid climates.

Mainka et al. [13] reported a study about homemade air purifiers, namely low-cost ozone generators to decrease the level of contaminants. In their experimental activity, they investigated the reduction of bacterial and fungal aerosol by using two available technological solutions of air purifiers. They observed a decrease by 78 % of bacteria concentration, after 20 min of ozone generation. The authors of also point out that ozone has the property of destroying the coronavirus, at the basis of the global SARS-CoV-2 pandemic. This aspect could justify a particular interest in the proposed device.

Kuru et al. [14] investigated the possible transfer of multifunctionality in nature into biomimetic strategies for engineered systems, by creating an effective framework to develop multifunctional biomimetic adaptive building skins. Such a framework may become a systemic collection of biological information. Numerical simulations proposed a comparison between the base-case building and a case study, demonstrating a decrease of discomfort hours by 23.18 %. The authors proposed biomimetic design as a suitable approach for future buildings.

Roy et al. [15] proposed a review paper dealing with perovskite solar cells, whose rapid worldwide diffusion in scientific research has attracted the attention of industries and governments. The authors investigated the main issues about building integration of perovskite-based solar cells, showing their main features, opportunities but also critical issues and current limits, which perhaps will soon be overcome, thanks to future scientific efforts.

Franco et al. [16] investigated the relevant role of ventilation rate, to achieve air quality in public buildings, with special attention to the required compromise between indoor air quality and the control of energy consumption. The authors proposed a method for the definition of optimal values of air exchange rates and found lower air flow exchange rates, compared to current Technical Standards, sensitive to occupation profiles. The authors claim that their adaptive solution represents a considerable progress, being based on simple measurements of carbon dioxide concentration, by means of commercial sensors.

Ćorić et al. [17] proposed a research article dealing with the application of seismic energy dissipaters based on a cost-effective copper-based shape memory alloy. The ausforming process was suitably optimized by controlling the parameters, achieving both adequate strength and suitable transformation behaviour.

Ciampi et al. [18] reported their experimental tests on extruded Acrylonitrile–Butadiene–Styrene, used in form of panels as a second-skin layer in a ventilated building façade, to be applied in eight refurbishment cases, analysed by means of a numerical simulation software platform. In comparison to the reference case study, the proposed system allowed reduction of energy demand for heating (−6.9 %) and for cooling (−3.1 %).

Zsembinszki et al. [19] studied a Deep Reinforcement Learning architecture, able to solving complex control problems and applied it to the control of novel hybrid energy storage systems. They found that a suitable strategy may reduce the system operating costs of cooling by more than 50 %, in residential buildings located in the Mediterranean climate.

In their work, Morano et al. [20] focussed their attention on one of the most relevant environmental factors, affecting human health: noise pollution. Its effects on the real estate market were investigated, by analysing functional relationships between noise pollution and selling prices in the city of Bari, in Apulia (Italy).

Hwang et al. [21] investigated the energy saving potential due to the integration of Phase Change Materials in rooftops, so as to reduce indoor temperatures in schools of Taiwan. The ideal melting temperature was found to be 29 °C, whereas the effective thickness of the material was 20 mm, for the selected locations, in Northern Taiwan.

The guest editors of this Special Issue would like to acknowledge all the authors for their scientific support and for kindly sharing their knowledge, from different fields of investigation. Furthermore, the editors would like to express their gratitude to the peer reviewers, for their rigorous analysis of manuscripts - effectively contributing to achieve the publication of the Special Issue - and the managing editors of *Buildings*, for their continuous support.

Author Contributions: All authors contributed to every part of this editorial. All authors have read and agree to the published version of the manuscript.

Funding: This research activity did not receive any funding.

Acknowledgments: The authors gratefully acknowledge Managing Editor of Buildings (MDPI) for her precious, professional support.

Conflicts of Interest: The authors declare no conflict of interest.

References

1. Lampert, C.M. Chromogenic smart materials. *Mater. Today* **2004**, *7*, 28–35. [CrossRef]
2. Cannavale, A. Chromogenic Technologies for Energy Saving. *Clean Technol.* **2020**, *2*, 29. [CrossRef]
3. Granqvist, C.G. Oxide-based chromogenic coatings and devices for energy efficient fenestration: Brief survey and update on thermochromics and electrochromics. *J. Vac. Sci. Technol. B* **2014**, *32*, 060801. [CrossRef]

4. Hörantner, M.T.; Nayak, P.K.; Mukhopadhyay, S.; Wojciechowski, K.; Beck, C.; McMeekin, D.; Kamino, B.; Eperon, G.E.; Snaith, H.J. Shunt-Blocking Layers for Semitransparent Perovskite Solar Cells. *Adv. Mater. Interfaces* **2016**, *3*, 1500837. [CrossRef]
5. Della Gaspera, E.; Peng, Y.; Hou, Q.; Spiccia, L.; Bach, U.; Jasieniak, J.J.; Cheng, Y.B. Ultra-thin high efficiency semitransparent perovskite solar cells. *Nano Energy* **2015**, *13*, 249–257. [CrossRef]
6. Chen, C.C.; Dou, L.; Zhu, R.; Chung, C.H.; Song, T.B.; Zheng, Y.B.; Hawks, S.; Li, G.; Weiss, P.S.; Yang, Y. Visibly transparent polymer solar cells produced by solution processing. *ACS Nano* **2012**, *6*, 7185–7190. [CrossRef] [PubMed]
7. Cannavale, A.; Martellotta, F.; Fiorito, F.; Ayr, U. The challenge for building integration of highly transparent photovoltaics and photoelectrochromic devices. *Energies* **2020**, *13*, 1929. [CrossRef]
8. Berardi, U. The development of a monolithic aerogel glazed window for an energy retrofitting project. *Appl. Energy* **2015**, *154*, 603–615. [CrossRef]
9. Jelle, B.P. Traditional, state-of-the-art and future thermal building insulation materials and solutions—Properties, requirements and possibilities. *Energy Build.* **2011**, *43*, 2549–2563. [CrossRef]
10. Baetens, R.; Petter, B.; Gustavsen, A. Phase change materials for building applications: A state-of-the-art review. *Energy Build.* **2012**, *42*, 1361–1368. [CrossRef]
11. De Matteis, V.; Cannavale, A.; Martellotta, F.; Rinaldi, R.; Calcagnile, P.; Ferrari, F.; Ayr, U.; Fiorito, F. Nano-encapsulation of phase change materials: From design to thermal performance, simulations and toxicological assessment. *Energy Build.* **2019**, *188–189*, 1–11. [CrossRef]
12. Cannavale, A.; Martellotta, F.; Berardi, U.; Rubino, C.; Liuzzi, S.; Carlo, V.D.; Ayr, U. Modeling of an Aerogel-Based "Thermal Break" for Super-Insulated Window Frames. *Buildings* **2020**, *10*, 60. [CrossRef]
13. Mainka, A.; Mucha, W.; Janoszek, A. Non-commercial Air Purifier—The E ffectiveness and Safety. *Buildings* **2020**, *10*, 104. [CrossRef]
14. Kuru, A.; Oldfield, P.; Bonser, S.; Fiorito, F. A framework to achieve multifunctionality in biomimetic adaptive building skins. *Buildings* **2020**, *10*, 114. [CrossRef]
15. Roy, A.; Ghosh, A.; Bhandari, S.; Sundaram, S.; Mallick, T.K. Perovskite solar cells for bipv application: A review. *Buildings* **2020**, *10*, 129. [CrossRef]
16. Franco, A.; Schito, E. Comfort and Energy Use in Indoor Spaces Using CO_2. *Buildings* **2020**, *10*, 135. [CrossRef]
17. Ćorić, D.; Žmak, I. Influence of ausforming treatment on super elasticity of cu-zn-al shape memory alloy for seismic energy dissipaters. *Buildings* **2021**, *11*, 22. [CrossRef]
18. Ciampi, G.; Spanodimitriou, Y.; Scorpio, M.; Rosato, A.; Sibilio, S. Energy performances assessment of extruded and 3d printed polymers integrated into building envelopes for a south Italian case study. *Buildings* **2021**, *11*, 141. [CrossRef]
19. Zsembinszki, G.; David, V.; Cabeza, L.F. Deep Learning Optimal Control for a Complex Hybrid Energy Storage System. *Buildings* **2021**, *11*, 194. [CrossRef]
20. Morano, P.; Tajani, F.; Di Liddo, F.; Darò, M. Economic evaluation of the indoor environmental quality of buildings: The noise pollution effects on housing prices in the city of Bari (Italy). *Buildings* **2021**, *11*, 213. [CrossRef]
21. Hwang, R.-L.; Chen, B.-L.; Chen, W.-A. Analysis of Incorporating a Phase Change Material in a Roof for the Thermal Management of School Buildings in Hot-Humid Climates. *Buildings* **2021**, *11*, 248. [CrossRef]

buildings

MDPI

Article

Analysis of Incorporating a Phase Change Material in a Roof for the Thermal Management of School Buildings in Hot-Humid Climates

Ruey-Lung Hwang, Bi-Lian Chen and Wei-An Chen *

Department of Industrial Technology Education, National Kaohsiung Normal University, Kaohsiung 82444, Taiwan; rueylung@nknu.edu.tw (R.-L.H.); ant88313@gmail.com (B.-L.C.)
* Correspondence: B0522@mail.nknu.edu.tw or vivian199038@gmail.com

Abstract: Strategies to reduce energy consumption are presently experiencing vigorous development. Phase change materials (PCMs) are novel materials that can reduce indoor temperatures via the change in material phase. Regarding the situation in Taiwan, there is no practical utilization of PCMs in school buildings at present, especially in combination with rooftops. In this paper, we discuss the feasibility and utilization potential of installing PCMs in the rooftops of school buildings. School buildings located in northern and southern Taiwan (Taipei and Kaohsiung) were selected to analyze the energy-saving potential and optimization of indoor thermal comfort by installing PCMs with different properties in rooftops over two time periods, including the air conditioning (AC) and natural ventilation (NV) seasons. Based on the simulation results, the feasible patterns of PCM simultaneity are found to be appropriate for improved indoor comfort and energy saving during the different seasons. Specifically, the efficient phase change temperature (PCT) for different PCM thicknesses is clarified to be 29 °C. The economic thickness of PCM was clarified to be 20 mm for Taipei and Kaohsiung. Through the recommendations proposed in this study, it is expected that the PCMs may be efficiently implemented in school buildings to realize the goal of energy conservation and improve thermal comfort.

Keywords: phase change materials; hybrid ventilated school building; indoor thermal comfort; thermal management; energy conservation

check for updates

Citation: Hwang, R.-L.; Chen, B.-L.; Chen, W.-A. Analysis of Incorporating a Phase Change Material in a Roof for the Thermal Management of School Buildings in Hot-Humid Climates. *Buildings* **2021**, *11*, 248. https://doi.org/10.3390/buildings11060248

Academic Editors: Alessandro Cannavale, Francesco Martellotta and Francesco Fiorito

Received: 12 May 2021
Accepted: 1 June 2021
Published: 9 June 2021

Publisher's Note: MDPI stays neutral with regard to jurisdictional claims in published maps and institutional affiliations.

Copyright: © 2021 by the authors. Licensee MDPI, Basel, Switzerland. This article is an open access article distributed under the terms and conditions of the Creative Commons Attribution (CC BY) license (https://creativecommons.org/licenses/by/4.0/).

1. Introduction

When considering the heat insulation of buildings, the great amount of heat gain from façades and rooftops cannot be neglected. Previous studies attempted to explore various materials and techniques for improving the heat insulation of building envelopes or structures and discuss opportunities and challenges. Furthermore, researchers have suggested feasible utilization strategies [1,2]. Roofs are exposed to direct sunlight for a long period of time and thereby significantly influence the energy consumption of a building when controlling the resulting indoor temperature, especially in hot climate areas, such as Taiwan. One of the most common measures for reducing roof heat gain is to reduce the overall heat transfer coefficient, i.e., the U-value, by adding insulative materials. Phase change materials (PCMs) allow incidental heat gain to be used to change the phase of the material and thus perform heat storage, and such materials may store a significant amount of thermal energy. PCMs are considered to be an innovative technology and an effective method for improving the thermal mass of buildings, owing to the large thermal capacity within a limited temperature range, which is similar to an isothermal energy tank [3]. Incorporating PCMs into roofs can significantly enhance the heat storage ability of a structure and disperse the heat gains from the peak hours of cooling demand, consequently reducing energy consumption and improving indoor thermal comfort.

Most of the previous studies focused on the use of PCMs have primarily considered walls and have neglected the potential for roofs, even though roofs represent significant heat gain. Although most attention has been paid to the installation of PCMs in walls, the ratio of studies discussing PCMs in roofs to those in walls is approximately 1:3 [4]; however, some studies have used experimental or numerical simulation methods to investigate the energy-saving potential of installing PCMs in roofs or floors to discuss the performance of heating systems [5–7]. Studies related to the installation of PCMs can be divided into two types, including reducing energy consumption in air-conditioned buildings and improving thermal comfort in non-air-conditioned buildings. In addition, the most common parameters of PCMs that have been analyzed in previous studies are the phase-change temperature (PCT), thickness, and position of the PCM layer. Kim et al. analyzed the energy consumption of a residential building with a PCM with a specific melting temperature under different air conditioning conditions and suggested suitable PCM properties for various cooling and heating conditions in different seasons [8]. Tokuç et al. [9] evaluated roofs with and without PCMs in terms of the cooling load for the summer season in four weather zones of Turkey. They indicated that the reduction in the cooling load in May was more significant when installing thicker PCM layers. Specifically, a PCM with a thickness of 50 mm reduced the cooling load by 48.2%, 56.6%, 78.7%, and 99.1% in Izmir, Istanbul, Ankara, and Erzurum, respectively.

Regarding studies related to the combination of PCMs and rooftops, Alawadhi and Alqallaf [10] carried out numerical and experimental studies of a roof structure with PCMs. It was found that the heat transferred into the indoor space could be reduced by 39%. Mushtaq et al. [11] experimentally studied the effect of incorporating a PCM with a melting temperature of 37 °C into a roof structure with a cooling system under summer weather conditions in Baghdad, Iraq, and found that a roof structure with a PCM reduced heat transfer by 46.71% when compared to the reference roof structure. Zwanzig et al. [12] evaluated the energy-saving potential of installing a PCM layer at different positions inside the roof structure of residential buildings located in three climatic zones in the USA (Minneapolis, Louisville, and Miami). The results showed that installing a PCM in the interior surface resulted in better energy-saving potential, as the cooling loads were reduced by up to 11.4%, 8.0%, and 4.1%, respectively. Xamán et al. [13] developed alternatives with different PCM types and thicknesses for warm weather conditions in Mexico by applying PCMs on the surfaces of roof components that were closer to the indoor environment.

Previous studies also discussed the application of PCMs for promoting thermal comfort. For instance, Mourid and Alami [14] evaluated the effectiveness of combining PCMs with roof structures to improve the summer thermal comfort in lightweight buildings and investigated the influence of the thickness and position of the PCM layer on its effectiveness. Piselli et al. [15] demonstrated that a combination of cool roofing membranes and PCMs is practical and effective at lowering surface temperatures and can mitigate indoor overheating in peak summer conditions without drawbacks in the cold season. Li et al. [16] conducted an experimental study to explore the influence of the melting temperature and thickness of a PCM on the thermal performance of a glazed roof. The results showed that the peak temperature decreased by approximately 16.3 °C, and the energy consumption reduced by 47.5% when utilizing a PCM with a melting temperature of 32 °C. With an increasing thickness of the PCM layer, the energy consumption and dissatisfaction rate for an indoor thermal environment gradually declined. Parametric analysis of the thermal behavior of PCMs with different melting temperatures, layer thicknesses, and numbers of layers incorporated into building roofs was carried out in the humid tropical weather of Chennai, India, by Reddy et al. [17]. Analysis has also been carried out in the cold northeastern area of China by Dong et al. [18] and over a range of climates across the United States by Kibria et al. [19]. Besides, Yu et al. [20] have numerically investigated the phase transition temperature and thickness of a PCM layer in a roof of typical design under climatic conditions in five regions in China. Their results showed that the optimal thickness of PCM is 30 mm and that the optimum phase transition temperature increases linearly with the increase in average outdoor solar air temperature in summer in different climatic regions. David and Javier [21] analyzed the selection of PCMs for building wall-boards and roofs via a comparison between multi-criteria decision methods and building energy simulations. The results demonstrated the importance of environment variables in appropriately assessing the performance of PCMs, and indicated that PCMs have different thermal behaviors depending on the climatic conditions. Accordingly, it is essential to assess an extensive evaluation of the climatic condition before applying this strategy.

The previous studies mentioned above analyzed the energy-saving potential and the effect of improving thermal comfort resulted from the utilization of PCM; however, most of these studies are conducted under a specific season. Different from the previous research, this study aims at suggesting a feasible and economic pattern of PCMs for school buildings that can simultaneously fulfill the conditions during the AC and NV seasons to reduce energy consumption and provide a better thermal comfort condition. In view of the fact that the Taiwanese government intends to complete the installation of air-conditioning in 103,000 classrooms across the entire island by the summer of 2022, it is considered that the utilization of PCMs is helpful for raising the effect of energy-saving potential and indoor thermal condition. However, there are no existed practical cases of combining the PCMs and rooftop structure in Taiwanese school buildings. Thus, this study acts as a pioneer study to discuss the effect and feasibility of utilizing PCMs on the rooftop of Taiwanese school buildings and attempts to suggest recommendations for the utilization strategies that the energy-saving potential and indoor thermal comfort are both taken into consideration. The investigated cases in this paper were set to analyze the reducing cooling load during the air-conditioning season and improving the thermal comfort of real classrooms during the ventilation season. Based on the simulation results, it is expected to suggest the appropriate properties of PCMs for Taiwanese school buildings, including thickness and PCT. In addition, a feasible pattern of PCM, which is appropriate to be utilized throughout the entire year, is proposed to fulfill the indoor comfort and energy-saving potential.

2. Materials and Methods

2.1. Building Prototype and Simulations Procedure

The top-floor classrooms of a typical high-school building (Figure 1) located in Taiwan are selected as a model for simulation. The setting conditions of the building model are listed in Table 1. Besides, a 2.0 m deep outdoor corridor is used as a horizontal shade on the south side. Regarding the air-conditioning operation and natural ventilation period of this school building model, the air-conditioning is available for use during the weekdays in the hot-humid period from May to October. Based on the regulation, it is allowed to turn on the air-conditioning when the outdoor temperature is higher than 27 °C. Besides, the windows are always closed during the occupancy period and out of the occupancy period at night and on weekends during the AC season. The natural ventilation season is from November to April.

In order to optimize the thermal design for the classroom, including cutting down the cooling load from May to October and improving the thermal environment from November to April, the case studies of installing PCM at the inner side of the existing roof were proposed. Due to the sol-air temperature on the external roof surface varies greatly between day and night, the inner side of the roof was chosen to install the PCM, which is particularly suitable for the application of phase-change energy-storage technology. Moreover, previous studies [22–26] have confirmed that placing PCM on the inner surface of a wall can better regulate the indoor temperature and save more energy than deploying PCM on the outer surface of a wall. The PCM considered in this paper is BioPCM™, which is generally used by other previous studies; besides, the database of the BioPCM™ is complete that can be regarded as a convincing material [27–32]. BioPCM™ is non-toxic, non-corrosive, biodegradable, and has a useful life of over 100 years. In addition, BioPCM™ enables users to reduce their carbon footprint and reduce the stress on HVAC systems in buildings, data centers, and telecom shelters [33]. In order to investigate the climate and seasonal adaptability of PCM, six types of phase-change temperature (PCT) were considered, including 23 °C (22–25 °C), 25 °C (24–27 °C), 27 °C (26–29 °C), 29 °C (28–31 °C), 31 °C (30–33 °C), and 33 °C (32–35 °C). Furthermore, four different thickness types of PCM layer were set, including 10 mm, 20 mm, 30 mm, and 40 mm. The latent heat of BioPCM is 219 kJ/kg. Figure 2 depicts the enthalpy–temperature curves for BioPCM [34].

Figure 1. Image and floor plan of a typical high-school building.

Table 1. Setting condition of school building model.

Building	Orientation	Face Toward the South
	Story	3 Stories
Classroom	Length × width × height	9.0 m × 8.1 m × 3.6 m
	Lighting density	12 W/m^2
	Room capacity	30 people (including teachers and students)
	Occupancy period	8:00–17:00 (weekdays)
Material of facade and roof structure	Window-wall ratio	North: 29%; South: 37%
	Window glass	6 mm thick clear glass
	Exterior wall	Reinforced concrete (RC) U value: 2.3 W/m^{2}K
	Roof	RC roof with 25 mm polystyrene (PS) insulation board (U value: 1.0 W/m^{2}K)

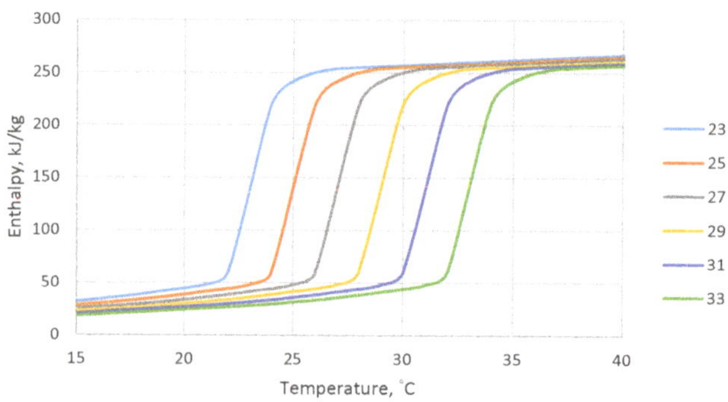

Figure 2. Enthalpy–temperature curves for BioPCM.

The building energy and thermal load simulation software EnergyPlus version 9.4 was used in this study to conduct the simulation, and the weather data of a typical meteorological year (TMY3) of two major cities in Taiwan: Taipei (25.03° N, 121.50° E) and Kaohsiung (22.57° N, 120.30° E) were used. Specifically, the climatic conditions, including the outdoor temperature, cooling degree hour (CDH), and horizontal solar radiation of each month for the two selected study cities, are shown in Figure 3. Taiwan is located in the sub-tropical zone that the weather features are hot and humid. Referring to Figure 3, it is known that the outdoor temperature in Taipei during winter is much lower than that of Kaohsiung, although Taipei and Kaohsiung are both located in a hot-humid climate zone; as for the outdoor temperature in Kaohsiung, it remains at an extremely high temperature during the entire year. In addition, the heat gains from the rooftop are directly affected by the solar air temperature. As shown in Figure 4, there is a wide range of the variation of daily solar air temperature, which means the utilization potential of PCM in rooftop structure is expectable in both Taipei and Kaohsiung.

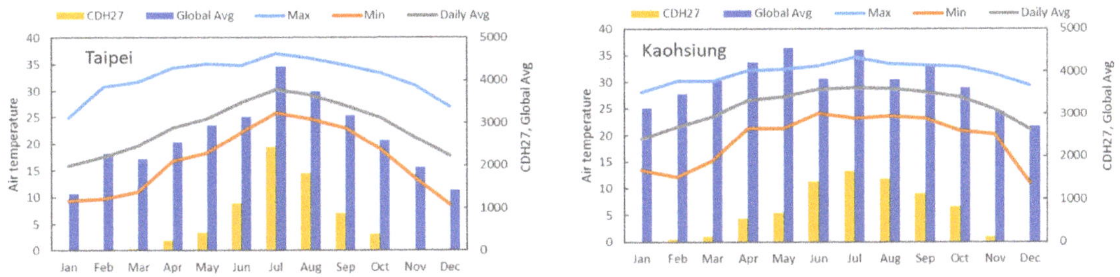

Figure 3. Climatic conditions of selected cities.

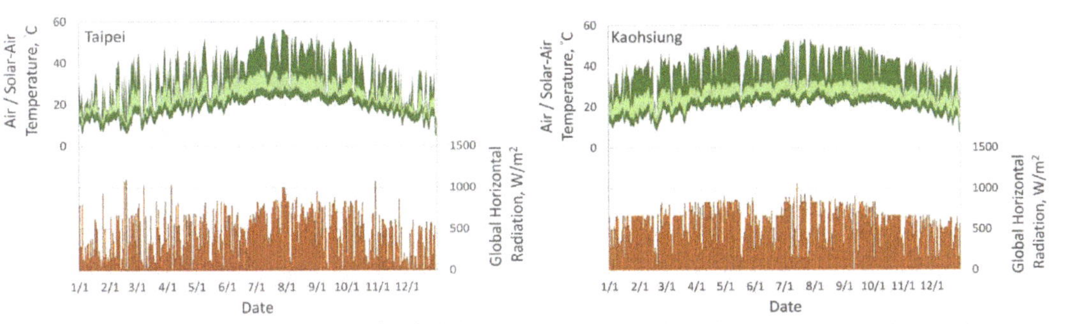

Figure 4. Daily range of air/solar-air temperature and daily peak radiation.

CDH 27 is a statistic value in the weather data (TMY3). Besides, owing to this value is closed to the standard value that is regulated to turn on the air-conditioning (27–28 °C); therefore, CDH 27 was set for the simulation in this study. All simulations were conducted using the conduction finite difference (CondFD) approach. During the months of May to October, air-conditioning equipment was activated to analyzing the cooling load of the classrooms; during the months of April to November, the thermal conditions of classrooms with natural ventilation opening areas such as windows were simulated (mainly for the operating temperature) through the airflow network model in EnergyPlus. The series of simulations mainly focuses on cases with PCM-installed roofs and night ventilation. Comparing to the traditional reference cases without installing PCM in the roof structure, the cooling load reduction and thermal comfort improvement resulted from the application of PCM in the roofs could be identified.

2.2. Selection of Performance Indicators

Through combining PCMs with a building's heating, ventilation, and air-conditioning (HVAC) system, it can utilize low nighttime temperatures to store or release energy and make use of the energy transfer for "peak shaving and valley filling", and thereby providing all or part of the heating or cooling required during the day. In order to take better advantage of PCMs in buildings, it is essential to evaluate the usage efficiency of PCMs based on latent heat. In addition to the usage efficiency of PCMs, other indicators must be considered to evaluate the effectiveness in terms of improving indoor thermal comfort and energy savings. Therefore, three performance indicators, including the daytime heating efficiency coefficient (heat storage), the nighttime cooling efficiency coefficient (heat release), and the effective latent storage, were selected to assess the usage efficiency of a PCM roof in energy storage/releasing and the effectiveness in terms of improving the indoor thermal environment and energy saving. Each of the evaluation indicators is respectively elaborated in the following section.

2.3. Usage Efficiency of PCM

There are two main factors that influence the usage efficiency of PCM: operational period and latent storing and releasing capability of heat. Specifically, regarding the operational period, the PCM layer can reach a completely solidified or completely melted stage within a relatively short period of time; in other words, instead of operating all day, the PCM layer must be in a partially melted state for a long period time, which means that it operates effectively in daily cycles. Concerning the latent storing and releasing capability of heat, it is unpopular to utilize the PCM layer with worse performance than storing or releasing a small amount of energy during the daily cycles. Scilicet, the better performance of PCMs is that the heat-storing at night is equivalent to the heat releasing during daytime. Aforementioned, the usage efficiency of a PCM can be analyzed through its latent storing and releasing capability of heat, as well as its operational period within one daily cycle.

Ramakrishnan et al. [28] developed a method to determine the latent storing and releasing capability of heat by combining the factors mentioned above, and thereby providing more accurate predictions of the daytime heating efficiency coefficient (heat storage) and the nighttime cooling efficiency coefficient (heat release) for a PCM, named HE and CE (Equations (1) and (2)). HE and CE are used to respectively describe the operating effectiveness of the PCM during the storing and releasing latent heat period and are separately calculated for the air-conditioning season from May to October and for the natural ventilation season from November to April. Referring to Equations (1) and (2), L_C and L_{DC} provide information regarding the potential energy that is effectively stored and released in the PCM layer via the melting and solidification processes. In this paper, the releasing time of heat during the daytime was 600 min (08:00–17:00), and the storing time of heat at nighttime was 840 min (17:00–08:00). Besides, T_C and T_{DC} can be used to identify the length of time that the PCM layer is activated during the heat-storage/releasing period; a longer time period in which the PCM layer is activated can provide longer isothermal energy storage.

$$\mathrm{HE} = \sqrt{L_C \times T_C} \tag{1}$$

$$\mathrm{CE} = \sqrt{L_{DC} \times T_{DC}} \tag{2}$$

where L_C and L_{DC} are the ratios of latent heat stored/released by the PCM during daytime/nighttime over the total latent heat capacity of the PCM; T_C and T_{DC} are the ratios of the actual operating time of the PCM during daytime/nighttime over the storing/releasing time during daytime/nighttime; T_C/T_{DC} is the interval between occurrence times of the maximum and the minimum value.

The effective latent storage (η) of the PCM layer at a specific moment is calculated by Equation (3). The value of η = 1 means that the PCM layer is completely solidified; η = 0 means that the PCM layer is completely melted.

$$\eta = \frac{\sum_{i=1}^{n}\{H(T_i) - H(T_0)\}}{\Delta H} \tag{3}$$

where ΔH is the phase-change enthalpy value of PCM; $H(T_i)$ and $H(T_0)$ are the specific enthalpies of PCM at the temperature T_i and the melting onset temperature T_0, respectively. The node temperature of the PCM can be confirmed from the simulation results of EnergyPlus, and the specific enthalpy can be obtained from the enthalpy–temperature curves in.

2.4. Evaluation of Thermal Comfort and Energy-Saving Potential

A long-term evaluation of building performance and the criteria for determining the performance should consider the different potential susceptibilities of occupants to overheating [35]. In addition, the thermal evaluation and the design corresponding to the occupants should apply the thermal comfort standard that adapted to occupants' thermal perception [36]. Hence, the adaptive comfort model for students in Taiwan [37] is used to evaluate the thermal environment of classrooms during the natural ventilation season from November to April. Specifically, this model is based on the experimental data of a thermal comfort survey in elementary schools and high schools. The upper limit temperature of thermal comfort (T_{max}) is defined by Equation (4):

$$T_{max} = 0.62 \times T_{om} + 14.5 \tag{4}$$

where T_{max} is the upper limit temperature of thermal comfort; T_{om} is the monthly moving average or the daily average of the outdoor air temperature for the past 30 days.

Regarding the long-term evaluation of indoor thermal comfort, Annex H of ISO 7730 [38] listed two types of measures for thermal discomfort to simultaneously consider the frequency and level of discomfort severity; besides, all of these two measures use the positive difference between the current operating temperature and the upper limit of thermal comfort range to evaluate the overheating discomfort. Briefly reviewing the two measures, the first measure is the number of hours in which the upper limit of thermal comfort range is exceeded; the second measure is the level of overheating severity that reflects the continuous duration and level of overheating, which is also called weighted overheating hours. The weighted measure indicates the level of thermal dissatisfaction among subjects and takes discomfort to be proportional to the non-linear ratio of the discomfort curve ($\mathrm{PD_h}$) [37]. The two measures are defined by Equations (5) and (6). Eventually, the energy-saving potential of PCM from May to October is defined as the percentage of energy savings based on the cooling load (Equation (7)):

$$I_i = \int wf_i(\tau)\cdot d\tau \ \ if \ T_{op}(\tau) \geq T_{max}, \ (during\ occupied\ hours) \tag{5}$$

$$PD_h(\tau) = \frac{e^{(0.6802\Delta T - 3.7690)}}{\left[1 + e^{(0.6802\Delta T - 3.7690)}\right]} \quad \Delta t = T_{op}(\tau) - T_{max} \tag{6}$$

$$Energy - saving\ potential\ of\ PCM = \frac{Cooling\ load\ without\ PCM\ roof - Cooling\ load\ wth\ PCM\ roof}{Cooling\ load\ without\ PCM\ roof} \times 100\% \tag{7}$$

where, I is the measures for thermal discomfort; for the first measure, $wf_1(\tau)$ = 1.0; for the second measure, $wf_2(\tau) = \mathrm{PD_h}(\tau)/0.2$; T_{op} is the operating temperature; PD_h is the discomfort curve, defined by Equation (6).

3. Results and Discussion

3.1. Cooling Load and Thermal Environment with No PCM Roof

The simulation results of monthly cooling loads (CL), the number of overheating hours (I1), and the number of weighted overheating hours (I2), which represent the thermal discomfort level of the classrooms without a PCM in the roof structure in Taipei and Kaohsiung are shown in Figure 5. According to Figure 5, under the weather conditions of CDH 27, it is known that the annual cooling load is 59.0 GJ and the thermal discomfort levels of I1 and I2 are respectively 234 h and 519 h for the top-floor classrooms in Taipei. Regarding the results of Kaohsiung, the cooling load is 79.8 GJ, and the thermal discomfort levels of I1 and I2 are respectively 344 h and 651 h. Owing to there is a higher CDH 27 and horizontal solar radiation, it is predictable that the cooling load of Kaohsiung will be higher than that of Taipei. For projects aimed at the improvement of roof structures, the ideal outcome is to reduce the cooling load and thermal discomfort in the top-floor classrooms to be the same level as in non-top-floor classrooms at the same time. Therefore, the simulation results of the non-top-floor classrooms, which had the same identical configuration with top-floor classrooms, are also shown in Figure 5 for comparing the difference. Specifically, the differences between the cooling loads and thermal discomfort levels of top-floor and non-top-floor classrooms can be regarded as being caused by the heat transferred from the roof.

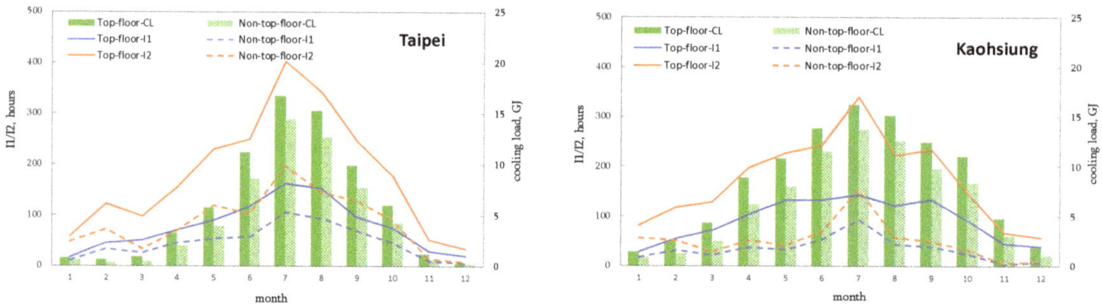

Figure 5. Comparison of cooling load, overheating hours, and weighted overheating hours between top-floor and non-top floor classrooms.

The contribution of heat transferred from the roof to the cooling loads and thermal discomfort levels of classrooms without PCM in the roof structure is shown in Table 2. The results show that the heat transferred from the roof accounts for 20.4% of the cooling load during the air-conditioning season in Taipei and 19.7% in Kaohsiung. The proportion of the overheating hours (I1) contributed by a roof in the ventilation season is 44.4% in Taipei and 64.2% in Kaohsiung; the proportion of the weighted overheating hours (I2) contributed by the roof is respectively 52.8% and 68.5% in Taipei and Kaohsiung. According to the results mentioned above, the importance and urgency of improving the thermal performance of the roof structure are confirmed.

Table 2. The contribution of heat transferred from the roof to the cooling loads and thermal discomfort levels of classrooms without PCM in the roof structure.

	May to October (Air-Conditioning Season)		November to April (Ventilation Season)			
	Cooling load (GJ)	Ratio (%)	I1 (hours)	Ratio (%)	I2 (hours)	Ratio (%)
Taipei	13.2	20.4%	104.0	44.4%	274.1	52.8%
Kaohsiung	15.7	19.7%	221.0	64.2%	445.9	68.5%

3.2. *Influence of PCM on The Energy Consumption of the Air-Conditioning Season*

In this paper, 30 cases of different properties of PCM are set to analyze the parameters which influence thermal performance. Specifically, six kinds of PCTs (23 °C, 25 °C, 27 °C, 29 °C, 31 °C, and 33 °C) and five types of PCM thickness (10–50 mm) were set. Figure 6 shows the energy-saving effects achieved by coupling different PCTs and PCM thicknesses during the AC season from May to October. Both PCT and thickness of PCM significantly influenced the reduction percentage of cooling load during the air-conditioning season. Within the scope of study cases, the energy-saving potential of a PCM layer on the cooling load was 1.2–7.8% in Taipei, which is equivalent to a cutback proportion of 6.3–41.6% cooling load from the roof; and 0.9–6.2% in Kaohsiung, which is equivalent to a cutback proportion of 4.8–33.1% of the cooling load from the roof. Under the circumstance of the same PCM layer thickness during the air-conditioning season, the analysis results of the six selected PCTs show that the energy-saving effect initially increased with the increasing PCT; however, after reaching the peak of energy-saving effect, the energy-saving effect decreases as the PCT increases. In addition, it is important to match the PCT with the outdoor air temperature at night to make the PCM layer completely release the latent heat absorbed and stored during the daytime; meanwhile, it can avoid the heat storage capacity of the PCM layer becoming smaller on the next day. In other words, a higher PCT restricts the heat that can be absorbed and stored by the PCM layer during daytime and lead to a poorer energy-saving effect. Nevertheless, a lower PCT cannot release enough heat at nighttime and results in the poor ability to absorb and store heat. Therefore, it had the same effect of reducing cooling load as higher PCT.

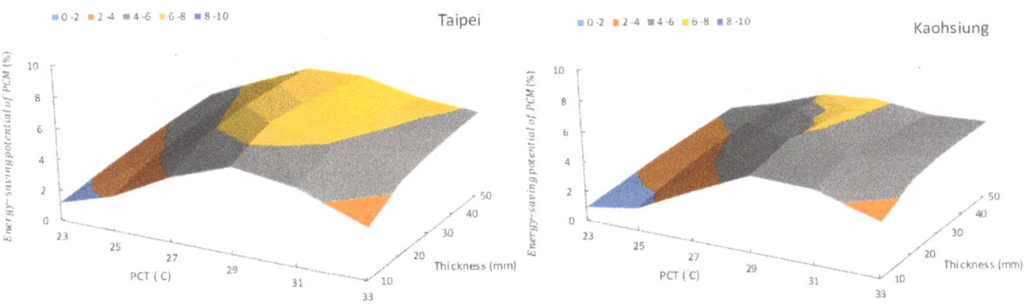

Figure 6. Energy-saving effects of different PCM roofs during the air-conditioning season from May to October.

For achieving the maximum energy-saving effect, the feasible PCTs for an air-conditioning season under different thicknesses of PCM are confirmed. Specifically, from the aspect of cost and efficiency, there is still an upper limitation that when increasing the PCT or thickness may not achieve the corresponding effect in a linear relationship. According to the results shown in Table 3, it is worth noting that only the case with a PCM thickness of 50 mm showed a better PCT at 27 °C in Taipei, all of the beneficial PCTs of a PCM layer in the air-conditioning season occurred at 29 °C both in Taipei or Kaohsiung. This result is in accordance with the previous literature reviews [39–41], which point out that it is not recommended to apply PCM with PCT > 30 °C as passive cooling for buildings.

Table 3. Feasible PCTs under different PCM layer thicknesses.

	May to October (Air-Conditioning Season)					November to April (Ventilation Season)				
Thicknesses of PCM (mm)	10	20	30	40	50	10	20	30	40	50
PCT (°C, Taipei)	29	29	29	29	27	27	25	25	25	25
PCT (°C, Kaohsiung)	29	29	29	29	29	31	29	29	29	29

As shown in Figure 6, when the thickness of the PCM layer in Taipei was 50 mm, the energy-saving effect of PCT = 29 °C or the beneficial PCT = 27 °C was almost the same that the differences between these two cases were negligible. Therefore, the beneficial PCT of a PCM roof in the air-conditioning season could be regarded as 29 °C. In addition, the results also show that though the solar radiation in Kaohsiung was higher than in Taipei; however, there is no need to apply the PCM layer with higher PCT in Kaohsiung to create a better energy-saving effect.

Figure 6 also shows the expected trends: the energy-saving effect increases with the increasing thickness of the PCM layer, and the increasing trend became moderate when the thickness of the PCM layer decreased. Taking a beneficial PCT as an example, according to the simulation results, when the thickness of the PCM layer increases from 10 mm to 20 mm, the energy-saving effect increased from 5.21% to 6.86% in Taipei, which was an increment of 1.65%; increased from 4.74% to 5.91% in Kaohsiung, with an increment of 1.16%. Regarding the thickness of the PCM layer increases from 40 mm to 50 mm, the energy-saving effect increases from 7.45% to 7.6% in Taipei, for which the increment was 0.15%; as for the results of Kaohsiung, the energy-saving effect increased from 6.17% to 6.22%, for which the increment was barely 0.05%. It is worth noting that the thickness of the two circumstances increased the same by 10 mm, but the increment of the energy-saving effect differs by nearly 11 times and 24 times. It is because of the low thermal conductivity of PCM (0.2 W/m·K) and a thicker PCM layer, the required time for complete melting of PCM is prolonged. If the time required for melting lasts too long, part of the PCM may remain in a solid state at the end of the daytime period of the endothermic process. Under the circumstance of a thicker PCM layer, the full capacity of PCM has not yet been utilized; therefore, the relationship between the energy-saving effect and the thickness of PCM was not linearly related.

3.3. *Influence of PCM on Thermal Comfort in the Ventilation Season*

The roof structure with PCM creates a more comfortable environment during the ventilation season from November to April, which can reduce the number of overheating hours and weighted overheating hours under the situation of classroom operating temperatures exceeding 80% of the acceptable upper limitation. Figure 7 shows the difference in the number of overheating hours (I1) between the baseline case of a roof without PCM and the case of a roof with different kinds of PCM. In Figure 7, the overheating hours (I1) are represented by a dashed line, and the weighted overheating hours (I2) are represented by a solid line. The reduction percentage in the number of overheating hours was 3.7–45.5% in Taipei and 8.1–47.4% in Kaohsiung, which were equivalent to 9–110 h and 28–163 h less than the case of not using PCM. As shown in Figure 7, among all the analyzed cases, the decrements in weighted overheating hours (I2) were higher than those in the number of overheating hours (I1).

Figure 8 explains how the PCM layer enhances the level of thermal comfort during the ventilation season. Specifically, it shows the operating temperatures of classrooms using and not using PCM in Taipei from 22–24 April. In this case, a PCM layer with a melting point of 25 °C and a thickness of 40 mm was set for simulation. As revealed in Figure 8, PCM begins to solidify by releasing heat at nighttime, and since the nighttime temperature of the ventilation season is low, most of the PCM layer was solidified by early morning. As the temperature rises during the daytime, the PCM begins to melt and store the heat transferred to the roof as latent heat, preventing it from being transferred into the interior space and thereby inhibiting the rise in indoor operating temperature. Moreover, there was no extremely high temperature during the ventilation season, so the PCM layer would melt slowly and makes the operating temperature during the day close to the comfort upper limit of 29 °C or within the comfort range. On average, the operating temperature during daytime was reduced by 1.0 °C, and the PD_h was decreased by 0.11, while the temperature of the hottest noon period (12:00–14:00) was reduced more by 1.8 °C, and the PD_h decreased by 0.23. Although in some high-temperature periods at noon, the PCM layer was not able to reduce the operating temperature to a comfortable range or reduce the number of overheating hours, it did reduce the PD_h significantly, which would help to improve the comfort level of classrooms. Furthermore, within the three days shown in Figure 8, the length of time periods in which the operating temperature was outside the comfort zone had also been decreased by installing PCM in the roof structure. Under the situation without PCM, the temperature exceeded the comfort zone for 14 h; while under the situation of using PCM, the continuous time was shortened to 8 h. Simultaneously, referring to the result of the previous night of 23 April, it showed that when the temperature was low enough for a complete solidification of the PCM, the greatest improvement in indoor comfort could be achieved.

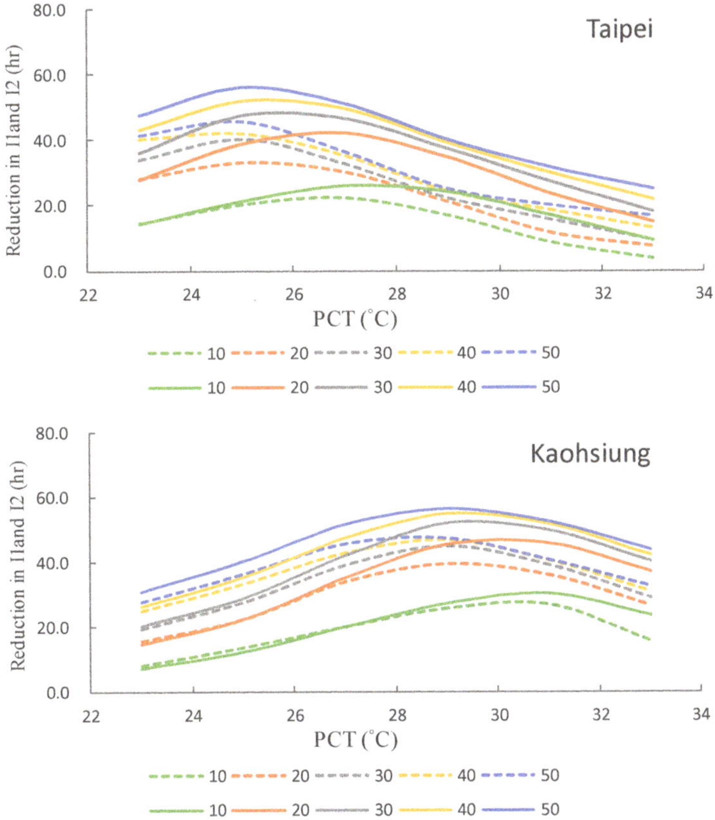

Figure 7. Effects on thermal comfort of different thicknesses of PCM roofs during the ventilation season from November to April.

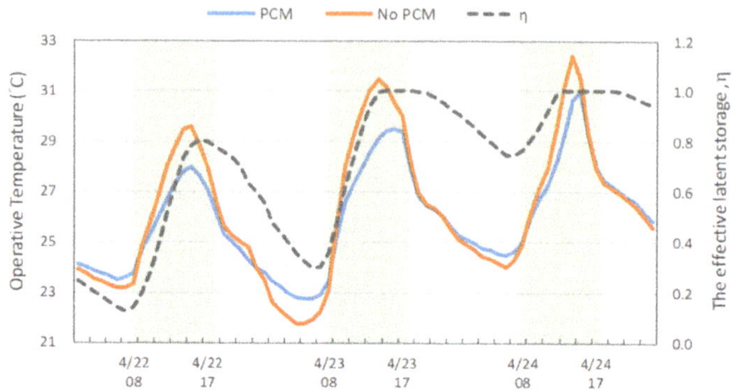

Figure 8. Comparison of simulating operative temperatures of classrooms with and without PCM roofs during the ventilation season.

According to the results shown in Figure 6 and Table 3, it is worth noting that for achieving the maximum comfort effect in the ventilation season under different thicknesses of PCM, it did not matter if the analysis was based on the number of overheating hours or the weighted overheating hours, the beneficial PCTs were identical. For instance, when the thickness of PCM was 10 mm, the beneficial PCTs were 27 °C in Taipei and 31 °C in Kaohsiung; when the thickness of PCM was over 20 mm, the beneficial PCTs were 25 °C in Taipei and 29 °C in Kaohsiung. In addition, the beneficial PCTs for the ventilation and air-conditioning seasons were identical in Kaohsiung; but in Taipei, the beneficial PCT for the air-conditioning season was 4 °C higher than that for the ventilation season. The main reason for this difference is because the outdoor air temperature of the air-conditioning season was higher than that of the ventilation season; ideally, dual-layer PCM roofs with different PCTs would be expected to be selected according to the outdoor climate features of the ventilation and air-conditioning seasons. Specifically, the outer PCM layer, which possesses a higher melting temperature, would be active in summer, and the inner PCM layer, which possesses lower melting temperatures, would be active in winter. Through this measure, the energy-saving effect and indoor thermal comfort of a building could be improved in both summer and winter [42,43]; however, it may increase construction costs and construction difficulties.

3.4. *Efficiency of PCM*

The corresponding efficiency coefficients for the PCM roof with the closest feasible PCT for applying to school buildings in Taiwan are listed in Table 4. According to the table, it could be concluded that the efficiency coefficients of CE (daytime heating) and HE (nighttime cooling) tended to be low for the selected PCM layers of various thicknesses in the two seasons. Regarding the air-conditioning season, CE and HE in Taipei, respectively, lay within 0.58–0.39 and 0.50–0.38, and those in Kaohsiung were within 0.45–0.73 and 0.43–0.55. For the ventilation season, CE and HE in Taipei, respectively, lay within 0.33–0.24 and 0.35–0.27, and those in Kaohsiung were within 0.35–0.48 and 0.38–0.53. It is predictable that the CE and HE of both cities in the two seasons were close to each other. In addition, it is worth mentioning that CE in the air-conditioning season was slightly higher than HE, particularly for the thinner PCM layer. This is because during the air-conditioning season, the outer surface temperature of the roof at 7:00–8:00 in the morning led to the PCM to start melting instead of continuously solidifying, offsetting part of the cold energy charged at nighttime. This could also be corroborated by the lower nighttime latent heat-storing ratio (L_C), as compared to the daytime latent heat-releasing ratio (L_{DC}). The reasons for the low CE and HE could be clarified by observing the latent heat charging ratio and operating time ratio of the PCM layer. Taking the simulation results of Taipei as examples, for various selected thicknesses of PCM, it was found that the nighttime operating time ratio (T_C)

and latent heat charging ratio (L_C) from May to October were respectively 0.62–0.74 and 0.21–0.57.

Figure 9 reveals the efficiency of PCM for the four days in May. A high T_C indicates that the PCM layer was solidifying and releasing heat at night for most of the time. On the other hand, a low L_C indicates that the PCM layer had not completely solidified before sunrise; as shown in Figure 9, the latent heat capacity could not be fully utilized. The daytime latent heat-releasing ratio (L_{DC}) of the PCM layer was usually close to the nighttime value L_C, but the daytime operating time ratio lay within 0.44–0.67, which means that they only operated for half of the time during the daytime, even becoming entirely liquid by noon.

According to the analysis mentioned above that the utilization efficiency barely lay within 0.44–0.67, which means that although the roof with PCM significantly improves indoor thermal comfort from November to April, it does not accompany by high efficiency. In the same manner, we also evaluated the operating time ratio (T_C) and latent heat-storing ratio (L_{DC}) for the operation of the PCM layer through the entire ventilation season (November to April). Unlike the air-conditioning season, the nighttime operating time ratios are not high (0.30–0.42 for Taipei, 0.37–0.54 for Kaohsiung) from November to April. Moreover, as revealed by comparing Figures 9 and 10, on those four days of February, the daytime latent heat discharging ratio (L_{DC}) of the PCM was 0.0 at 08:00 and did not reach 1.0 until 17:00; besides, the PCM layer even started solidification after 15:00. According to the results, it means that the lack of demand was the main reason for the PCM not operating for the entire day and could not make full utilization of the latent heat effect, and accordingly lead to the poor CE or HE. Overall, the lack of demand during the daytime was the main reason for the low usage efficiency of the PCM layer in the ventilation season. This result was different for the situation during the air-conditioning season as the PCM was not able to completely solidify at nighttime. However, for the better utilization of storage capacity, the PCM must be fully melted during daytime and fully solidified during nighttime; in other words, the PCM must complete its daily phase change cycle to reach greater performance.

Table 4. Corresponding efficiency coefficients for PCM roofs for school buildings in Taiwan.

Thickness (mm)	May to October (Air-Conditioning Season)							November to April (Ventilation Season)						
	PCT (°C)	Storing			Releasing			PCT (°C)	Storing			Releasing		
		T_C	L_C	CE	T_{DC}	L_{DC}	HE		T_C	L_C	CE	T_{DC}	L_{DC}	HE
Taipei														
10	29	0.62	0.57	0.58	0.44	0.62	0.50	25	0.30	0.38	0.33	0.33	0.43	0.35
20	29	0.71	0.42	0.53	0.57	0.45	0.49	25	0.37	0.29	0.32	0.40	0.32	0.34
30	29	0.73	0.31	0.47	0.61	0.34	0.44	25	0.39	0.21	0.29	0.44	0.25	0.32
40	29	0.74	0.25	0.42	0.65	0.27	0.41	25	0.41	0.17	0.26	0.46	0.20	0.29
50	29	0.74	0.21	0.39	0.67	0.23	0.38	25	0.42	0.14	0.24	0.47	0.17	0.27
Kaohsiung														
10	29	0.79	0.69	0.73	0.46	0.72	0.55	29	0.37	0.62	0.47	0.45	0.68	0.53
20	29	0.85	0.49	0.64	0.60	0.51	0.54	29	0.48	0.49	0.48	0.56	0.51	0.52
30	29	0.86	0.37	0.56	0.67	0.39	0.49	29	0.51	0.36	0.43	0.60	0.38	0.46
40	29	0.86	0.30	0.50	0.71	0.31	0.45	29	0.53	0.28	0.38	0.62	0.30	0.42
50	29	0.86	0.25	0.45	0.75	0.26	0.43	29	0.54	0.23	0.35	0.63	0.24	0.38

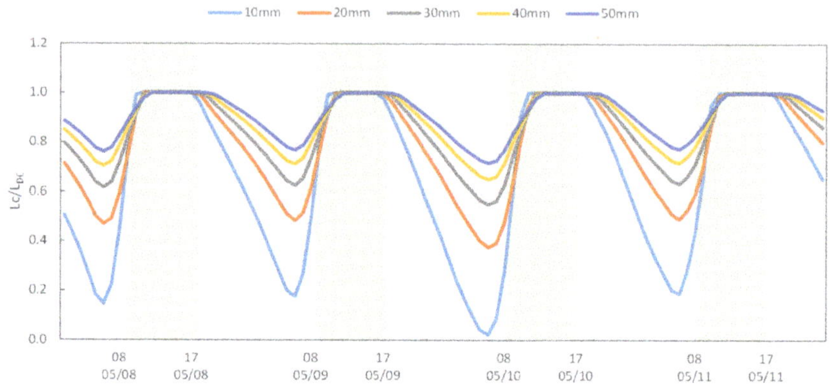

Figure 9. Operating efficiency of the PCM during the air-conditioning season.

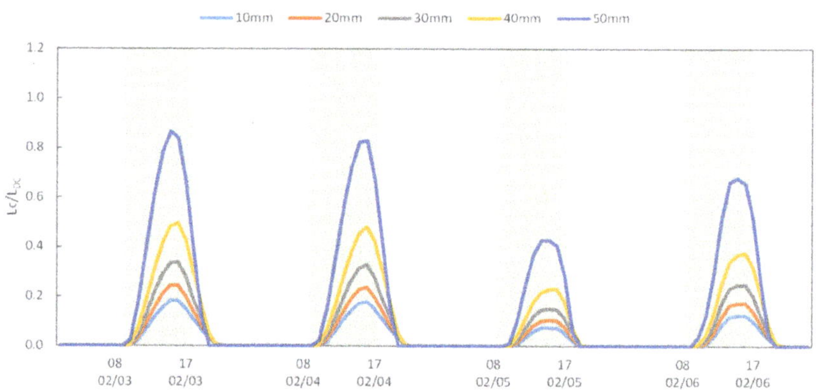

Figure 10. Operating efficiency of the PCM during the ventilation season.

3.5. Selection of PCM

After discussing the feasible PCTs, the proper thickness of the PCM layer for achieving a better economic outcome was analyzed in this section. Under the situation of the advantageous PCT, the relationship between energy-saving effect, thermal comfort, as well as HE, and the thickness of PCM is shown in Figure 11. According to the results, when the thickness was greater than 20 mm, the rates of increase in both energy-savings during the air-conditioning season and thermal comfort enhancement during the ventilation season became significantly more moderate as the PCM layer became thicker. Simultaneously, the decreasing rate of HE became rapid. Therefore, it was more economical to select the PCM with a thickness of 20 mm.

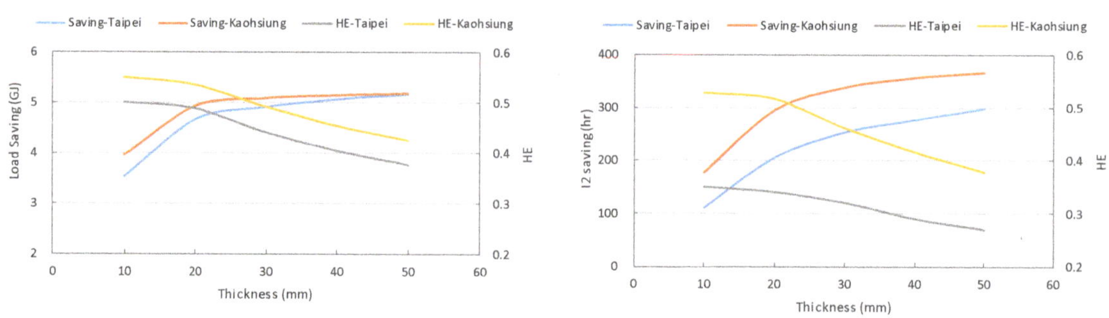

Figure 11. The energy-saving and thermal comfort efficacy of different thicknesses of PCM layers under the beneficial PCT.

As revealed in the analysis above, the beneficial PCT during the air-conditioning season in Taipei was 4 °C higher than the PCT during the ventilation season. Under the thermal management approach of schools in Taiwan, the feasible selection of PCM roofs must simultaneously consider the energy consumption during the air-conditioning season and the annual benefit of thermal comfort during the ventilation season. The issue of whether to prioritize thermal comfort during the ventilation season or to prioritize energy consumption during the air-conditioning season may not alter by the 4 °C temperature difference when selecting a feasible PCT. Under the economic PCM thickness of 20 mm, the benefits of a PCT of 25 °C or 29 °C in the air-conditioning season and the ventilation season can be observed according to Table 5. Specifically, when 29 °C is selected as the feasible PCT, only an additional 29 h of overheating and 22 h of weighted overheating hours are added in the ventilation season, which does not significantly increase the level of thermal discomfort. Besides, it can achieve an additional 2.5 GJ (3.7%) energy-saving potential of cooling load compared with selecting 25 °C as the beneficial PCT. Even if the air conditioner is turned on for an additional 22 h during the ventilation season, the percentage of increasing cooling load is much less than 3.7%. Thus, we can confirm that the most economical PCM layer in Taipei was PCT = 29 °C and a thickness of 20 mm, which was the same as that in Kaohsiung.

Table 5. Comparison of annual efficacy of a PCT of 25 °C or 29 °C in Taipei.

PCT (°C)	Overheating Hours, I1 (hr)	Weighted Overheating Hours, I2 (hr)	Energy-Saving Potential (GJ)
25	80	206	2.5 (3.7%)
29	51	184	4.7 (6.9%)
Difference	29	22	−2.2 (−3.2%)

4. Conclusions

Owing to the undergoing policy of broadly installing the air-conditioning in school buildings in Taiwan, it is essential to draw up energy-saving strategies that not only reduce energy consumption but create better indoor comfort. This paper selected the school buildings located in Taipei and Kaohsiung as target buildings to discuss their cooling load in air-conditioning season and thermal discomfort in natural ventilation season under the circumstance of installing PCM on rooftops. In order to confirm that combining the rooftop structure and PCM brings benefits to building energy-saving, this study firstly revealed the importance of heat insulation from the rooftop by simulating the incoming heat from the rooftop. Secondly, thirty cases of different setting parameters of PCM, including thickness and PCT, were set to simulate their energy-saving effect and thermal discomfort. Eventually, the recommended feasible utilization patterns of PCM for better indoor comfort and energy-saving potential was proposed.

According to the simulation results of incoming heat from a rooftop in AC season, it is known that the ratio of cooling load resulting from the rooftop is significant and cannot be ignored. Regarding the NV season, the ratio of overheating hours resulting from the rooftop should not be overlooked either. As aforementioned, the urgency of heat insulation from the rooftop could be confirmed. The investigated cases of different PCT and thicknesses of PCM show that the features of PCM make noticeable influences on the reducing potential of cooling load. Specifically, regarding the relationship between different PCT and energy-saving effects in the AC season, when the thickness of PCM is fixed, it is worth noting that the energy-saving effect increases as the PCT increase. However, when reaching the peak of the energy-saving effect, the energy-saving effect decreased as the PCT increased. Furthermore, based on the simulation results, for achieving the maximum energy-saving potential, the beneficial PCT for all different thicknesses of PCM was clarified as 29 °C. Additionally, this study also analyzed the relationship between the thickness of PCM and the energy-saving effect. The results show that the energy-saving effect increased as the thickness of the PCM increased, whereas the increasing trend becomes moderate when the thickness of PCM keeps increasing. Briefly, the relationship between the energy-saving effect and the thickness of PCM was not linear.

Concerning the relationship between PCM and thermal discomfort in the NV season, the results showed that installing the PCM on the rooftop apparently cut down the weighted overheating hours in both Taipei and Kaohsiung. Furthermore, the feasible PCT under the conditions of different thicknesses of PCM was suggested. In addition, it is worth noting that the beneficial PCT in AC season and NV season was the same value in Kaohsiung; as for Taipei, the beneficial PCT in the air-conditioning season was 4 °C higher than in the NV season. After confirming the feasible PCT, the most economical thickness of PCM for improving the energy-saving effect and thermal comfort is clarified as 20 mm.

According to the aforementioned, this study suggests the feasible selection of PCM, including its thickness and PCT for the school buildings during different seasons. In addition, based on the simulation conducted respectively in the AC season and NV season, the feasible pattern of PCM through the year is proposed to fulfill the whole year indoor comfort and energy-saving potential. Through the recommendations proposed in this study, it is expected to efficiently utilize the PCM in school buildings to reduce the additional energy consumption and eventually realize the goal of energy conservation.

Author Contributions: Conceptualization, R.-L.H. and B.-L.C.; Methodology, R.-L.H. and W.-A.C.; Formal Analysis, R.-L.H. and W.-A.C.; Investigation, B.-L.C.; Resources, R.-L.H.; Data Curation, R.-L.H. and B.-L.C.; Writing—Original Draft Preparation, R.-L.H. and W.-A.C.; Writing—Review and Editing, W.-A.C.; Supervision, R.-L.H.; Project Administration, W.-A.C. All authors have read and agreed to the published version of the manuscript.

Funding: This research received no external funding.

Institutional Review Board Statement: Not applicable.

Informed Consent Statement: Not applicable.

Data Availability Statement: The data presented in this study are available on request from the corresponding author.

Conflicts of Interest: The authors declare no potential conflict of interest with respect to the research, authorship, and/or publication of this article.

References

1. Wang, X.; Sun, X.; Yu, C.W.F. Building envelope with variable thermal performance: Opportunities and challenges. *Indoor Built Environ.* **2018**, *27*, 729–733. [CrossRef]
2. Zhou, Y.; Yu, C.W. The year-round thermal performance of a new ventilated Trombe wall integrated with phase change materials in the hot summer and cold winter region of China. *Indoor Built Environ.* **2019**, *28*, 195–216. [CrossRef]
3. Madad, A.; Mouhib, T.; Mouhsen, A. Phase Change Materials for Building Applications: A Thorough Review and New Perspectives. *Buildings* **2018**, *8*, 63. [CrossRef]
4. Al-Yasiri, Q.; Szabó, M. Incorporation of phase change materials into building envelope for thermal comfort and energy saving: A comprehensive analysis. *J. Build. Eng.* **2021**, *36*, 102122. [CrossRef]
5. Stamatiadou, M.E.; Katsourinis, D.I.; Founti, M.A. Computational assessment of a full-scale Mediterranean building incorporating wallboards with phase change materials. *Indoor Built Environ.* **2016**, *26*, 1429–1443. [CrossRef]
6. Ramakrishnan, S.; Sanjayan, J.; Wang, X. Experimental Research on Using Form-stable PCM-Integrated Cementitious Composite for Reducing Overheating in Buildings. *Buildings* **2019**, *9*, 57. [CrossRef]
7. Bravo, J.P.; Venegas, T.; Correa, E.; Álamos, A.; Sepúlveda, F.; Vasco, D.A.; Barreneche, C. Experimental and Computational Study of the Implementation of mPCM-Modified Gypsum Boards in a Test Enclosure. *Buildings* **2020**, *10*, 15. [CrossRef]
8. Kim, T.; Ahn, S.; Leigh, S.-B. Energy consumption analysis of a residential building with phase change materials under various cooling and heating conditions. *Indoor Built Environ.* **2014**, *23*, 730–741. [CrossRef]
9. Tokuç, A.; Yesügey, S.C.; Başaran, T. An evaluation methodology proposal for building envelopes containing phase change materials: The case of a flat roof in Turkey's climate zones. *Arch. Sci. Rev.* **2017**, *60*, 408–423. [CrossRef]
10. Alqallaf, H.J.; Alawadhi, E.M. Concrete roof with cylindrical holes containing PCM to reduce the heat gain. *Energy Build.* **2013**, *61*, 73–80. [CrossRef]

11. Mushtaq, T.H.; Ahmed, Q.M.; Hasanain, M.H. Experimental and Numerical Study of Thermal Performance of a Building Roof Including Phase Change Material (PCM) for Thermal Management. 2013. Available online: http://ijersonline.org/HTML_Papers/ResearchJournalofEngineeringandTechnology__PID__2013-4-3-7.html (accessed on 31 October 2013).
12. Zwanzig, S.D.; Lian, Y.; Brehob, E.G. Numerical simulation of phase change material composite wallboard in a multi-layered building envelope. *Energy Convers. Manag.* **2013**, *69*, 27–40. [CrossRef]
13. Xamán, J.; Rodriguez-Ake, A.; Zavala-Guillén, I.; Hernández-Pérez, I.; Arce, J.; Sauceda, D. Thermal performance analysis of a roof with a PCM-layer under Mexican weather conditions. *Renew. Energy* **2020**, *149*, 773–785. [CrossRef]
14. Mourid, A.; El Alami, M. Thermal Behavior of a Building Provided With Phase-Change Materials on the Roof and Exposed to Solar Radiation. *Sol. Energy Eng.* **2017**, *139*, 061012. Available online: https://asmedigitalcollection.asme.org/solarenergyengineering/article-abstract/139/6/061012/379777 (accessed on 24 December 2019). [CrossRef]
15. Piselli, C.; Castaldo, V.L.; Pisello, A.L. How to enhance thermal energy storage effect of PCM in roofs with varying solar reflectance: Experimental and numerical assessment of a new roof system for passive cooling in different climate conditions. *Sol. Energy* **2019**, *192*, 106–119. [CrossRef]
16. Beltrán, R.D.; Martínez-Gómez, J. Analysis of phase change materials (PCM) for building wallboards based on the effect of environment. *J. Build. Eng.* **2019**, *24*, 100726. [CrossRef]
17. Reddy, K.S.; Mudgal, V.; Mallick, T.K. Thermal Performance Analysis of Multi-Phase Change Material Layer-Integrated Building Roofs for Energy Efficiency in Built-Environment. *Energies* **2017**, *10*, 1367. [CrossRef]
18. Li, D.; Zheng, Y.; Liu, C.; Wu, G. Numerical analysis on thermal performance of roof contained PCM of a single residential building. *Energy Convers. Manag.* **2015**, *100*, 147–156. [CrossRef]
19. Roman, K.K.; O'Brien, T.; Alvey, J.; Woo, O. Simulating the effects of cool roof and PCM (phase change materials) based roof to mitigate UHI (urban heat island) in prominent US cities. *Energy* **2016**, *96*, 103–117. [CrossRef]
20. Yu, J.; Yang, Q.; Ye, H.; Luo, Y.; Huang, J.; Xu, X.; Gang, W.; Wang, J. Thermal performance evaluation and optimal design of building roof with outer-layer shape-stabilized PCM. *Renew. Energy* **2020**, *145*, 2538–2549. [CrossRef]
21. Li, D.; Wu, Y.; Zhang, G.; Arıcı, M.; Liu, C.; Wang, F. Influence of glazed roof containing phase change material on indoor thermal environment and energy consumption. *Appl. Energy* **2018**, *222*, 343–350. [CrossRef]
22. Jin, X.; Medina, M.A.; Zhang, X. Numerical analysis for the optimal location of a thin PCM layer in frame walls. *Appl. Therm. Eng.* **2016**, *103*, 1057–1063. [CrossRef]
23. Heim, D.; Wieprzkowicz, A. Positioning of an isothermal heat storage layer in a building wall exposed to the external environment. *J. Build. Perform. Simul.* **2015**, *9*, 542–554. [CrossRef]
24. Murathan, E.K.; Manioğlu, G. Evaluation of phase change materials used in building components for conservation of energy in buildings in hot dry climatic regions. *Renew. Energy* **2020**, *162*, 1919–1930. [CrossRef]
25. Yang, L.; Qiao, Y.; Liu, Y.; Zhang, X.; Zhang, C.; Liu, J. A kind of PCMs-based lightweight wallboards: Artificial controlled condition experiments and thermal design method investigation. *Build. Environ.* **2018**, *144*, 194–207. [CrossRef]
26. Cascone, Y.; Capozzoli, A.; Perino, M. Optimisation analysis of PCM-enhanced opaque building envelope components for the energy retrofitting of office buildings in Mediterranean climates. *Appl. Energy* **2018**, *211*, 929–953. [CrossRef]
27. Muruganantham, K. Application of Phase Change Material in Buildings: Field Data vs. EnergyPlus Simulation. Arizona State University, December 2010; p. 84. Available online: http://repository.asu.edu/attachments/56138/content/Muruganantham_asu_0010N_10151.pdf (accessed on 1 December 2010).
28. Jamil, H.; Alam, M.; Sanjayan, J.; Wilson, J.L. Investigation of PCM as retrofitting option to enhance occupant thermal comfort in a modern residential building. *Energy Build.* **2016**, *133*, 217–229. [CrossRef]
29. Liu, J.; Liu, Y.; Yang, L.; Liu, T.; Zhang, C.; Dong, H. Climatic and seasonal suitability of phase change materials coupled with night ventilation for office buildings in Western China. *Renew. Energy* **2020**, *147*, 356–373. [CrossRef]
30. Berardi, U.; Soudian, S. Benefits of latent thermal energy storage in the retrofit of Canadian high-rise residential buildings. *Build. Simul.* **2018**, *11*, 709–723. [CrossRef]
31. Sage-Lauck, J.; Sailor, D. Evaluation of phase change materials for improving thermal comfort in a super-insulated residential building. *Energy Build.* **2014**, *79*, 32–40. [CrossRef]
32. Ramakrishnan, S.; Wang, X.; Alam, M.; Sanjayan, J.; Wilson, J.L. Parametric analysis for performance enhancement of phase change materials in naturally ventilated buildings. *Energy Build.* **2016**, *124*, 35–45. [CrossRef]
33. Biopcm Webpage.Pdf. Available online: https://phasechange.com/biopcm/ (accessed on 12 May 2021).
34. What is BioPCM®? Available online: https://phasechange.com/biopcm/ (accessed on 12 May 2021).
35. Lomas, K.J.; Porritt, S.M. Overheating in buildings: Lessons from research. *Build. Res. Inf.* **2016**, *45*, 1–18. [CrossRef]
36. Teli, D.; Bourikas, L.; James, P.; Bahaj, A.S. Thermal Performance Evaluation of School Buildings using a Children-based Adaptive Comfort Model. *Procedia Environ. Sci.* **2017**, *38*, 844–851. [CrossRef]
37. Liang, H.-H.; Lin, T.-P.; Hwang, R.-L. Linking occupants' thermal perception and building thermal performance in naturally ventilated school buildings. *Appl. Energy* **2012**, *94*, 355–363. [CrossRef]
38. ISO. ISO 7730:2005 Ergonomics of the Thermal Environment—Analytical Determination and Interpretation of Thermal Comfort Using Calculation of the PMV and PPD Indices and Local Thermal Comfort Criteria. Available online: https://www.iso.org/standard/39155.html (accessed on 1 November 2005).

39. Pomianowski, M.; Heiselberg, P.; Zhang, Y. Review of thermal energy storage technologies based on PCM application in buildings. *Energy Build.* **2013**, *67*, 56–69. [CrossRef]
40. Waqas, A.; Din, Z.U. Phase change material (PCM) storage for free cooling of buildings—A review. *Renew. Sustain. Energy Rev.* **2013**, *18*, 607–625. [CrossRef]
41. Zhou, D.; Zhao, C.; Tian, Y. Review on thermal energy storage with phase change materials (PCMs) in building applications. *Appl. Energy* **2012**, *92*, 593–605. [CrossRef]
42. Zhu, N.; Liu, P.; Liu, F.; Hu, P.; Wu, M. Energy performance of double shape-stabilized phase change materials wallboards in office building. *Appl. Therm. Eng.* **2016**, *105*, 180–188. [CrossRef]
43. Zhu, N.; Hu, N.; Hu, P.; Lei, F.; Li, S. Experiment study on thermal performance of building integrated with double layers shape-stabilized phase change material wallboard. *Energy* **2019**, *167*, 1164–1180. [CrossRef]

buildings

MDPI

Article

Economic Evaluation of the Indoor Environmental Quality of Buildings: The Noise Pollution Effects on Housing Prices in the City of Bari (Italy)

Pierluigi Morano [1], Francesco Tajani [2], Felicia Di Liddo [1,*] and Michele Darò [3]

1 Department of Civil, Environmental, Land, Building Engineering and Chemistry (DICATECh), Polytechnic University of Bari, Via Orabona 4, 70125 Bari, Italy; pierluigi.morano@poliba.it
2 Department of Architecture and Design, Sapienza University of Rome, Via Flaminia 359, 00196 Rome, Italy; francesco.tajani@uniroma1.it
3 Mi.Da—Sound Design, Via Porro 9, 10064 Pinerolo, Italy; michele.daro@awn.it
* Correspondence: felicia.diliddo@poliba.it

Abstract: Among environmental factors, noise represents one of the most relevant determinants on human health and on the urban quality level and, consequently, on real estate values. Thus, the noise pollution issue plays a significant role in public urban policies aimed at increasing the acoustic comfort level and creating more sustainable and comfortable cities. The real estate market is highly sensitive to noise factor and the residential prices can be strongly influenced by a high acoustic pollution rate. The present research aims to analyze the functional relationships between noise pollution and selling prices in four municipal areas of the city of Bari (Southern Italy). For each area, a study sample constituted by two hundred residential properties sold in 2017–2019 was detected for the identification of the main influential factors on prices and the investigation of the contribution of noise on them. The implementation of an econometric technique was used to obtain four different models (one for each municipal area of the city of Bari) able to explain the specific impact of noise pollution level on selling prices. From the comparison of the results obtained for each area, the outputs confirm the expected phenomena in terms of a decrease of noise component influence on residential prices from the central area to the peripheral. For the suburban area of the city of Bari, the model obtained does not include the noise pollution factor, showing a lower (scarce) importance of the environmental factor among the buyer and seller bargaining phases.

Keywords: noise pollution; environmental factors; property prices; genetic algorithm; residential market

check for updates

Citation: Morano, P.; Tajani, F.; Di Liddo, F.; Darò, M. Economic Evaluation of the Indoor Environmental Quality of Buildings: The Noise Pollution Effects on Housing Prices in the City of Bari (Italy). *Buildings* **2021**, *11*, 213. https://doi.org/10.3390/buildings11050213

Academic Editors: Alessandro Cannavale, Francesco Martellotta and Francesco Fiorito

Received: 5 May 2021
Accepted: 16 May 2021
Published: 19 May 2021

Publisher's Note: MDPI stays neutral with regard to jurisdictional claims in published maps and institutional affiliations.

Copyright: © 2021 by the authors. Licensee MDPI, Basel, Switzerland. This article is an open access article distributed under the terms and conditions of the Creative Commons Attribution (CC BY) license (https://creativecommons.org/licenses/by/4.0/).

1. Introduction

Noise is one of the most significant environmental risks to health. It continues to be a growing question of concern discussed by policy decision-makers. Road traffic, public transportation systems, workplaces, industrial machinery, loud music and electronic equipment have a relevant impact on people's well-being. The correlations between some pathologies and the continuous exposure to the physical agent of noise are known; in the context of health protection in the workplace, for example, there have been specific laws and procedures that regulate the exposu1re of workers to noise, identifying limits of level and exposure time, and in some cases, suitable personal protective equipment to protect hearing. In the context of traffic noise, the Environmental Noise Directive of the European Union [1] is focused on the noise generated by this source and defines the assessment threshold; currently, at least 100 million EU people are affected by traffic noise.

Especially in urban areas, there are also additional sources of noise that weigh on the population, such as anthropic noise in crowded places and the so-called "nightlife" noise. This type of noise occurs in a punctual way around aggregation centers and changes during the day, avoiding analysis on the incidence when referring to large populations.

Sound emission becomes noise when it produces negative effects on the environment and on human health, that is, when it compromises quality of life.

Noise pollution, like air pollution, "directly" affects health [2–4], differing from water and soil pollution which damage the health only if water or soil products are used for food production.

In recent decades, European technological development has led to several lifestyle changes that have effects on urban systems. At the same time, appropriate measures aimed to reduce the noise of the working and home environment and to fully preserve the life quality level have not been adopted in all countries because of the high costs connected to the interventions for the introduction of acoustic isolation systems (acoustic glazing and acoustic window and façade units, as well as internal sound-insulating partitions).

Moreover, the economic influence of noise is growing. The World Health Organization (WHO) estimates that, as of 2012, at least 1.6 million healthy life years in Western Europe were being lost annually to environmental noise [5]. Other assessments put the external noise cost related to health issues between 0.3–0.4% of GDP in the European Union [6,7] and 0.2% of GDP in Japan [8]. Long-term exposure to noise has been shown to cause 12,000 premature deaths and contribute to 48,000 new cases of ischemic heart disease per year in Europe. Furthermore, it is estimated that 22 million people are affected by chronic discomfort and 6.5 million people are affected by chronic sleep disturbances [9].

The WHO Regional Office for Europe has developed noise guidelines [10] for (i) the definition of the noise level that causes significant harmful effects on health and must not be exceeded, (ii) the identification of the measures to reduce its effects as much as possible. The document constitutes an important reference for political decision-makers and expert technicians for law development in order to regulate the noise issue at local, national and international levels. The guidelines aim to raise awareness the public administrations and private investors of the harmful impacts on the local community, influencing urban planning choices to achieve the Goals of the Agenda 2030 for Sustainable Development [11]. In particular, the noise protection contributes to the achievement of (i) Goal 3 "Ensure healthy lives and promote well-being for all at all ages," as the noise exposure is a significant cause of health deterioration, (ii) Goal 9 "Build resilient infrastructure, promote inclusive and sustainable industrialization and foster innovation" through the development of low-cost sensors (MEMS (Micro Electro-Mechanical Systems) microphones) which, combined with the increase in the transmission and management of large amounts of data, could be an incentive to create widespread acoustic monitoring networks, (iii) Goal 11 "Make cities and human settlements inclusive, safe, resilient and sustainable," as the progressive expansion of the population in urban agglomerations determines an increase in the demand for mobility and an increase in social and entertainment activities, potentially causing disturbance.

The Directive 2002/49/EC [1] modifies the methods of assessing the harmful effects of noise pollution, taking into account: (i) ischemic heart disease; (ii) high annoyance; (iii) high sleep disturbance.

The adverse effects are separately calculated using the relative (or absolute) risk formulas and they must be independently assessed for each noise source. Currently, road, rail and airport traffic are only considered, and Member States are required to transpose the new Directive by 31 December 2021.

In the framework outlined, different studies in the U.S. and Europe have demonstrated the relationship between real estate markets and environmental noise, showing a decrease of housing prices per decibel (dB) of noise increase.

This approach has limitations, as the "noise level" parameter alone does not consider the typology and specific factors of the individual sources, but it can provide a first indication about the correlation between housing price and the quality of the acoustic environment.

Regarding the disturbance generated by multiple types of anthropic activities that are different from industrial noise and traffic noise, there are greater difficulties in framing the issue in a generalizable perspective, especially in urban areas, and therefore the definition of universally feasible strategies and solutions is more complex. The harmful noise, connoted as neighborhood disturbance, often derives from (i) discos, (ii) bars and squares frequented by young people and (iii) productive or commercial activities located in residential contexts. These sources are randomly distributed throughout the territory and they are sometimes concentrated in areas with particular vocations, such as the so-called nightlife "movida" which, on certain days and times, upsets the acoustic environment of a residential area. The "movida" noise originates from a plurality of sources, sometimes fixed, but often linked to the anthropic emissions of subjects which can be assimilated to punctual, temporary and mobile sound sources. The difficulty in determining and quantifying the disturbance generated by this type of noise makes it complicated to circumscribe and summarize in coherent data that can be suitable for correlation with others. In this sense, the different connotation given to the voices and bawl that are considered "festive" in a holiday location but "unbearable" if they are under the home windows should be taken into account. Therefore, it is difficult to correlate this type of environmental noise with its effects on property values. Furthermore, it should be added that the noise sources are often concentrated around an attraction pole, for example a bar, which can also be isolated and located in an acoustically quiet context. This determines an excess of noise that could literally cause the collapse of the market prices of the residential units located in its immediate closeness; on the other hand, this "disturbance" is only limited, as at a distance of a few tens of meters the situation varies and the property values are consistent with the area average values. In this sense, the discrimination of the entity and reliability of an economic indicator that takes into account specific elements is a complex operation.

Moreover, robust and extended data able to allow a first-level study of the correlation between environmental noise and property prices by focusing on traffic-induced noise are known:

1. The noise caused by traffic on infrastructures, in particular in urban areas, determines the highest impact on the population, both in quantitative terms and regarding the exposure.
2. European and national directives consider as a first priority the analysis, the mapping, the monitoring and the mitigation of transport infrastructures: this allows for immediate and updated availability of studies and checking data.
3. The "traffic" source is generated by vehicles (or trains) that move on predetermined and fixed paths (and in this way, to a certain extent, for aircraft). Therefore, this source is acoustically comparable to a linear-type source whose emission levels are directly proportional (i) to the number of vehicles in transit (i.e., vehicles/hour), (ii) to the average travel speed and (iii) to the percentage of heavy vehicles (or freight trains). Once a road axis has been acoustically characterized, it knows with good approximation the values of sound emission (and the consequent input at the receivers) by monitoring one or more correlated parameters (e.g., the number of transits).
4. The "length" of a road axis can be considered as a single homogeneous "traffic" source involving multiple properties, such as in an urban area where a stretch of road defined by typical acoustic factors has a minimum length equal to one block of properties. The road stretches with almost homogeneous traffic conditions, which determine the typical sound emission, are generally much longer and cross entire neighborhoods.
5. The vehicular traffic, as well as air and rail traffic, presents considerable differences in noise emissions between the day and night, allowing the influence of the exposure to noise phenomena for the residential units in the night period (where a sleep disturbance can occur) on the property market to be analyzed compared to the more relevant, but less complained about daytime disturbance.

2. Aim

The present research concerns the analysis of the influence of noise pollution, generally defined by the traffic on road infrastructure, on housing prices in the city of Bari (Southern

Italy). With reference to the four municipal areas in which the territorial context is divided, taking into account the geographical distribution developed by the Real Estate Market Observatory (OMI) of the Italian Revenue Agency [12], a study sample relating to two hundred properties sold in 2017–2019 for each area was considered. For each sample, the total market prices and the relevant influential factors for their formation were detected in order to assess the contribution of acoustic component on the local residential market. The implementation of an econometric technique allowed for obtaining four models (one for each municipal OMI area of the city of Bari) able to point out the functional relationships between the most influential factors selected and the selling prices. The study aims to examine the contribution of sound level calculated in the street where each property is located on the housing prices in order to (i) investigate the influence of noise pollution in each municipal OMI area of the city of Bari considering the homogeneous noise factors on the road axes considered; and (ii) compare the outputs obtained for defining the different impacts on residential market segments in the city.

This work is part of a wider research line aimed to assess the externalities deriving from different sound levels that characterize the acoustic environment of the different areas in terms of housing prices. The results could represent a useful tool for urban planning to guide the decision-making processes towards public strategic actions able to reduce noise pollution levels. In fact, the output of the analysis could constitute a reference for Public Administrations for orienting local initiatives aimed to decrease the noise pollution level and, in this way, to improve environmental quality in the city. In this sense, this research intends to highlight the benefits for the community from a decrease of noise pollution levels in terms of well-being and the economy. Firstly, it is well known that the main positive impact of living in a noiseless and acoustically comfortable environment is connected to human health. Secondly, from the economic point of view, the increase of the existing residential asset market value is strongly linked to housing factors, among which the acoustic comfort constitutes a fundamental factor for the reference market appreciation and the buyers' decisions.

Furthermore, in the contexts of civil trials, the practical outcome of the analysis could be a significant reference for experts to define the loss of property value caused by long exposure to noise; therefore, this work intends to provide appropriate indications, including charts, in order to support the official technical consultants with acoustic damage assessments.

The structure of the paper is summarized as follows. In Reference Literature (Section 3), the main international and national research that have examined the relationships between noise pollution and property prices are illustrated. In Case Study (Section 4), the application is introduced: the variables considered are listed and the main descriptive statistics related to the four study samples are analyzed. In Methodology (Section 5), the technique implemented and the model obtained for each study sample is explained. In particular, the specific statistic performances are reported, the empirical reliability of the functional correlations is checked, and the results are discussed. Finally, in Conclusions (Section 6), the findings of the work are outlined.

3. Reference Literature

A high noise pollution level has a relevant impact on rents, costs and values of residential properties. Among the factors taken into account in the choice processes (architectural aspects, distribution of internal spaces, quality of the materials used, thermal insulation, etc.), the acoustic comfort level constitutes an important element considered by buyers and sellers relating to house sales.

In this sense, the noise pollution is an essential indicator in the environmental quality definition that, in turn, influences the property's market value, as it could indicate the occurrence of a noise nuisance negatively impacting social relationships and people's well-being. The building's acoustic comfort depends on different factors: some of them are immediately perceived, e.g., the road traffic exposure, the proximity to airports or railway, the presence of industrial areas or commercial and recreational activities (restaurants or public places or clubs open until late at night, etc.), others are more "hidden" and are related to the building construction techniques, materials and technologies.

In recent years, new needs and services has led a continuous growth of the noise pollution level: the buildings are "invaded" by internal and external sound that generates sleeping and working disorders. The noise, in fact, determines health problems—widely documented in the reference literature [13,14]—including annoyance, tiredness, headache, anxiety and physiological and even cardiovascular stress effects. In this context, the inclusion of a property in an acoustically degraded area or with a lack of sound insulation can cause a depreciation of the property. It has been noted that a house or an office placed in a noisy environment has less market value compared to the same one, ceteris paribus, located in a peaceful and noiseless place.

With reference to the property acoustic quality, it is linked to (i) the acoustic insulation of both the internal and external dividing walls and of horizontal partitions that can be obtained through the introduction of acoustically performing materials and by adopting adequate technical solutions; (ii) the footfall sound insulation, which can be obtained through floating or resilient floors; (iii) the low emission level of sewage systems, which can be obtained through appropriate installation measures and choice of low intrinsic noise solutions; (iv) the airborne and structural noise of systems such as lift, heating, autoclaves, etc.; (v) the replacement of older paved roads with smoother asphalt; and (vi) a better management of traffic flows and the introduction of speed limits in urban contexts. Furthermore, in recent years strategic measures aimed at raising the awareness of people to change their behavior to use less noisy transport modes, e.g., cycling, walking or electric vehicles, have been promoted in several cities.

The correlations between noise pollution factors and housing prices have been studied in numerous research papers [15–28]. All the developed analyses intend to highlight the fundamental role played by the acoustic component on selling prices. Chang and Kim [29] demonstrated a drop in prices equal about to 0.5% per additional decibel emitted by the rail network in the city of Seoul (Korea). Furthermore, Bureau and Glachant [30] have shown that the selling prices of the residential units located in peaceful districts of the city of Paris (France) are 1.5% higher than those of properties in noisier areas. Similarly, the European Environment Agency have attested to a price decrease in the range of 0.2–1.5% for each decibel above 50–55 dB [31].

The analysis of the road noise influence on the prices of condominiums in the city of Hamburg (Germany) represents the main goal of Brandt and Maennig's research: on the basis of micro-level datasets capturing road traffic-noise exposure, price discounts in the amount of 0.23% following a 1 dB(A) increase in road noise were observed [32]. Most studies aimed to monetize noise have focused on road and air noise [33–38]. Andersson et al. [39] examined the market participants' preferences regarding improvements in environmental quality through the willingness to pay to reduce road and railway noise.

With reference to 292 single-family houses located in a suburb of the city of Stockholm, the empirical analysis carried out by Wilhelmsson [40] found an average noise discount of 0.6% of the house price per decibel, or a total discount of 30% of the price for a house in a noisy location compared to a house in a quiet one.

In the last decades, a steady growth in road traffic intensity in most urban areas has forced politicians and city planners to seriously consider the resulting environmental impact, such as traffic noise. The assessment of negative factors related to acoustic component is required in order to reveal the social benefit of infrastructure plans and to orient the public decisions toward efficient strategies to decrease noise pollution. For the Copenhagen region, Rich and Nielsen [41] analyzed the implicit costs of traffic noise, measured as the marginal percentage loss in property values with respect to the decibel traffic noise, highlighting a Noise Sensitivity Depreciation Index (NSDI)—defined as the marginal percentage depreciation in house prices with respect to dB noise—of 0.54 for houses and 0.47 for apartments. Some years later, Duarte and Tamez [42] demonstrated that the environmental noise has become a major issue in densely urbanized areas, defining a NDSI equal to 0.08% by differentiating according to city areas and to the nature of the noise source. The outputs obtained by Wen et al. [43] and based on data of 7590 multi-story and 4980 high-rise residential properties in the city of Hangzhou (China) in 2017 attested that the urban road traffic generates noise and air pollution, thereby resulting in a disamenity effect on surrounding residential property and subsequently affecting the willingness to pay from homebuyers. The results confirmed a significant moderating effect of floor level on the impact of road traffic.

In order to identify the relationships between traffic noise levels and the prices of apartments located in three residential districts in the city of Olsztyn in north-eastern Poland, Szczepańska et al. [44] implemented a linear correlation analysis. The negative relationship obtained confirms that a building's location relative to the road network and, consequently, the level of traffic noise, is one of the most significant factors in the potential real estate buyers' choice of property.

The functional link between rental prices and noise levels in the city of Munich (Germany) was determined with the implementation of the hedonic price regression model in the studies of Kuehnel and Moeckel [45]. In particular, the output indicated a rental price discount equal to 0.4% per dB(A), on average. Similar negative correlations between housing prices and noise levels have been reported by other authors [46–48].

In the framework outlined, the present research analyzes the correlations between the noise pollution and the housing prices in a city located in the Southern Italy. Compared to similar research in international and national contexts, this aim was carried out by borrowing (i) the urban acoustic maps developed by an Italian public entity, (ii) the classification of the sound pressure level through different acoustic ratings and (iii) an econometric technique that implements a genetic algorithm for the assessment of the contribution of the influential factors on selling prices.

4. Case Study

This case study concerns four samples of two hundred residential properties located in the city of Bari (Italy). Bari is the main city of the Apulia region in Southern Italy. It covers an area of about 120 km^2 and it has a population of over 325,000 inhabitants.

Each residential unit of the four study samples had been sold between the second half of 2017 and the first half of 2019 and is located in one of the four municipal areas of the city of Bari, taking into account the geographical distribution developed by OMI of the Italian Revenue Agency. In particular, the definition provided by OMI of the municipal trade areas is the following: "Aggregation of homogeneous contiguous zones. It represents a territorial area with specific geographical position, and it reflects a consolidated urbanistic location. The municipal context is divided into the following OMI areas: central, semi-central, peripheral, suburban, extraurban".

The central area identifies the municipality portion corresponding to the city urban center, i.e., the distinguishable building aggregate able to attract a larger settlement. The semi-central area identifies the municipal zone immediately adjacent to the city urban center and directly connected to it for services, transport system, infrastructure. This area presupposes the presence of the central and peripheral ones, being located between the two. Therefore, the peripheral area identifies the portion of the municipal territory that is bound by built settlement external limiting. The suburban area indicates the territory zone with the urbanized portions separated from the urban agglomeration from an undeveloped area, i.e., a natural or artificial barrier. The extra-urban area represents the municipal zone where the main activity is agricultural with rural buildings or where the residential properties are almost or completely absent. In this sense, the extra-urban area identifies the remaining part of the municipal context not included in the previous areas and it is bound by the administrative city border; therefore, it is one residual municipal region [49].

With reference to medium-sized municipalities, the following areas are normally identified: old town and central zones (OMI central area), city semi-central zones (OMI semi-central area), residential expansion zones and artisan-industrial zones (OMI peripheral and/or suburban area), agricultural and rural zones (OMI extra-urban area). Each of the municipal areas may require an articulation in multiple different areas.

For the present analysis, OMI areas considered were the central, the semi-central, the peripheral and the suburban. The extra-urban area was excluded as the building is scarcely present or mainly rural.

For each property considered in the analysis, the selling price and the most influential intrinsic and extrinsic characteristics taken into account by buyers and sellers in the negotiation phases were considered. With reference to the indications provided by the market operators in the four municipal OMI areas of the city of Bari, the influential factors in the corresponding market segment of each property were analyzed. Figures 1–4 show the localization of the four study samples in the related municipal OMI areas of the city of Bari. In Figure 5, the localization of the eight hundred properties considered for the case study in the entire municipal territorial context of the city of Bari is reported.

Figure 1. Localization of the study sample properties in the OMI central area of the city of Bari.

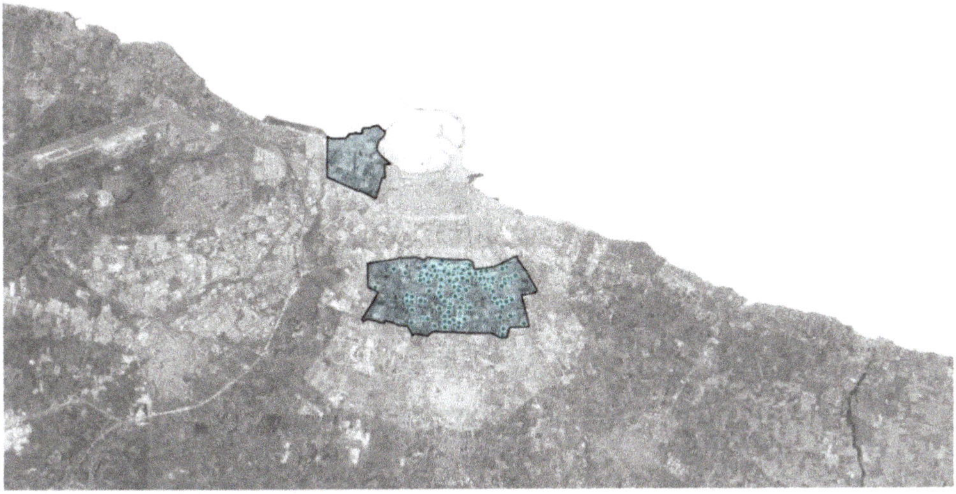

Figure 2. Localization of the study sample properties in the OMI semi-central area of the city of Bari.

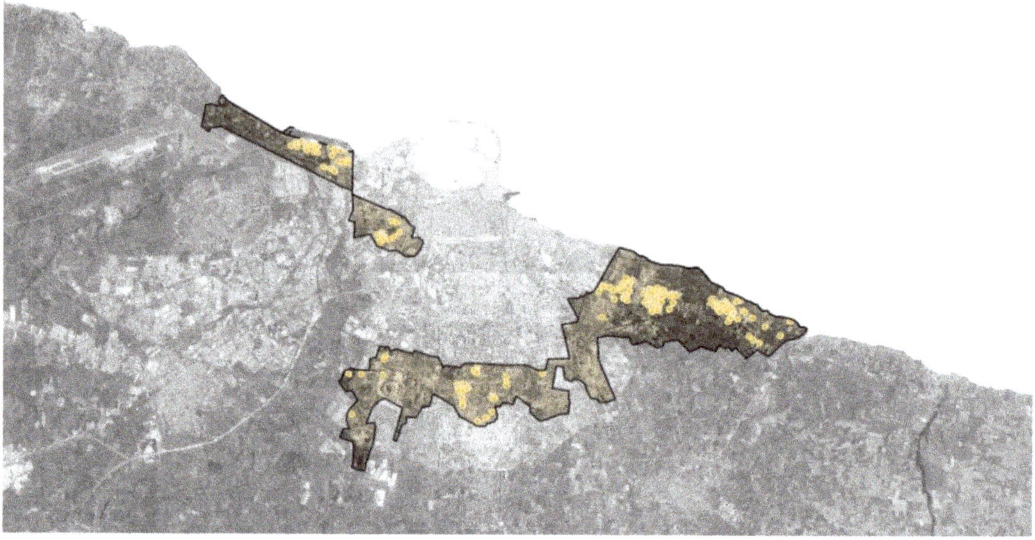

Figure 3. Localization of the study sample properties in the OMI peripheral area of the city of Bari.

Figure 4. Localization of the study sample properties in the OMI suburban area of the city of Bari.

Figure 5. Localization of the study sample properties in the city of Bari.

Variables

With reference to the model developed in this research, the dependent variable is represented by the total selling price (Pr), expressed in euro. For each residential unit, the factors considered by local operators—buyers and sellers—in the bargaining phases [50] were detected. In particular, the variables were selected on the basis of (i) the indications given by the market operators in the area, (ii) the data generated by OMI and (iii) research found in the reference literature [51–54]. In fact, several scientific papers pointed out the unavoidable tradeoff between bias from omitted factors and increased sampling variance related to the collinearity that is involved in this step [55], even if a relative agreement on the major influencing factors is observed [56–58].

Among the variables analyzed, the acoustic one selected by local real estate agents confirms the importance assumed by this factor in the context of the Italian market. The recent fiscal incentive measures taken within the residential asset redevelopment and ex-novo realizations framework aim to enhance the housing comfort improvement and, generally, to reach a high housing quality.

The explanatory variables considered are described below:

- the size of the property (S) in square meters of gross floor area;
- the number of bathrooms in the property (B);
- the floor level on which the property is located (L);
- the presence of the lift in the building where the property is located (A). In the model, the factor is considered as a dummy variable, for which the presence of the service is represented by the value "one," whereas the absence of the service is indicated with the value "zero";
- the presence of the parking space in the building where the property is situated (P);
- the maintenance conditions of the property (Sc), assumed as a qualitative variable and differentiated, through a synthetic evaluation, by the scores 1, 3 and 5, respectively corresponding to the categories "to be restructured," "fit for habitation" and "restructured." Following the logic of the dummy variables, the score "one" is assigned to the category that defines the specific quality of each property, and the score "zero" for the remaining two categories. In particular, the "to be restructured" state refers to properties that require significant refurbishment interventions, because the functionality and the livability of the property are not good due to the inappropriate conservative state

of the elements that compose it; the "good" state indicates properties whose maintenance conditions are acceptable and whose functions can be carried out without heavy refurbishment interventions. Finally, the "excellent" state refers to properties characterized by high construction and aesthetic quality, possibly affected by recent redevelopment and renovation initiatives;

- the distance of the property from the Araldo di Crollalanza waterfront of the city of Bari (Dl), identified as the landmark for the local community, according to different surveys carried out. The variable is measured in kilometers it takes to walk to it;
- the maintenance conditions of the public spaces adjacent to the property (Sa), assessed through a scale of scores (1, 3, 5) attributed by panels of experts (sociologists, landscape architects, etc.), where the score "1" indicates a bad maintenance condition of the public spaces, the score "3" is a good state and the score "5" is an excellent state. In particular, fixed furniture suitable for equipping public spaces, such as public lighting lamps, waste baskets, benches, planters, parking canopies, display boards for billboards, etc., are included in the category of street furniture;
- the property distance from the nearest food market or grocery shop (Dm), calculated in kilometers it takes to walk to it. The category "grocery shop" includes the self-service retail shops of consumer products (supermarkets and hypermarkets) present in the municipal area;
- the maintenance conditions of the building facades adjacent to the property (Sf), assessed through a scale of scores (1, 3, 5) attributed by panels of experts (sociologists, landscape architects, etc.): the score "1" indicates bad maintenance conditions of the facades' conservative state, the score "3" is a good state and the score "5" is an excellent state.
- the road private and public vehicular traffic (buses) level (T), evaluated by a team of experts (sociologists, landscapers, architects, engineers, etc.) through a scale of scores defined as follows: score "1" indicates a road characterized by high traffic intensity, score "3" indicates a medium traffic intensity, score "5" indicates a road characterized by low traffic congestion;
- the property distance from the nearest public green space (Dv), calculated in the kilometers it takes to walk to it;
- the property distance from the nearest highway (Dt), measured in the kilometers it takes to get there by car;
- the distance of the property from the nearest railway station (Ds), measured in the kilometers it takes to walk to it;
- the perceived environmental quality level of the property area (Qn), assessed by assigning a numerical score from "1" (disagreement with the item) to "5" (agreement with the item), given by a sample of users sufficiently representative of the urban area. The items considered are:
  - this neighborhood is generally not polluted,
  - this is a quiet neighborhood,
  - residents' health is threatened by pollution,
  - the heavy traffic in this neighborhood is very annoying,
  - there are green areas for relaxing,
  - going to a park means travelling to other parts of the city,
  - the green areas are well-equipped.
- the sound level (Ld), expressed in decibels dB(A), measured on day, evening and night intervals, in the street where the residential unit is located. The data are borrowed by the Strategic Noise Map of the Bari agglomeration, published in June 2017 by the Scientific Directorate of the Regional Agency for Environmental Prevention and Protection of the Puglia Region (ARPA Puglia) and reported in Figure 6 [59]. The sound pressure level expressed in decibels are shown in the map for each city road axis and it is divided into seven classes of ratings as follows:
  (i) Rating 1: $\leq$40 dB(A);

(ii) Rating 2: >40 dB(A) and ≤50 dB(A);
(iii) Rating 3: >50 dB(A) and ≤55 dB(A);
(iv) Rating 4: >55 dB(A) and ≤60 dB(A);
(v) Rating 5: >60 dB(A) and ≤65 dB(A);
(vi) Rating 6: >65 dB(A) and ≤70 dB(A);
(vii) Rating 7: >70 dB(A) and ≤75 dB(A).

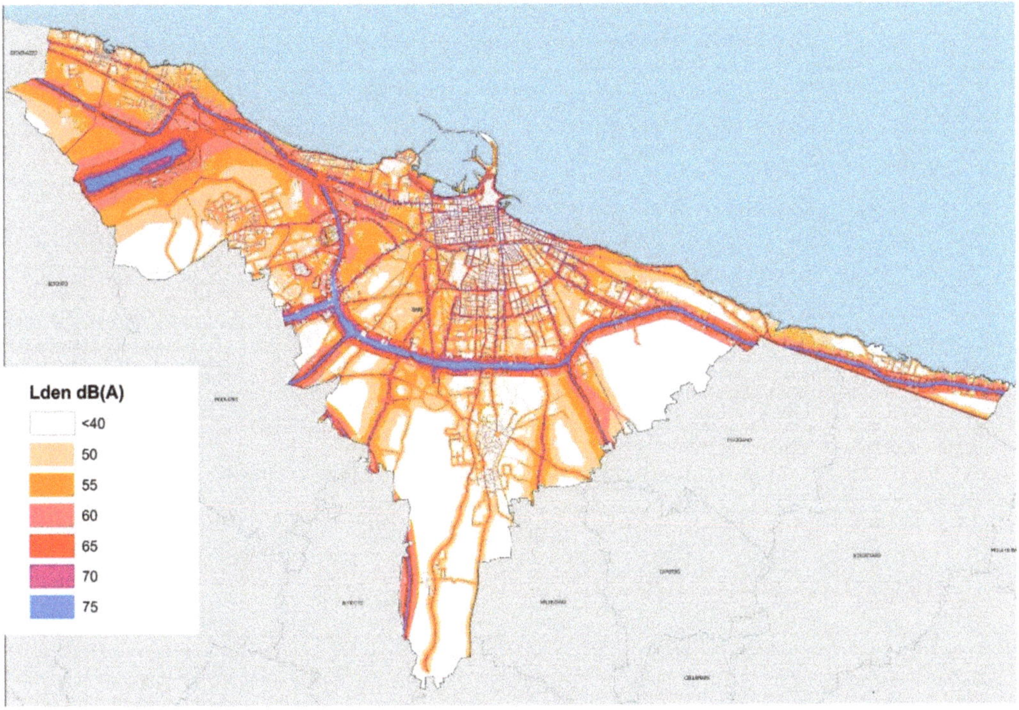

Figure 6. Strategic Noise Map of the city of Bari (Source: Arpa Puglia—Scientific Directorate, adapted from ref. [59]).

It should be noted that some variables considered in the present research are directly or indirectly connected to the noise and to its effects on selling prices. In this sense, the floor level on which the property is located may influence the acoustic pollution, as the road traffic or leisure activity noises are mainly perceived on lower floors. Furthermore, the presence of the parking spaces in the building where the property is located might cause different noise sources, e.g., a car's engine or horn. At the same time, the property distance from the nearest railway station or from the nearest highway influences the noise pollution level since there are different noisy elements that characterize these places, e.g., the trains' passage or the announcements for the trains' information or, in the case of highway proximity, the traffic congestion. It is evident that the variable "road, private and public vehicular traffic (buses) level of the building area" is strongly linked to the noise pollution level recorded for each property, as the vehicular traffic is one of the most relevant noise sources. Finally, it is expected that the maintenance conditions of the property have an impact on selling prices, as an "excellent" state presupposes recent refurbishment initiatives, e.g., the realization of acoustic insulation interventions. Contrarily, the "to be restructured" state is reserved for properties that are characterized by bad maintenance conditions and require significant refurbishment interventions, among which acoustic improvement operations are included.

The data, i.e., the values of the selected explanatory variables relating to each property of the four study samples, were subsequently homogenized through a normalization operation in order to obtain numerical terms of the same size order and to facilitate the comparison between them.

In Table 1, a summary of the variables considered in the analysis is shown. The acronym, the type (dummy or cardinal or discrete) and the measurement unit for each variable is specified.

Table 1. Summary of the variables selected in the analysis.

Variable	Acronym	Typology	Measurement Unit
Total selling price	Pr	Cardinal	€
Size	S	Cardinal	m^2
Bathroom number	B	Cardinal	number
Floor level on which the property is located	L	Cardinal	number
Presence of the lift in the building where the property is located	A	Dummy	1—presence, 0—absence
Presence of the parking space in the building where the property is located	P	Dummy	1—presence, 0—absence
Quality of the property maintenance conditions	Sc	Discrete	1—"to be restructured" property, 3—"fit for habitation" property, 5—"restructured" property.
Property distance from the Araldo di Crollalanza waterfront of the city of Bari	Dl	Cardinal	kilometers by walking
Maintenance conditions of the public spaces adjacent to the property	Sa	Discrete	1—bad maintenance, 3—good state, 5—excellent state.
Property distance from the nearest food market or grocery shop	Dm	Cardinal	Kilometers by walking
Maintenance conditions of the building facades adjacent to the property	Sf	Discrete	1—bad maintenance, 3—good state, 5—excellent state.
Road, private and public vehicular traffic (buses) level of the building area	T	Discrete	1—high traffic intensity, 3—medium traffic intensity, 5—low traffic congestion
Property distance from the nearest public green space	Dv	Cardinal	Kilometers by walking
Distance from the nearest highway	Dt	Cardinal	Kilometers by car
Distance from the nearest railway station	Ds	Cardinal	Kilometers by walking
Perceived environmental quality level of the property area	Qn	Discrete	Scores scale from 1—disagreement with the item to 5—agreement with the item
Sound level in the street where the property is located	Ld	Discrete	Rating 1: ≤40 dB(A); Rating 2: >40 dB(A) and ≤50 dB(A); Rating 3: >50 dB(A) and ≤55 dB(A); Rating 4: >55 dB(A) and ≤60 dB(A); Rating 5: >60 dB(A) and ≤65 dB(A); Rating 6: >65 dB(A) and ≤70 dB(A); Rating 7: >70 dB(A) and ≤75 dB(A).

Tables S1–S4 in Supplementary Materials report the main descriptive statistics of the selling prices and values of the explanatory variables for the four municipal OMI areas of the city of Bari. The analysis has allowed a global reading of the phenomenon through the chosen samples. For each characteristic, the recorded average value, the standard deviation, the different levels or intervals that each variable can be divided into and the frequencies with which each is verified for the properties of the study samples are indicated.

5. Methodology

The methodologic approach used is the Evolutionary Polynomial Regression (EPR). In particular, a recent version of EPR [60], called EPR-MOGA, was implemented. The EPR-MOGA applies a multi-objective evolutionary genetic algorithm as an optimization strategy based on the Pareto frontier and it allows to simultaneously reach different objective functions, such as to define an optimal Pareto frontier of the fixed conflictual objectives that provide for (i) the maximization of the statistical model accuracy; (ii) the maximization of the model parsimony, by means the minimization of the number of coefficients (a_i) of the equation; (iii) the minimization of the number of explanatory variables (X_j) for reducing the complexity of the model.

From the starting database that includes the set of variables detected according to the data available, the technique allows for extrapolating the selling price function, making explicit the factors that influence the values. In particular, EPR-MOGA is a hybrid data-driven technique that combines numerical and symbolic regression methods through polynomial structures [61,62]. The implementation procedure consists of two main phases. In the first step, the search is performed to identify the model structure by generating a set of polynomial expressions. In the second one, the classical (numerical) regression method is used in order to estimate the polynomial coefficient values. In this sense, the main goal of EPR-MOGA is to find the best price function as a combination of independent variable vectors by implementing a regression with the least squares method aimed at generating the value of the coefficients of each variable, elevated by the proper exponents [63,64].

The generic expression of the polynomial equation generated by EPR is represented by Equation (1), which describes a generic nonlinear model structure:

$$Y = a_0 + \sum_{i=1}^{n} [a_i \cdot (X_1)^{(i,1)} \cdot \ldots \cdot (X_j)^{(i,j)} \cdot f((X_1)^{(i,j+1)} \cdot \ldots \cdot (X_j)^{(i,2j)})] \quad (1)$$

Each equation element is explained below: n is the number of additive terms, a_0 represents the constant additive term, a_i are numerical coefficients to be assessed, X_j are candidate explanatory variables, (i, l)—with $l = (1, \ldots, 2j)$—is the exponent of the l-th input within the i-th term in Equation (1), f is a function selected by the user. The exponents (i, l) are also selected by the user from a set of real numbers.

The EPR-MOGA technique allows to generate a set of models; the quantity and the complexity of the solutions depend on the maximum terms number and on the possible exponents defined by the user.

Each polynomial expression generated by the EPR-MOGA technique is characterized by a specific statistical performance level. In this sense, the statistical accuracy of each model returned following the EPR-MOGA implementation is checked through its Coefficient of Determination (COD), which ranges between 0 and 1, reported in Equation (2):

$$COD = 1 - \frac{N-1}{N} \cdot \frac{\sum_N (y_{estimated} - y_{detected})^2}{\sum_N (y_{detected} - mean(y_{detected}))^2} \quad (2)$$

where $y_{estimated}$ are the values of the dependent variable assessed by the method, $y_{detected}$ are the collected values of the dependent variable, N is the sample size in analysis.

The model statistical accuracy is higher when the COD is close to the value 100% and, therefore, the user will choose a model characterized by a high COD level. In this regard, it should be outlined that the user choice is linked not only to the statistical performance of the model, but also to the empirical evidence of the functional relationships between explanatory variables (independent variables) and the selling prices (dependent variables).

Application of the Method

With reference to the four samples related to the four municipal areas in which the city of Bari is divided according to OMI, in the present research the EPR-MOGA technique implemented the generic model mathematical structure shown in Equation (1) with no function f considered, so that the technique chose the best one and the algebraic expression was not excessively complicated.

Taking into account several study outputs [65,66], Pr (the dependent variable) is represented by the natural logarithm of the total selling price (Y = ln(Pr)).

The log-linear model was used as it is characterized by many attractive advantages [67,68]. Moreover, the log-linear form partially mitigates a common form of heteroskedasticity.

Each additive monomial term of the polynomial expressions generated by EPR-MOGA is a combination of the selected explanatory variables (inputs of the model) raised to the proper exponents.

For each EPR-MOGA application, the maximum number *n* of additive terms is equal to eight in order to avoid a high complexity of model functional form and to allow an immediate interpretation of the results. It is evident that a larger number of terms in the equation complicates the analysis' functional correlations and the verification of the empirical evidence with the expected outputs. The candidate exponents belong to the range (0; 0.5; 1; 2) in order to have a sufficiently wide set of solutions able to describe the phenomenon of the formation of selling prices and, at the same time, to avoid an excessive complexity of the mathematical forms. In fact, the exponent equal to zero value nullifies the variable influence and the exponent equal to unit value corresponds to the "pure" explanatory variable, whereas the exponents equal to 0.5 and 2 allow for considering functional quadratic-type forms.

The implementation of the EPR-MOGA technique was carried out four times (one for each study sample) including the same initial conditions (maximum number of additive terms, range candidate exponents) in order to obtain a model for each municipal OMI area of the city of Bari and to compare the outputs.

For each elaboration of EPR-MOGA, several models were generated. Table 2 shows the four equations selected (a model for the OMI central area, one for the semi-central area, one for the peripheral area and, finally, one for the suburban area). For each of them, the statistical accuracy of the outputs in terms of COD is indicated in Table 2.

Table 2. Models generated by EPR-MOGA implementation for each municipal OMI areas of the city of Bari.

Municipal OMI Area	Model	COD (%)
Central	$- 2.1364 \cdot Dl^{0.5} + 2.0268 \cdot Sc \cdot Dl^{0.5} \cdot Ld + 3.2642 \cdot A \cdot Ds +$ $+ 1.555 \cdot L^{0.5} + 4.4836 \cdot S^{0.5} - 6.9297 \cdot S^{0.5} \cdot L^{0.5} \cdot Ds^{0.5} \cdot Ld^{0.5} + 9.4094$	86.73
Semi-central	$+ 33.7673 \cdot Sc^{0.5} \cdot Dl^{2} \cdot Ds^{0.5} \cdot Qn^{2} + 0.34205 \cdot P^{0.5} \cdot Sf^{0.5} -$ $+ 28.409 \cdot P^{0.5} \cdot Dl^{2} \cdot Dv^{0.5} \cdot Qn^{2} \cdot Ld^{0.5} + 2.0779 \cdot L^{0.5} -$ $+ 2.0641 \cdot L \cdot Ld^{0.5} + 0.56217 \cdot B^{0.5} \cdot A^{2} \cdot Sc^{2} \cdot T^{0.5} + 3.6085 \cdot S^{0.5} -$ $+ 4.9704 \cdot S^{2} \cdot B \cdot Sc \cdot Dl^{0.5} \cdot Dt^{0.5} + 9.2473$	83.15
Peripheral	$+ 2.2214 \cdot Dv^{0.5} \cdot Ds \cdot Qn + 0.54754 \cdot Sc^{0.5} \cdot Sa^{0.5} + 0.83222 \cdot P^{0.5} \cdot$ $\cdot Ds + 0.18301 \cdot A^{0.5} + 4.5316 \cdot L^{0.5} \cdot Sf^{0.5} \cdot Dt^{0.5} \cdot Qn \cdot Ld^{2} -$ $+ 2.5508 \cdot L^{0.5} \cdot P^{0.5} \cdot Sf^{0.5} \cdot Dt + 5.2405 \cdot S^{0.5} - 2.0766 \cdot S \cdot Ld + 8.6718$	83.14
Suburban	$+ 1.6114 \cdot Dl^{0.5} + 2.1545 \cdot Dl \cdot Dv + 1.8803 \cdot Sc \cdot Dv + 0.19514 \cdot$ $A^{0.5} + 8.3442 \cdot S^{0.5} - 5.2521 \cdot S + 7.5065$	81.71

It should be noted that each model is characterized by a high statistical accuracy level (COD = +86.73% for the OMI central area model, +83.15% for the semi-central one, +83.14% for the OMI peripheral area equation and +81.71% for the model related to the suburban area) and includes a large number of influential factors (seven for the central area study sample, fourteen for the semi-central area sample, twelve for the peripheral one, five for the suburban area study sample).

Furthermore, for all models the algebraic structure of the equations, i.e., the complexity of the terms that compose them, does not allow an immediate interpretation of the functional relationships among the variables. In fact, each term of the equations is a combination of different variables that occur several times in the same model.

Therefore, the functional correlations of the *i*-th independent explanatory variable with the variation of the selling prices were explicated by means of a simplified approach that considers the other variable values equal to the average values of the starting database (value 1 or 0 if the factor is a dummy variable). Moreover, the approach provides the analysis of the changes in value of the assessed variations of selling prices in correspondence of each *i*-th variable in the admissible range of its corresponding sample values.

In particular, for each municipal OMI area Tables 3–6 show the functional relationships typology (direct or negative) found for each explanatory variable selected by the four models. For each variable, the marginal influence on selling prices in terms of average percentage variation obtained in correspondence of the passage from a level to the next one is reported. From the analysis of functional correlations between influential factors selected by the models and market prices, the empirical evidence of the coefficient signs is verified for all municipal OMI areas of the city of Bari. Thus, for each explanatory variable the consistency of the links with the expected phenomena is verified.

Table 3. Total selling price average percentage variation for the variables chosen by the model for the OMI central area of the city of Bari.

Variable Denomination	Variable Acronym	Functional Correlation	Average Percentage Variation (%)
Distance from the waterfront	Dl	negative	−3.56
Maintenance conditions	Sc	direct	20.22
Sound level in the street where the property is located	Ld	negative	−3.31
Lift	A	direct	30.95
Distance from the nearest railway station	Ds	negative	−3.01
Floor level	L	direct	2.67
Size	S	direct	22.80

Table 4. Total selling price average percentage variation for the variables chosen by the model for the OMI semi-central area of the city of Bari.

Variable Denomination	Variable Acronym	Functional Correlation	Average Percentage Variation (%)
Maintenance conditions	Sc	direct	19.49
Distance from the waterfront	Dl	direct	0.63
Distance from the nearest railway station	Ds	direct	2.48
Perceived environmental quality level	Qn	direct	0.52
Presence of parking	P	direct	4.3
Maintenance conditions of the building facades adjacent to the property	Sf	direct	9.01
Distance from the nearest public green space	Dv	negative	−1.14
Sound level in the street where the property is located	Ld	negative	−3.46
Floor level	L	parabolic	0.77
Bathroom number	B	direct	0.09
Lift	A	direct	8.85
Road private and public vehicular traffic level	T	direct	3.08
Size	S	direct	19.91
Distance from the nearest highway	Dt	negative	−1.77

Table 5. Total selling price average percentage variation for the variables chosen by the model for the OMI peripheral area of the city of Bari.

Variable Denomination	Variables Acronym	Functional Correlation	Average Percentage Variation (%)
Distance from the nearest public green space	Dv	direct	3.73
Distance from the nearest railway station	Ds	direct	7.11
Perceived environmental quality level	Qn	direct	4.73
Maintenance conditions	Sc	direct	8.69
Maintenance conditions of the public spaces adjacent to the property	Sa	direct	11.26
Presence of the parking	P	direct	3.59
Lift	A	direct	16.72
Floor level	L	direct	1.38
Maintenance conditions of the building facades adjacent to the property	Sf	direct	2.52
Distance from the nearest highway	Dt	parabolic	−2.85
Sound level in the street where the property is located	Ld	negative	−2.47
Size	S	direct	21.57

Table 6. Total selling prices average percentage variation for the variables chosen by the model for the OMI suburban area of the city of Bari.

Variable Denomination	Variables Acronym	Functional Correlation	Average Percentage Variation (%)
Distance from the waterfront	Dl	direct	5.52
Distance from the nearest public green space	Dv	direct	1.2
Maintenance conditions	Sc	direct	17.74
Lift	A	direct	17.73
Size	S	direct	0.88

It should be pointed out that for the suburban area the variable Ld was not selected by the model as an influencing factor on selling prices. Therefore, the following considerations related to the noise pollution incidence on property prices exclusively refer to the remaining three OMI areas (central, semi-central, peripheral) of the city of Bari.

Thus, with reference to the sound level detected in the street where the property is located (Ld), the graphs in Figures 7–9 show a negative correlation between the unit selling prices and the Ld label for the central, semi-central and peripheral municipal areas. Furthermore, in each Figure the percentage variations of the total selling prices for the noise pollution class rating for the three OMI areas of the city of Bari are reported.

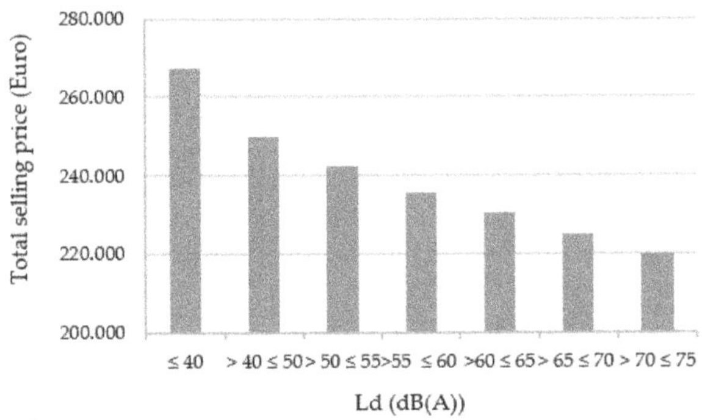

Rating	Percentage variations of the total selling prices
≤ 40	
>40 - ≤50	-6.88%
>50 - ≤55	-3.15%
>55 - ≤60	-2.87%
>60 - ≤65	-2.26%
>65 - ≤70	-2.45%
>70 - ≤75	-2.25%

Figure 7. Functional relationship between the Ld label and the prices and percentage variations for the noise pollution class rating for the OMI central area of the city of Bari.

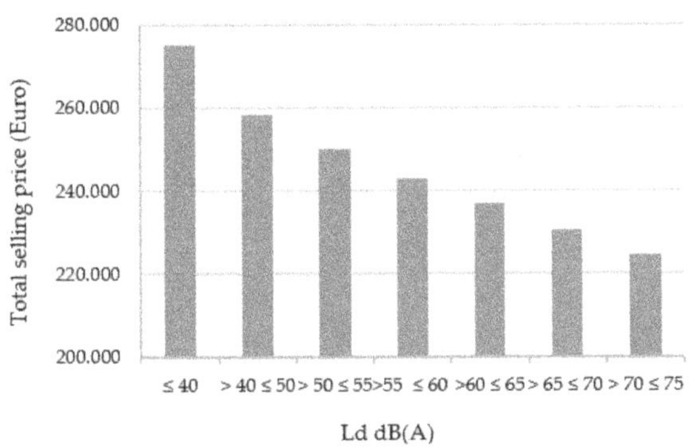

Rating	Percentage variations of the total selling prices
≤ 40	
>40 - ≤50	-6.51%
>50 - ≤55	-3.19%
>55 - ≤60	-3.04%
>60 - ≤65	-2.49%
>65 - ≤70	-2.81%
>70 - ≤75	-2.70%

Figure 8. Functional relationship between the Ld label and the prices and percentage variations for the noise pollution class rating for the OMI semi-central area of the city of Bari.

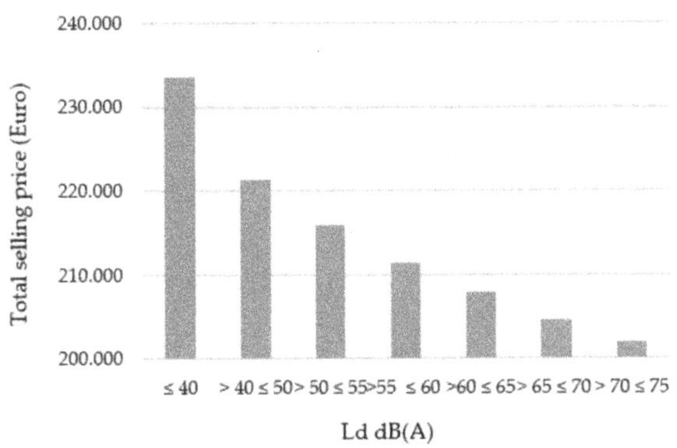

Rating	Percentage variations of the total selling prices
≤ 40	
>40 - ≤50	-5.49%
>50 - ≤55	-2.49%
>55 - ≤60	-2.19%
>60 - ≤65	-1.64%
>65 - ≤70	-1.64%
>70 - ≤75	-1.34%

Figure 9. Functional relationship between the Ld label and the prices and percentage variations for the noise pollution class rating for the OMI peripheral area of the city of Bari.

With regard to the aim of the study, the results obtained for the variable Ld—the sound level detected in the street where the property is located—allow for obtaining interesting considerations.

Firstly, as already mentioned, the variable Ld is included in the models related to the OMI central area, semi-central and peripheral municipal areas of the city of Bari, whereas it is excluded from the variables selected by the EPR-MOGA technique in the OMI suburban area.

This result confirms an expected phenomenon: the central and semi-central municipal OMI areas are affected by a higher noise pollution level compared to the peripheral and suburban trade areas, for which a lower noise pollution is detected. The Strategic Noise Map of the city of Bari, reported in Figure 6, attests that, excluding the airport area characterized by a relevant noise pollution level, the other portions of the peripheral and suburban of city of Bari show a sound level (Ld) between <40 dB(A) and 55dB(A). In this sense, the central and semi-central municipal areas are characterized by a larger number of leisure and commercial activities that cause higher loudness level, especially in the evening hours, instead of peripheral and suburban areas in which there are a greater amount of green or empty urban spaces that determine lower noise pollution levels.

With reference to the outputs obtained, the average percentage decrease in the selling price corresponding to a variation from one Ld level to the next one is almost constant for the OMI central area (−3.31%) and for the OMI semi-central area (−3.46%) of the city of Bari. Conversely, for the peripheral area, the average percentage decrease in the housing price corresponding to a passage from a Ld level to the next one is lower and equal to −2.47%. This confirms the lower influence given by the factor related to the noise pollution on selling prices in the municipal OMI area in which the pollution is minor, due to a less levels of road traffic and industrial traffic and less of a preference for recreational activities. For the suburban municipal area, the model generated by the EPR-MOGA technique implementation does not include the variable Ld among the influencing factors on selling prices. Thus, it is evident that a lower noise pollution level (average sound level Ld for the central area is equal to 64.8 dB(A), whereas the sound level Ld for the suburban area is equal to 57.02 dB(A)) corresponds to a less significant effect of the acoustic factor in the selling price formation.

For each study sample, in the Table 7 the percentage variations in the total selling residential prices, from the best situation—i.e., rating 1 (<40 dB(A))—to the worst one—rating 7 (>70 and ≤75 dB(A))—are reported.

Table 7. Percentage variation in the market prices from rating 1 to rating 7 for the four municipal OMI areas of the city of Bari.

Municipal OMI Area	Average Percentage Variation from Rating 1 to Rating 7 (%)
Central	−21.50
Semi-central	−22.56
Peripheral	−15.67
Suburban	-

Finally, the average unitary contribution of the explanatory variable Ld on the housing prices for the four OMI areas of the city of Bari and in percentage terms is shown in Table 8.

Table 8. Average unitary influence of the variable Ld for the four municipal OMI areas of the city of Bari.

Municipal OMI Area	Average Unitary Contribution of Ld (%)
Central	−0.56
Semi-central	−0.58
Peripheral	−0.42
Suburban	-

6. Conclusions

The increase of public and private road traffic, loud recreation activities and industrial activities located in residential areas in European and U.S. cities has led to a relevant growth of the noise pollution level. In particular, the road networks cover entire cities, undergoing and connecting the residential, commercial and directional districts. It is noted that the road noise is one of the greatest sources of discomfort [69] because an increasing congestion level characterizes the traffic on the city street. Therefore, currently the environmental aspects play a fundamental role in the urban policies aimed to define strategies for the reduction of noise pollution and, at the same time, for the development of more sustainable cities characterized by a high urban quality level [70–72]. In this sense, the noise levels may directly and strongly influence the market residential prices.

With reference to four study samples composed of two hundred residential properties sold in 2017–2019 and located in the four municipal OMI areas in which the city of Bari (Southern Italy) is divided (central, semi-central, peripheral and suburban), the aim of the present paper was double: (i) to investigate the presence and, eventually, the influence of noise among the most influencing factors on property prices and (ii) to compare the contribution of acoustic components on selling prices in the four municipal OMI areas according to the different features of each in terms of the traffic congestion level and presence of loud activities. For each study sample, the total market prices and the relevant influential factors for their formation were detected. An econometric technique was implemented for determining the contribution of each factor selected and for analyzing the influence of noise pollution factor on property prices. For each municipal OMI area, a model was selected among those generated by the technique, and the functional correlations between the influential factors and the market prices were analyzed. With reference to the acoustic component, the models for the central, semi-central and peripheral areas included the variable related to the sound level calculated in the street where the property was located, whereas for the suburban area this variable was not selected.

Furthermore, for the central, semi-central and peripheral areas, a negative link between the noise level and the housing prices was noted, which confirms that a building's location characterized by a noise high level causes a reduction in residential market values. In particular, the average percentage decrease of prices given by a high noise pollution level was similar for the central and OMI semi-central areas of the city of Bari, and it was lower for the peripheral one. Thus, the results of the study confirm that the impact of noise on the decisions made by real estate buyers is less relevant in the urban area where the acoustic pollution is weaker.

This analysis could represent a useful tool for public administration to orient specific urban interventions aimed to create acoustic comfort in urban spaces and to promote measures for the reduction of the noise pollution, e.g., introduction of limited traffic or pedestrian areas, planning of ecological days, provision of public services for sustainable and slow mobility and diffusion of bike sharing systems.

The present research has intended to focalize the relevance of the issue. In recent years a higher attention has been focused on the housing aspects able to increase the well-being of inhabitants. Furthermore, strong awareness-raising actions and information campaigns have been developed in order to promote the realization of domestic spaces characterized by high quality levels. In this sense, the main three topics sponsored concern energy saving, seismic safety and noise reduction. The definition of incentives—also in fiscal and economic terms—aimed at stimulating the housing comfort improvement contributes to the awareness growth toward residential property factors redevelopment and ex-novo realization. In the Italian context, the variables associated with the acoustic component vary among different urban territories because of urban morphology (e.g., building typology, density of the property asset, houses proximity, green barriers, etc.). In particular, with reference to the present analysis, the factors considered were selected taking into account the indications of the local market operators that reported the acoustic factor as one of the most influencing in the buyers' decision process. This attests to the prominence of domestic space comfort aspects in the current specific market of the city of Bari. It should be added that the Covid-19 pandemic lockdown has raised the attention for indoor noise even more due to (i) the smart working spread, (ii) the functional adaptation of housing spaces for a longer duration of home living, (iii) the temporary reduction of external noise (cars, trains, airplanes, people, etc.). Therefore, this contingence has determined a higher appreciation for specific residential property factors that probably in the past were not so influential.

Further insight into the research may address the implementation of the econometric technique to other national or international territorial contexts in order to analyze the influence of noise pollution on selling prices in different urban areas and to compare the results.

Moreover, the study may be developed through the investigation of the relationship between the environmental factors related to the acoustic component and the floor level on which the property is located. In this sense, the floor level could have a significant moderating effect on the external effect of road traffic and, at the same time, the sound pressure deriving from the road traffic could drop significantly with the increase of floor level [73]. It could be interesting to implement the econometric technique using the noise level as a dependent variable and the floor level as one of the factors that influences its value in order to define a direct functional link between the two characteristics.

Supplementary Materials: The following are available online at https://www.mdpi.com/article/10.3390/buildings11050213/s1, Table S1. Descriptive statistics of the variables for the municipal OMI central area; Table S2. Descriptive statistics of the variables for the municipal OMI semi-central area; Table S3. Descriptive statistics of the variables for the municipal OMI peripheral area; Table S4. Descriptive statistics of the variables for the municipal OMI suburban area.

Author Contributions: The contribution is the result of the joint work of the authors. In particular: conceptualization, P.M., F.T., M.D.; methodology, P.M., F.D.L., F.T.; validation, P.M.; formal analysis, F.T.; investigation, M.D.; data curation, F.D.L.; writing—review and editing, F.D.L., F.T.; supervision, P.M. All authors have read and agreed to the published version of the manuscript.

Funding: This research received no external funding.

Institutional Review Board Statement: Not applicable.

Informed Consent Statement: Not applicable.

Data Availability Statement: The data is contained within the article.

Conflicts of Interest: The authors declare no conflict of interest.

References

1. European Commission. Directive 2002/49/EC of the European Parliament and of the Council of 25 June 2002 Relating to the Assessment and Management of Environmental Noise-Declaration by the Commission in the Conciliation Committee on the Directive Relating to the Assessment and Management of Environmental Noise. *Off. J. Eur. Communities* **2002**, *189*, 12–25.
2. Ozdemir, B.; Bayramoglu, E.; Demirel, O. Noise Pollution and Human Health in Trabzon Parks. *Stud. Ethno Med.* **2014**, *8*, 127–134. [CrossRef]
3. Önder, S.; Akay, A. Reduction of Traffic Noise Pollution Effects by Using Vegetation, Turkey'Sample. *J. Eng. Econ. Dev.* **2015**, *2*, 23.
4. Oroke, A.M.; Sarihan, E.; Ogbuene, E.B.; Ani, C. Assessment of Noise Pollution Level in Abakaliki Metropolis, Ebonyi State Nigeria. *Theor. Empir. Res. Urban Manag.* **2020**, *15*, 75–83.
5. World Health Organization. *Burden of Disease from Environmental Noise: Quantification of Healthy Life Years Lost in Europe*; World Health Organization, Regional Office for Europe: Copenhagen, Denmark, 2011.
6. Bello, J.P.; Silva, C.; Nov, O.; DuBois, R.L.; Arora, A.; Salamon, J.; Doraiswamy, H. SONYC: A system for the monitoring, analysis and mitigation of urban noise pollution. *Commun. ACM* **2019**, *62*, 68–77. [CrossRef]
7. Maibach, M.; Schreyer, C.; Sutter, D.; Van Essen, H.P.; Boon, B.H.; Smokers, R.; Bak, M. Report Produced within the Study Internalisation Measures and Policies for All External Cost of Transport (IMPACT). In *Handbook on Estimation of External Costs in the Transport Sector*; Version 1.1; European Commission DG TREN: Delft, CE, The Netherlands, 2008; Available online: http://www.europadecentraal.nl/documents/dossiers/Transport/2008_01_15_handbook_external_cost_en.pdf (accessed on 10 March 2020).
8. Mizutani, F.; Suzuki, Y.; Sakai, H. Estimation of Social Costs of Transport in Japan. *Urban Stud.* **2011**, *48*, 3537–3559. [CrossRef]

9. European Environment Agency Number of Europeans Exposed to Harmful Noise Pollution Expected to Increase. 2020. Available online: www.eea.europa.eu (accessed on 27 February 2020).
10. World Health Organization. Environmental Noise Guidelines for the European Region. 2018. Available online: www.euro.who.int (accessed on 15 March 2020).
11. UN. Transforming our World: The 2030 Agenda for Sustainable Development. Available online: https://www.un.org/ga/search/view_doc.asp?symbol=A/RES/70/1&Lang=E (accessed on 17 July 2020).
12. Agenzia Delle Entrate. Available online: http://www.agenziaentrate.gov.it/ (accessed on 25 January 2020).
13. Jariwala, H.J.; Syed, H.S.; Pandya, M.J.; Gajera, Y.M. Noise Pollution & Human Health: A Review. *Indoor Built Environ.* **2017**, 1–4. Available online: https://www.researchgate.net/profile/Hiral-Jariwala/publication/319329633_Noise_Pollution_Human_Health_A_Review/links/59a54434a6fdcc773a3b1c49/Noise-Pollution-Human-Health-A-Review.pdf (accessed on 10 April 2020).
14. Aluko, E.; Nna, V. Impact of Noise Pollution on Human Cardiovascular System. *Int. J. Trop. Dis. Health* **2015**, *6*, 35–43. [CrossRef]
15. Ozdenerol, E.; Huang, Y.; Javadnejad, F.; Antipova, A. The Impact of Traffic Noise on Housing Values. *J. Real Estate Pract. Educ.* **2015**, *18*, 35–54. [CrossRef]
16. Franck, M.; Eyckmans, J.; De Jaeger, S.; Rousseau, S. Comparing the impact of road noise on property prices in two separated markets. *J. Environ. Econ. Policy* **2015**, *4*, 15–44. [CrossRef]
17. Beimer, W.; Maennig, W. Noise effects and real estate prices: A simultaneous analysis of different noise sources. *Transp. Res. Part D Transp. Environ.* **2017**, *54*, 282–286. [CrossRef]
18. Li, W.; Saphores, J.D. Assessing impacts of freeway truck traffic on residential property values: Southern California Case Study. *Transp. Res. Rec.* **2012**, *2288*, 48–56. [CrossRef]
19. Guijarro, F. Assessing the Impact of Road Traffic Externalities on Residential Price Values: A Case Study in Madrid, Spain. *Int. J. Environ. Res. Public Health* **2019**, *16*, 5149. [CrossRef] [PubMed]
20. Hogan, C.M. Analysis of highway noise. *Water Air Soil Pollut.* **1973**, *2*, 387–392. [CrossRef]
21. Swoboda, A.; Nega, T.; Timm, M. Hedonic analysis over time and space: The case of house prices and traffic noise. *J. Reg. Sci.* **2015**, *55*, 644–670. [CrossRef]
22. Mínguez, R.; Montero, J.-M.; Fernández-Avilés, G. Measuring the impact of pollution on property prices in Madrid: Objective versus subjective pollution indicators in spatial models. *J. Geogr. Syst.* **2013**, *15*, 169–191. [CrossRef]
23. Del Giudice, V.; De Paola, P. The Effects of Noise Pollution Produced by Road Traffic of Naples Beltway on Residential Real Estate Values. *Appl. Mech. Mater.* **2014**, *587–589*, 2176–2182. [CrossRef]
24. Amrusch, P. The impact of urban traffic and environmental conditions on the housing market: An analysis of Italian and Slovenian urban areas. *WIT Trans. Built Environ.* **2005**, *77*. [CrossRef]
25. Ottoz, E.; Pavese, P.; Sella, L. The effect of Movida on residential property prices: An example from Turin. In *Proceedings of the Pedestrians, Urban Spaces and Health*; Apple Academic Press: Palm Bay, FL, USA, 2020; pp. 170–174.
26. D'Arcangelo, F.M.; Percoco, M. Housing rent and road pricing in Milan: Evidence from a geographical discontinuity approach. *Transp. Policy* **2015**, *44*, 108–116. [CrossRef]
27. Blanco, J.C.; Flindell, I. Property prices in urban areas affected by road traffic noise. *Appl. Acoust.* **2011**, *72*, 133–141. [CrossRef]
28. Paunović, K.; Belojević, G.; JakovljeviĆ, B. Noise annoyance is related to the presence of urban public transport. *Sci. Total. Environ.* **2014**, *481*, 479–487. [CrossRef] [PubMed]
29. Chang, J.S.; Kim, D.-J. Hedonic estimates of rail noise in Seoul. *Transp. Res. Part D Transp. Environ.* **2013**, *19*, 1–4. [CrossRef]
30. Bureau, B.; Glachant, M. Évaluation de l'impact des politiques. *Econ. Prévis.* **2010**, *192*, 27–44. [CrossRef]
31. EEA, European Environment Agency. Technical Report No 11/2010 EEA. Good Practice Guide on Noise Exposure and Potential Health Effects. 2010. Available online: http://www.eea.europa.eu/publications/goodpractice-guide-on-noise (accessed on 7 February 2020).
32. Brandt, S.; Maennig, W. Road noise exposure and residential property prices: Evidence from Hamburg. *Transp. Res. Part D Transp. Environ.* **2011**, *16*, 23–30. [CrossRef]
33. Arsenio, E.; Bristow, A.L.; Wardman, M. Stated choice valuations of traffic related noise. *Transp. Res. Part D Transp. Environ.* **2006**, *11*, 15–31. [CrossRef]
34. Bateman, I.; Day, B.; Lake, I.; Lovett, A. *The Effects of Road Traffic on Residential Property Values: A Literature Review and Hedonic Pricing Study*; Technical Report; University of East Anglia: Norwich, UK; Economic & Social Research Council: Swindon, UK; University College London: London, UK, 2001.
35. Garrod, G.D.; Scarpa, R.; Willis, K.G. Estimating The Benefits of Traffic Calming on Through Routes: A Choice Experiment Approach. *SSRN Electron. J.* **2003**, *36*, 211–231. [CrossRef]
36. Navrud, S. The economic value of noise within the European Union—A review and analysis of studies. *Acústica* **2004**, 14–17. Available online: http://citeseerx.ist.psu.edu/viewdoc/download?doi=10.1.1.582.6811&rep=rep1&type=pdf (accessed on 26 April 2020).
37. Nelson, J.P. Highway noise and property values: A survey of recent evidence. *J. Transp. Econ. Policy* **1982**, *16*, 117–138.
38. Nelson, J.P. Meta-analysis of airport noise and hedonic property values. *J. Transp. Econ. Policy* **2004**, *38*, 1–27.
39. Andersson, H.; Jonsson, L.; Ögren, M. Property Prices and Exposure to Multiple Noise Sources: Hedonic Regression with Road and Railway Noise. *Environ. Resour. Econ.* **2009**, *45*, 73–89. [CrossRef]

40. Wilhelmsson, M. The Impact of Traffic Noise on the Values of Single-family Houses. *J. Environ. Plan. Manag.* **2000**, *43*, 799–815. [CrossRef]
41. Rich, J.H.; Nielsen, O.A. Assessment of traffic noise impacts. *Int. J. Environ. Stud.* **2004**, *61*, 19–29. [CrossRef]
42. Duarte, C.M.; Tamez, C.G. Does noise have a stationary impact on residential values? *J. Eur. Real Estate Res.* **2009**, *2*, 259–279. [CrossRef]
43. Wen, H.; Gui, Z.; Zhang, L.; Hui, E.C. An empirical study of the impact of vehicular traffic and floor level on property price. *Habitat Int.* **2020**, *97*, 102132. [CrossRef]
44. Szczepańska, A.; Senetra, A.; Wasilewicz-Pszczółkowska, M. The Influence of Traffic Noise on Apartment Prices on the Example of a European Urban Agglomeration. *Sustainability* **2020**, *12*, 801. [CrossRef]
45. Kuehnel, N.; Moeckel, R. Impact of simulation-based traffic noise on rent prices. *Transp. Res. Part D: Transp. Environ.* **2020**, *78*, 102191. [CrossRef]
46. Szopińska, K.; Krajewska, M. Prices of Apartments in Relation to Noise Level in Poland. *J. Civ. Eng. Arch.* **2013**, *7*, 1189–1195. [CrossRef]
47. Ligus, M.; Peternek, P. Measuring Structural, Location and Environmental Effects: A Hedonic Analysis of Housing Market in Wroclaw, Poland. *Procedia-Soc. Behav. Sci.* **2016**, *220*, 251–260. [CrossRef]
48. Cellmer, R. Spatial analysis of the effect of noise on the prices and value of residential real estates. *Geomat. Environ. Eng.* **2011**, *5*, 13–28.
49. Iovine, A. *La Revisione delle Rendite Delle Unità Immobiliari Urbane*; Sistemi Editoriali: Naples, Italy, 2006; ISBN 9788851303464.
50. Bourassa, S.C.; Hoesli, M.; Peng, V.S. Do housing submarkets really matter? *J. Hous. Econ.* **2003**, *12*, 12–28. [CrossRef]
51. Grether, D.; Mieszkowski, P. Determinants of real estate values. *J. Urban Econ.* **1974**, *1*, 127–145. [CrossRef]
52. Vogiazas, S.; Alexiou, C. Determinants of Housing Prices and Bubble Detection: Evidence from Seven Advanced Economies. *Atl. Econ. J.* **2017**, *45*, 119–131. [CrossRef]
53. Égert, B.; Mihaljek, D. Determinants of House Prices in Central and Eastern Europe. *Comp. Econ. Stud.* **2007**, *49*, 367–388. [CrossRef]
54. Dziauddin, M.F.; Ismail, K.; Othman, Z. Analysing the local geography of the relationship between residential property prices and its determinants. *Bull. Geogr. Socio-Econ. Ser.* **2015**, *28*, 21–35. [CrossRef]
55. Zoppi, C.; Argiolas, M.; Lai, S. Factors influencing the value of houses: Estimates for the city of Cagliari, Italy. *Land Use Policy* **2015**, *42*, 367–380. [CrossRef]
56. Mbachu, J.I.; Lenono, N. Factors influencing market values of residential properties. In Proceedings of the Queensland University of Technology Research Week International Conference, QUT Research Week 2005, Brisbane, QLD, Australia, 4–8 July 2005.
57. Cui, N.; Gu, H.; Shen, T.; Feng, C. The Impact of Micro-Level Influencing Factors on Home Value: A Housing Price-Rent Comparison. *Sustainability* **2018**, *10*, 4343. [CrossRef]
58. Bourassa, S.; Cantoni, E.; Hoesli, M. Predicting House Prices with Spatial Dependence: A Comparison of Alternative Methods. *J. Real Estate Res.* **2010**, *32*, 139–160. [CrossRef]
59. ARPA Puglia. Available online: www.arpa.puglia.it (accessed on 14 December 2019).
60. Giustolisi, O.; Savic, D. Advances in data-driven analyses and modelling using EPR-MOGA. *J. Hydroinform.* **2009**, *11*, 225–236. [CrossRef]
61. Rezania, M.; Javadi, A.A.; Giustolisi, O. An evolutionary-based data mining technique for assessment of civil engineering systems. *Eng. Comput.* **2008**, *25*, 500–517. [CrossRef]
62. Morano, P.; Tajani, F.; Di Liddo, F.; Anelli, D. A Feasibility Analysis of The Refurbishment Investments in The Italian Residential Market. *Sustainability* **2020**, *12*, 2503. [CrossRef]
63. Tajani, F.; Morano, P.; Torre, C.M.; Di Liddo, F. An Analysis of the Influence of Property Tax on Housing Prices in the Apulia Region (Italy). *Buildings* **2017**, *7*, 67. [CrossRef]
64. Morano, P.; Rosato, P.; Tajani, F.; Manganelli, B.; Di Liddo, F. Contextualized Property Market Models vs. Generalized Mass Appraisals: An Innovative Approach. *Sustainability* **2019**, *11*, 4896. [CrossRef]
65. Cassel, E.; Mendelsohn, R. The choice of functional forms for hedonic price equations: Comment. *J. Urban Econ.* **1985**, *18*, 135–142. [CrossRef]
66. Lynch, A.K.; Rasmussen, D.W. Proximity, Neighbourhood and the Efficacy of Exclusion. *Urban Stud.* **2004**, *41*, 285–298. [CrossRef]
67. Malpezzi, S.; Chun, G.H.; Green, R.K. New Place-to-Place Housing Price Indexes for U.S. Metropolitan Areas, and Their Determinants. *Real Estate Econ.* **1998**, *26*, 235–274. [CrossRef]
68. Selim, H. Determinants of house prices in Turkey: Hedonic regression versus artificial neural network. *Expert Syst. Appl.* **2009**, *36*, 2843–2852. [CrossRef]
69. Liu, J.; Kang, J.; Luo, T.; Behm, H.; Coppack, T. Spatiotemporal variability of soundscapes in a multiple functional urban area. *Landsc. Urban Plan.* **2013**, *115*, 1–9. [CrossRef]
70. Guarini, M.R.; D'Addabbo, N.; Morano, P.; Tajani, F. Multi-Criteria Analysis in Compound Decision Processes: The AHP and the Architectural Competition for the Chamber of Deputies in Rome (Italy). *Buildings* **2017**, *7*, 38. [CrossRef]
71. Morano, P.; Tajani, F. The Transfer of Development Rights for the Regeneration of Brownfield Sites. *Appl. Mech. Mater.* **2013**, *409–410*, 971–978. [CrossRef]

72. Morano, P.; Tajani, F. Break Even Analysis for the Financial Verification of Urban Regeneration Projects. *Appl. Mech. Mater.* **2013**, *438–439*, 1830–1835. [CrossRef]
73. Mak, C.M.; Leung, W.; Jiang, G. Measurement and prediction of road traffic noise at different building floor levels in Hong Kong. *Build. Serv. Eng. Res. Technol.* **2010**, *31*, 131–139. [CrossRef]

buildings

MDPI

Article

Deep Learning Optimal Control for a Complex Hybrid Energy Storage System

Gabriel Zsembinszki, Cèsar Fernández, David Vérez and Luisa F. Cabeza *

GREiA Research Group, INSPIRES Research Centre, University of Lleida, 25001 Lleida, Spain; gabriel.zsembinszki@udl.cat (G.Z.); cesar.fernandez@udl.cat (C.F.); david.verez@udl.cat (D.V.)
* Correspondence: luisaf.cabeza@udl.cat

Abstract: Deep Reinforcement Learning (DRL) proved to be successful for solving complex control problems and has become a hot topic in the field of energy systems control, but for the particular case of thermal energy storage (TES) systems, only a few studies have been reported, all of them with a complexity degree of the TES system far below the one of this study. In this paper, we step forward through a DRL architecture able to deal with the complexity of an innovative hybrid energy storage system, devising appropriate high-level control operations (or policies) over its subsystems that result optimal from an energy or monetary point of view. The results show that a DRL policy in the system control can reduce the system operating costs by more than 50%, as compared to a rule-based control (RBC) policy, for cooling supply to a reference residential building in Mediterranean climate during a period of 18 days. Moreover, a robustness analysis was carried out, which showed that, even for large errors in the parameters of the system simulation models corresponding to an error multiplying factors up to 2, the average cost obtained with the original model deviates from the optimum value by less than 3%, demonstrating the robustness of the solution over a wide range of model errors.

Keywords: deep reinforcement learning; optimal control; optimization; HYBUILD; thermal energy storage; residential buildings

check for updates

Citation: Zsembinszki, G.; Fernández, C.; Vérez, D.; Cabeza, L.F. Deep Learning Optimal Control for a Complex Hybrid Energy Storage System. *Buildings* **2021**, *11*, 194. https://doi.org/10.3390/buildings11050194

Academic Editors: Alessandro Cannavale, Francesco Martellotta and Francesco Fiorito

Received: 29 March 2021
Accepted: 29 April 2021
Published: 3 May 2021

Publisher's Note: MDPI stays neutral with regard to jurisdictional claims in published maps and institutional affiliations.

Copyright: © 2021 by the authors. Licensee MDPI, Basel, Switzerland. This article is an open access article distributed under the terms and conditions of the Creative Commons Attribution (CC BY) license (https://creativecommons.org/licenses/by/4.0/).

1. Introduction

As building energy consumption accounts for a large percentage of the total energy consumption, an extensive work on new methods and strategies for more efficient control systems has been done. In this sense, many approaches have been proposed, from classical control theory to reinforcement learning, particularly related to heating, ventilation and air conditioning (HVAC) systems. The availability, ubiquity and performance of current digital systems, as well as their reduced cost, allow to devise control scenarios where many parameters can be easily monitored (i.e., batteries state, instant photovoltaic production, current consumption demand, etc.) and take real-time decisions according to different control techniques, always pursuing some predefined objectives such as lowest operating costs or better efficiencies, among others.

Even though the use of machine learning techniques is relatively recent, model predictive control (MPC) and all its flavors, produced a large number of publications in the field of optimal control for energy storage systems. A complete review on control of storage systems can be found in [1–4] with particular reference to MPC approaches. Even though MPC is able to reach optimal or quasi-optimal solutions, its implementation for complex systems is challenging. Aside from its computational requirements that can difficult a real-time control, MPC optimization problems usually require to be formulated as mixed integer non-linear programming (MINLP) problems [5,6], requiring specialized solvers to find optimal solutions as SCIP [7,8]. Current state-of-the-art solvers only deal with certain type of non-linearities, making it sometimes hard or impossible to express a complex system as a quasi-linear system. Not to mention its difficulty to adapting under uncertainty scenarios, especially arising from model inaccuracies.

In recent years, reinforcement learning (RL) has emerged as an efficient alternative to MPC. RL is based on a mathematical framework for experience-driven autonomous learning [9]. In essence, the learning process is established on a trial-and-error basis, interacting with either the real system or its model. Early RL algorithms, back to the 1980s, proved to solve a wide range of problems in different areas. Q-Learning was one of the most often recurred [10], being firstly used in the field of thermal storage control by Liu and Henze [11,12], and proving experimentally its feasibility. It was in that work where the RL drawbacks for optimal control were mentioned. An inaccurate model may lead to unexpected behavior and, as in all the RL approaches, the curse of dimensionality of the actions-state space arose, putting difficulties in future approaches for complex systems. It was at this point that neural networks came to the rescue, by substituting the time and memory consuming value tables in classical RL schemes and becoming deep reinforcement learning (DRL).

Since its appearance in 2013 [13], DRL has been applied successfully to many complex control problems and, particularly, was first used in HVAC control in [14]. A good review of RL for energy management can be found in [15–17], while [18] made an exhaustive analysis of DRL applications for HVAC systems. For the particular case of thermal energy storage (TES) systems, only a few studies were reported in [15,17]. Actually, [17] only identifies 6 publications related to the control of TES systems, all of them with a complexity degree of the TES system far below the one of this study. The study and experimentation by Liu and Henze [19] were the cornerstones of the application of RL to active and passive TES systems, proving the advantages of hybrid approaches that allow accelerating the learning phase by simulation. Later and distinct uses of RL are found in [20,21] describing the first use, to the authors knowledge, of RL techniques to phase change materials (PCM) storage.

The main contribution in this study is twofold. First, the use of DRL for optimal control under demand response in a complex and innovative system to reduce the energy demand for heating, cooling and domestic hot water of a standard single-family residential building is presented in detail. The system, proposed and developed within the H2020 research project, HYBUILD [22], integrates different subsystems such as photovoltaic (PV) panels, Fresnel solar thermal collector, a sorption chiller connected with a reversible heat pump and electrical and thermal energy storages. The application of DRL for optimal control of such a complex system is the first to the best of the authors knowledge. Second, a robustness analysis of the learning process was performed, showing that the learned model results are useful and accurate even for large deviations between the real and the simulated system, answering one of the open questions reported in [17] and proving that the model presented in this study and evaluated under a simulated scenario may fit the control requirements for the real test pilot plant.

2. Methodology

This section explains the details of the system and the approach used for system modelling and optimization.

2.1. System Description

The system considered in this study (Figure 1) was designed to ensure comfort indoor conditions and domestic hot water (DHW) in residential buildings, and specifically to reduce primary energy consumption of single-family houses located in Mediterranean climate regions. Therefore, the different system components were chosen and sized with the main purpose to meet most of the cooling demand using solar energy. To enhance the energy efficiency of the system and the share of renewable energy, it incorporates four different energy storage technologies: an electric battery connected to PV panels, a low-temperature phase change material (PCM) storage unit connected to the low-pressure side of the heat pump, a sorption chiller connected to the high-pressure side of the heat pump and a buffer water tank that stores the heat produced by the Fresnel solar collectors. The hot water stored in the buffer tank is used to drive the sorption chiller and also to contribute to heating and DHW supply.

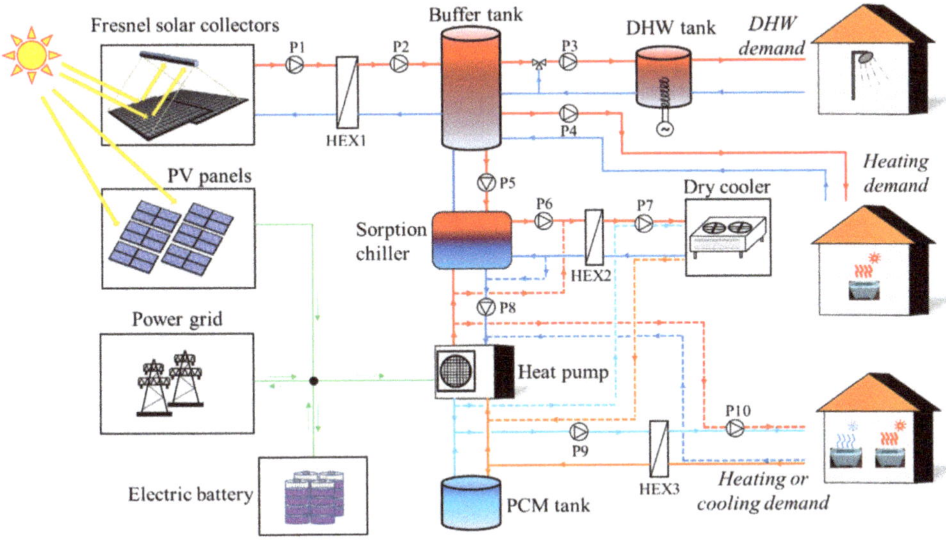

Figure 1. Schematic of the system and its main components.

The heat pump is fed with DC current by means of a DC-bus, which interconnects the PV panels, electric battery, heat pump and (by means of an AC/DC inverter) the power grid. Even though this type of connection enhances the complexity of the system, it also gives a high flexibility and improves the efficiency of the system by reducing the number of multiple stages of conversions from DC to AC and vice-versa.

To obtain the energy demand of the building, a single-family residential building located in Athens was considered as a reference building for Mediterranean climate regions. The building has a total surface of 100 m^2 distributed in two floors, each having a living surface area of 50 m^2, and it was assumed to be inhabited by four people. The ceiling/floor heights considered were 2.5 m/3.0 m, while the building width/depth were 6.5 m/8.0 m. The glazing ratio considered was of 20% on the south side, 10% on the north side and 12% on the east and west sides. The energy demand profile for cooling, heating and DHW of the building were obtained within the HYBUILD project [23] activities and it is out of the scope of this paper to present the details of energy demand calculations.

2.2. Components Models and Operating Modes Description

This subsection presents the main system components mathematical models along with the associated operating modes, which were implemented in the control strategies developed for the system.

2.2.1. Fresnel Collectors

The Fresnel collectors consist of flat mirrors that can rotate around a fixed horizontal axis oriented along the north-south direction. There are only two possible operating modes of this component: mode 1 (on) and mode 2 (off). In mode 1, the orientation of the mirrors is set by a controller in such a way that they focus the incident solar radiation to the receiver that is located on top of the mirrors to heat the water in the primary circuit up to 100 °C. The heat is transferred to the buffer tank by means of a heat exchanger (HEX1) and two circulation pumps (P1 and P2) installed in the primary and secondary circuits, respectively.

No heat losses were considered in HEX1 for simplicity. In mode 2, no heat is harvested by the solar collectors and pumps P1 and P2 are switched off.

When operating in mode 1, the thermal power generated by the Fresnel collectors ($\dot{Q}_{solar}$, in kW) is given by Equation (1), otherwise $\dot{Q}_{solar} = 0$.

$$\dot{Q}_{solar} = \left[\eta_{opt} \cdot \eta_{clean} \cdot DNI - \left(4.8703 - 0.0981 \cdot T_m + 9 \cdot 10^{-4} \cdot T_m^2\right) - \left(\frac{(T_m - T_{amb}) - 80}{4} + 29.043 + 1.0983 \cdot v_w + 0.4188 \cdot v_w^2 + 4 \cdot 10^{-5} \cdot v_w^3\right)\right] \cdot A_{Fres}/1000, \quad (1)$$

where η_{opt} is the optical efficiency of the receiver, $\eta_{clean} = 1$ is the mirror cleanness factor, DNI (in W/m^2) is the direct normal irradiance at the specified location [24], $T_m = 95$ °C is the mean receiver temperature, T_{amb} (in °C) is the ambient air temperature [24], v_w (in m/s) is the wind speed [24] and $A_{Fres} = 60$ m^2 is the total surface area of the solar collectors.

The values for the optical efficiency of the receiver (η_{opt}) depend on the month of the year and on the geographic coordinates of the location, and were provided by the manufacturer within HYBUILD activities [22].

The overall electricity consumption of Fresnel collectors is the sum of the consumption of the circulation pumps P1 (34 W) and P2 (34 W), when the Fresnel collectors operate in mode 1, otherwise the electricity consumption of this component is zero.

2.2.2. PV Panels

The PV panels were assumed to face south and have a tilt angle of 30° with respect to the horizontal plane. The net power generated by the PV panels (PV, in kW) is given by Equation (2):

$$PV = \eta_{PV} \cdot E_{POA} \cdot A_{PV}, \quad (2)$$

where $\eta_{PV} = 0.16$ [25] is the efficiency of the PV system, E_{POA} (in W/m^2) is the plan of array (POA) irradiance at the specified location and $A_{PV} = 20.9$ m^2 is the PV panels surface area. The efficiency of auxiliary components related to the PV system (DC/DC converter, connections, etc.) was assumed to be accounted for in η_{PV}.

The value of the solar irradiance incident to the PV surface (E_{POA}, in W/m^2) is the sum of three contributions, as shown in Equation (3):

$$E_{POA} = E_b + E_g + E_d, \quad (3)$$

where E_b (in W/m^2) is the POA beam component, E_g (in W/m^2) is the POA ground-reflected component and E_d (in W/m^2) is the POA sky-diffuse component.

The three contributions shown in the right-hand member of Equation (3) were obtained using Pysolar library [26] and Reindl model [27–29], and assuming an albedo of 0.2.

2.2.3. Heat Pump and PCM Tank

The heat pump (HP) is one of the core components of the system and it is mainly used to provide space cooling, although it can also provide space heating. On the one hand, the low-pressure circuit of the HP is connected to an innovative type of PCM tank, which can at the same time act as the evaporator of the heat pump. The main purpose of the PCM tank is to store the surplus of coolness produced by the HP during periods of low cooling demand and high PV production, when the electric battery is already completely charged. On the other hand, the high-pressure circuit of the HP (condenser) is connected, by means of a hydraulic loop, to the evaporator of a sorption chiller, so that the heat rejected by the HP condenser is absorbed by the evaporator of the sorption chiller. The objective of this connection is to increase the efficiency (EER) of the HP and reduce therefore the overall electricity consumption of the heat pump.

As shown in Figure 1, the hydraulic connections allow the HP and PCM tank to operate in different modes, either for cooling or heating purposes, as summarized in Table 1.

Table 1. Operating modes of the HP and PCM tank.

Mode	Description	Active Pumps	Dry Cooler	Fan-Coils
Cooling 1	PCM tank is charged by the heat pump, no cooling is provided to the building	P5–P8 (if sorption is on) P7 and P8 (if sorption is off)	On	Off
Cooling 2	PCM tank is discharging to provide cooling to the building	P9 and P10	Off	On
Cooling 3	Cooling is provided by the HP through the PCM tank	P5–P10 (if sorption is on) P7–P10 (if sorption is off)	On	On
Cooling 4	Cooling is provided by the HP through the standard evaporator	P5–P10 (if sorption is on) P7–P10 (if sorption is off)	On	On
Heating	Heating is provided by the HP	P7 and P10	On	On
0	No cooling or heating is provided	None	Off	Off

The PCM tank consists of a compact three-fluids (refrigerant-PCM-water) heat exchanger, in which PCM is placed in an array of parallel channels containing aluminum fins, sandwiched between refrigerant and water channels in an alternating sequence. This configuration allows for efficient heat transfer between the three fluids in the same container, also made of aluminum, which allows for easy charging and discharging of the PCM, as well as direct heat transfer between the supply and demand circuits. An amount of 160 kg of the commercial RT4 PCM, which is a paraffin that melts around 5 °C, was considered in the PCM tank. A complete description of the HP and PCM tank model can be found in [30]. A slightly improved model for the PCM tank was used in this study to also consider the sensible contribution to the overall energy stored in the PCM tank, as well as energy losses to the ambient. The updated relation between the charging level of the PCM tank ($E_{PCM,t}$, in kJ) and the PCM temperature ($T_{PCM,t}$, in °C) at time t is shown in Equation (4):

$$E_{PCM,t} = \begin{cases} 43186.8,\ if\ T_{PCM,t} < -2\ °C \\ -211.75 \cdot T_{PCM,t}{}^2 - 2110.7 \cdot T_{PCM,t} + 39812,\ if -2\ °C \le T_{PCM,t} \le 3\ °C \\ -1270.5 \cdot T_{PCM,t}{}^2 + 3183 \cdot T_{PCM,t} + 33460,\ if\ 3\ °C < T_{PCM,t} < 6\ °C \\ -1136.7 \cdot T_{PCM,t} + 13640,\ if\ 6\ °C\& \le T_{PCM,t} \le 12\ °C \\ 0,\ if\ T_{PCM,t} > 12\ °C \end{cases}, \quad (4)$$

Regardless the operating mode, the change in the energy stored in the PCM tank at time t is calculated from the charging level at the previous time slot ($E_{PCM,t-1}$, in kJ) and the net rate of coolness transfer to the PCM in the time interval Δt (in seconds), as shown in Equation (5):

$$E_{PCM,t} = E_{PCM,t-\Delta t} + \left(\dot{Q}_{PCM} - \dot{Q}_{losses}\right) \cdot \Delta t, \quad (5)$$

where $\dot{Q}_{PCM}$ (in kW) is the rate of coolness transfer to the PCM and $\dot{Q}_{losses} = (T_{amb,t} - T_{PCM,t-1})/R_{PCM}$ (in kW) are the coolness losses from the PCM tank to the ambient air at temperature $T_{amb,t}$ (in °C). The thermal resistance of the PCM tank (R_{PCM}) was estimated to be equal to 424.5 K/kW.

The rate of coolness transfer to the PCM ($\dot{Q}_{PCM}$) depends on the operating mode. When operating in cooling mode 1, the entire energy (coolness) generated by the HP ($\dot{Q}_{evap}$, in kW) [30] is transferred to the PCM, so that $\dot{Q}_{PCM} = \dot{Q}_{evap}$. In cooling mode 2, the PCM is discharged by the heat transfer fluid (HTF) of the building cooling circuit, and it was assumed to be equal (in absolute value) to the cooling demand, i.e., $\dot{Q}_{PCM} = -\dot{Q}_{cool,demand}$. In cooling mode 3, an energy balance is needed to determine the net rate of coolness transferred to the PCM because, on the one hand, the PCM is cooled down by the refrigerant and, on the other hand, it is heated up by the HTF. Therefore, in cooling mode 3, the PCM tank can actually be charging or discharging, depending on the charge level and the cooling demand. In cooling mode 4, the heat pump operates with the standard evaporator and the PCM tank is by-passed, and the same occurs when the heat pump operates in the heating mode. Therefore, $\dot{Q}_{PCM} = 0$ in cooling mode 4 and in heating mode.

When the heat pump operates in heating mode, the sorption chiller is always off. The heat required by the building is taken from the ambient air through the dry cooler by activating pump P7 and by-passing HEX2, and it is delivered to the building heating loop connected to the condenser of the heat pump by activating P10 and by-passing HEX3.

Once the charging level of the PCM tank at time slot t ($E_{PCM,t}$) is calculated, the PCM temperature and the water temperature at condenser outlet ($T_{wc,out,t}$) can be updated. Moreover, the electricity consumption of the compressor of the heat pump ($\dot{Q}_{comp}$), as well as the electricity consumption of all auxiliary equipment (pumps, dry cooler, fan-coils), can be calculated by taking into account what components are active in each operating mode according to Table 1. Only the compressor of the heat pump is driven by the DC-bus, while all other equipment uses electricity directly from the grid.

2.2.4. Sorption Chiller

The sorption chiller consists of two adsorbers based on a silica gel/water system, which switch periodically between adsorption and desorption operation in counter phase, a condenser and an evaporator. There are only two possible operating modes for the sorption chiller: mode 1 in which the sorption chiller is on and mode 2 in which it is off. In mode 1, the adsorption cycle is activated thanks to the hot water provided by the buffer tank. To work properly, the temperature of the hot water provided by the buffer tank ($T_{HT,in}$) should lie between 65 °C and 95 °C. At the evaporator side of the sorption chiller, heat is taken from the condenser of the HP. The waste heat produced by the sorption chiller is drained by the dry cooler to the ambient air at temperature T_{amb}.

The thermal coefficient of performance (COP_{th}) of the sorption chiller is defined as $COP_{th} = \dot{Q}_{LT}/\dot{Q}_{HT}$, where $\dot{Q}_{LT}$ (in kW) is the cooling power (heat taken from the condenser of the HP) and $\dot{Q}_{HT}$ (in kW) is the thermal power extracted from the buffer tank. Experimental tests performed in the lab showed that COP_{th} can be considered constant and equal to 0.55 for a large range of operating conditions.

The cooling power of the sorption module ($\dot{Q}_{LT}$) is a function of the water temperature at the evaporator inlet ($T_{LT,in}$, in °C), the water temperature that returns from the dry cooler ($T_{MT,in}$, in °C) and the water temperature that returns from the buffer tank ($T_{HT,in}$, in °C) [31], as shown in Equation (6):

$$\dot{Q}_{LT} = 4.559 + 1.36245 \cdot T_{LT,in} - 1.64553 \cdot T_{MT,in} + 0.47773 \cdot T_{HT,in}, \quad (6)$$

The return water temperature from the dry cooler ($T_{MT,in}$) was assumed to be 5 K above the ambient temperature, i.e., $T_{MT,in} = T_{amb} + 5$. The water temperature at the evaporator inlet ($T_{LT,in}$) was assumed to be equal to the water temperature at the outlet of the condenser of the HP evaluated at the previous time slot, i.e., $T_{LT,in,t} = T_{wc,out,t-1}$.

Therefore, the thermal power extracted from the buffer tank ($\dot{Q}_{HT}$) can be calculated according to Equation (7) [31]:

$$\dot{Q}_{HT} = \frac{\dot{Q}_{LT}}{COP_{th}} = \frac{\dot{Q}_{LT}}{0.55}, \quad (7)$$

Water temperature at the outlet of the adsorption module ($T_{HT,out,t}$, in °C) can be obtained using an energy balance as shown in Equation (8):

$$T_{HT,out,t} = T_{HT,in,t} - \frac{\dot{Q}_{HT}}{\dot{m}_{ad} \cdot c_{p,w}}, \quad (8)$$

where $\dot{m}_{ad} = 0.694$ kg/s is the mass flow rate of the water in the loop that connects the buffer tank with the sorption chiller and $c_{p,w} = 4.18$ kJ/(kg·K) is the specific heat capacity of the water.

Finally, the water temperature at the evaporator outlet ($T_{LT,out,t}$, in °C) can be obtained using an energy balance as shown in Equation (9):

$$T_{LT,out,t} = T_{LT,in,t} - \frac{\dot{Q}_{LT}}{\dot{m}_{wc} \cdot c_{p,w}}, \quad (9)$$

where $\dot{m}_{wc} = 1.417$ kg/s is the mass flow rate of the water in the loop that connects the condenser of the HP with the evaporator of the sorption chiller.

In mode 2, the sorption chiller is off and the following values were assumed for the main variables related to the sorption chiller: $\dot{m}_{ad} = \dot{Q}_{HT} = \dot{Q}_{LT} = 0$, $T_{HT,out,t} = T_{HT,in,t} = T_{buffer,top,t-1}$, $T_{LT,in,t} = T_{amb} + 5$ and $T_{LT,out,t} = T_{LT,in,t}$, where $T_{buffer,top,t-1}$ (in °C) is the temperature of the water at the top part of the buffer tank at the previous time slot.

The overall electricity consumption of the sorption chiller in mode 1 is the sum of the electricity consumption of the dry cooler, pumps P5–P8 and the actuators of the hydraulic system and controller (around 200 W). In mode 2, the electricity consumption of the sorption chiller is zero. The electricity needed to feed the sorption module is taken from the grid.

2.2.5. Dry Cooler

The dry cooler switches on whenever there is a need to reject heat from the system to the ambient air, i.e., when the sorption chiller and/or the heat pump are on. The electricity consumption of the dry cooler ($\dot{W}_{dc}$, in kW) depends on the part load of the dry cooler (PL_{dc}) and it is given in Equation (10) [32]:

$$\dot{W}_{dc} = 0.0176 - 0.1622 \cdot PL_{dc} + 0.8781 \cdot PL_{dc}^2, \quad (10)$$

The part load (PL_{dc}) is defined as the actual thermal power to be rejected or absorbed by the dry cooler divided by its nominal thermal power (40 kW), i.e., $PL_{dc} = \dot{Q}_{dry\ cooler}/40$. The actual thermal power ($\dot{Q}_{dry\ cooler}$, in kW) depends on the operating modes of both the sorption chiller and the HP, as shown in Equation (11):

$$\dot{Q}_{dry\ cooler} = \begin{cases} \dot{Q}_{HT} + \dot{Q}_{LT}, \text{ if sorption chiller is on} \\ \dot{Q}_{cond}, \text{ if sorption chiller is off and HP operates in cooling mode} \\ \dot{Q}_{evap}, \text{ if HP operates in heating mode} \end{cases}, \quad (11)$$

where $\dot{Q}_{cond}$ (in kW) is the rate of heat rejected by the HP condenser and $\dot{Q}_{evap}$ (in kW) is the heat absorbed by the dry cooler from the ambient air.

2.2.6. DHW Tank

The DHW tank is used in the system to store a sufficient amount of hot water able to meet the DHW demand of the building at any moment. Therefore, the water stored in the tank should always be kept above a minimum temperature level. To achieve it, the DHW tank should be heated with hot water from the buffer tank. An electric heater can also be used as a backup in case the temperature in the buffer tank is not high enough to be able to charge the DHW tank. In case the water temperature inside the DHW tank lies within the required temperature range, no heat is provided to the DHW tank. This means that there are three possible operating modes for the DHW tank: mode 1, in which the DHW tank is heated by the buffer tank, mode 2, in which it is heated by the electric heater and mode 3, when no heat is provided to the DHW tank.

In mode 1, pump P3 is activated to circulate hot water from the top part of the buffer tank to heat the DHW tank. This mode can be activated whenever the temperature inside the DHW tank (T_{DHW}, in °C) is below a lower threshold ($T_{DHW} < T_{set,DHW} - 5$) and the temperature of the water at the top part of the buffer tank ($T_{buffer,top}$, in °C) is above a given threshold ($T_{buffer,top} \geq T_{set,DHW} + 10$). The DHW tank is heated by the water from the buffer tank until the water temperature inside the DHW reaches the upper threshold of the set-point temperature ($T_{DHW} \geq T_{set,DHW} + 5$) or the buffer tank temperature is lower than the required threshold ($T_{buffer,top} < T_{set,DHW} + 10$), whichever occurs first. The value considered for the set-point temperature of the DHW tank is $T_{set,DHW} = 50$ °C.

Mode 2 is activated when the temperature inside the DHW tank is below the set-point range ($T_{DHW} < T_{set,DHW} - 5$) and the DHW tank is heated by the electric heater (instead of the buffer tank). Similar to mode 1, the electric heater is switched off when the water temperature reaches the upper threshold of the set-point temperature ($T_{DHW} \geq T_{set,DHW} + 5$). In mode 3, water temperature inside the DHW tank lies within the required temperature range ($T_{set,DHW} - 5 < T_{DHW} < T_{set,DHW} + 5$), so no heat is supplied to the DHW tank, although heat can be discharged from the DHW tank to meet the demand.

In all three modes, the temperature distribution inside the tank is considered homogeneous, and it can be calculated at any time t by means of an energy balance given in Equation (12):

$$a_1 \cdot \dot{Q}_{DHW,buffer} + a_2 \cdot \dot{Q}_{DHW,el} - \dot{Q}_{DHW,demand} - \dot{Q}_{loss,DHW} = \frac{M_{DHW} \cdot c_{p,w} \cdot (T_{DHW,t} - T_{DHW,t-1})}{\Delta t}, \tag{12}$$

where $a_1 = (1, 0, 0)$ and $a_2 = (0, 1, 0)$ for DHW mode $= (1, 2, 3)$, respectively, $\dot{Q}_{DHW,buffer}$ (in kW) is the heat extracted from the buffer tank, $\dot{Q}_{DHW,el} = 2$ kW is the thermal power supplied by the electric heater, $\dot{Q}_{DHW,demand}$ (in kW) is the DHW demand, $\dot{Q}_{loss,DHW}$ (in kW) are the heat losses from the DHW tank to the ambient air, $M_{DHW} = 250$ kg is the mass of the water inside the DHW tank and $T_{DHW,t-1}$ (in °C) is the temperature of the water inside the DHW tank calculated in the previous time slot.

The heat extracted from the buffer tank ($\dot{Q}_{DHW,buffer}$) depends on the temperature inside the DHW tank ($T_{DHW,t}$) as shown in Equation (13):

$$\dot{Q}_{DHW,buffer} = \dot{m}_{DHW} \cdot c_{p,w} \cdot (T_{DHW,in} - T_{DHW,t}), \tag{13}$$

where $\dot{m}_{DHW} = 0.556$ kg/s is the water mass flow rate (displaced by pump P3) in the loop that charges the DHW tank and $T_{DHW,in} = 60$ °C is the set-point of water temperature at the DHW tank inlet.

Heat losses from the DHW tank to the ambient air ($\dot{Q}_{loss,DHW}$) are calculated using Equation (14):

$$\dot{Q}_{loss,DHW} = \frac{T_{DHW,t} - T_{amb,t}}{R_{DHW}}, \tag{14}$$

where $R_{DHW} = 830.8$ K/kW is the overall thermal resistance of the DHW tank.

The electricity consumption of the DHW tank from the grid is associated to the circulating pump P3 (only in mode 1) and the electric heater (only in mode 2). There is no electricity consumption in mode 3.

2.2.7. Buffer Tank

The buffer tank is modelled considering three different regions (volumes) and assuming a uniform water temperature distribution inside each volume (Figure 2) [33,34]. The temperature of the buffer tank at time slot t is calculated by applying an energy balance to each of the three different volumes of the tank. Heat transfer by conduction or natural convection between two adjacent regions is neglected and the only heat transfer mechanism considered is through mass transfer. The buffer tank charging is assumed to be done with hot water at constant inlet temperature of 95 °C coming from the Fresnel solar field. The heat generated by the solar collectors ($\dot{Q}_{solar}$, in kW) is transferred to the buffer tank by means of a heat exchanger placed between the solar field loop and the buffer tank loop (HEX1 in Figure 1). Heat losses between the solar field and the buffer tank were neglected for simplicity. The mass flow rate of the water in the buffer tank loop ($\dot{m}_{solar}$, in kg/s) is variable to maintain a constant water temperature at the buffer tank inlet.

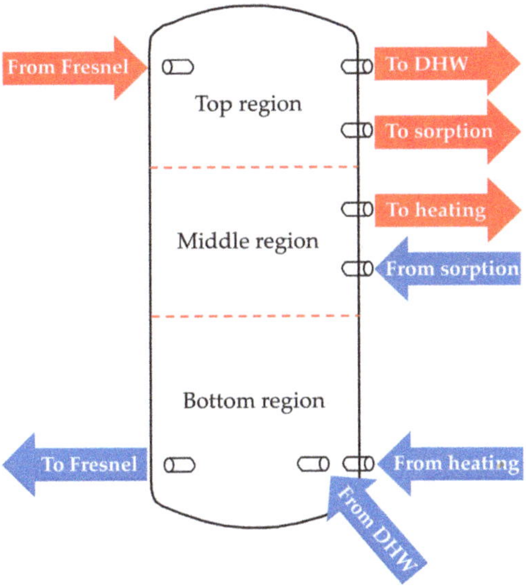

Figure 2. Schematic of the different inlets and outlets of the buffer tank.

For the top region, the energy balance is shown in Equation (15):

$$\dot{m}_{solar}\cdot c_{p,w}\cdot\left(T_{solar,in}-T_{buffer,top,t-1}\right)+\dot{m}\prime_{DHW}\cdot c_{p,w}\cdot\left(T_{buffer,mid,t-1}-T_{buffer,top,t-1}\right)+\dot{m}_{ad}\cdot c_{p,w}\cdot\left(T_{buffer,mid,t-1}-T_{buffer,top,t-1}\right)-\frac{\left(T_{buffer,top,t-1}-T_{amb}\right)}{R_{buffer,top}}=f_{top}\cdot M_{buffer}\cdot c_{p,w}\cdot\frac{T_{buffer,top,t}-T_{buffer,top,t-1}}{\Delta t},\quad(15)$$

where M_{buffer} = 800 kg is the mass of the water inside the buffer tank, f_{top} = 0.3 is the mass fraction of the top part of the buffer tank, $T_{solar,in}$ = 95 °C is the inlet temperature of the water flow coming from the Fresnel collectors, $T_{buffer,top,t-1}$ (in °C) and $T_{buffer,mid,t-1}$ (in °C) are the temperatures of water at the top and middle parts of the buffer tank in the previous time slot, respectively, and Δt is the time step (in seconds). If sorption module is off, $\dot{m}_{ad}$ = 0. The sorption module is automatically switched off when $T_{buffer,top,t} < 65$ °C and it may be switched on again when $T_{buffer,top,t} \geq 68$ °C (if the high-level controller decides it is best to do it, and whenever the heat pump is working in one of the cooling modes 1, 3 or 4).

The mass flow rate of the loop that connects the buffer tank with the solar field ($\dot{m}_{solar}$, in kg/s) is given by Equation (16):

$$\dot{m}_{solar}=\frac{\dot{Q}_{solar}}{c_{p,w}\cdot\left(T_{solar,in}-T_{buffer,bot,t-1}\right)},\quad(16)$$

where $T_{buffer,bot,t-1}$ (in °C) is the water temperature at the bottom part of the buffer tank evaluated at the previous time slot. When the water temperature at the top of the buffer tank reaches 94 °C during charging, the solar field is switched off and the charging of the buffer tank stops ($\dot{m}_{solar}$ = 0) until the water temperature at the top of the buffer tank decreases to 90 °C, when it may be switched on again if $\dot{Q}_{solar} > 0$.

The water mass flow rate at the buffer tank outlet towards the DHW tank charging circuit ($\dot{m}\prime_{DHW}$, in kg/s) is given by Equation (17):

$$\dot{m}\prime_{DHW} = \frac{\dot{m}_{DHW}\cdot\left(T_{DHW,in} - T_{DHW,t-1}\right)}{\left(T_{buffer,top,t-1} - T_{DHW,t-1}\right)}, \tag{17}$$

where $T_{DHW,t-1}$ (in °C) is the temperature inside the DHW tank at the previous time slot. Equation (17) only applies if the DHW tank works in mode 1 (charging with heat supplied from the buffer tank), otherwise $\dot{m}\prime_{DHW} = 0$.

Heat losses from the top part of the buffer tank to the ambient air depend on the thermal resistance of this part of the tank ($R_{buffer,top}$, in K/kW), which can be calculated using Equation (18):

$$R_{buffer,top} = \frac{R_{buffer}\cdot\left(A_{edge} + 2\cdot A_{base}\right)}{f_{top}\cdot A_{edge} + A_{base}}, \tag{18}$$

where $R_{buffer} = 430.3$ K/kW is the overall thermal resistance of the buffer tank, $A_{edge} = 4.095$ m^2 is the surface area of the buffer tank edge (lateral surface area) and $A_{base} = 0.62$ m^2 is the surface area of the base of the buffer tank.

For the middle part of the buffer tank, the energy balance is shown in Equation (19):

$$\begin{gathered}\dot{m}_{solar}\cdot c_{p,w}\cdot\left(T_{buffer,top,t-1} - T_{buffer,mid,t-1}\right) + \dot{m}\prime_{DHW}\cdot c_{p,w}\cdot\\ \left(T_{buffer,bot,t-1} - T_{buffer,mid,t-1}\right) + \dot{m}_{ad}\cdot c_{p,w}\cdot\left(T_{HT,out,t-1} - T_{buffer,mid,t-1}\right) +\\ \dot{m}_{heat}\cdot c_{p,w}\cdot\left(T_{buffer,bot,t-1} - T_{buffer,mid,t-1}\right) - \frac{\left(T_{buffer,mid,t-1} - T_{amb}\right)}{R_{buffer,mid}} = f_{mid}\\ \cdot M_{buffer}\cdot c_{p,w}\cdot\frac{T_{buffer,mid,t} - T_{buffer,mid,t-1}}{\Delta t},\end{gathered} \tag{19}$$

where $\dot{m}_{heat} = 0.63$ kg/s is the mass flow rate of the building heating loop (circulated by pump P4), $f_{mid} = 0.3$ is the mass fraction of the middle part of the buffer tank and $T_{HT,out,t-1}$ (in °C) is the temperature of the water returning from the adsorption module. If there is no heating demand ($\dot{Q}_{heat,demand} = 0$) or heat is provided to the building by the heat pump working in heating mode, $\dot{m}_{heat} = 0$.

Heat losses from the middle part of the buffer tank to the ambient air depend on the thermal resistance of this part of the tank ($R_{buffer,mid}$, in K/kW), which can be calculated using Equation (20):

$$R_{buffer,mid} = \frac{R_{buffer}\cdot\left(A_{edge} + 2\cdot A_{base}\right)}{f_{mid}\cdot A_{edge}}, \tag{20}$$

For the bottom region of the buffer tank, the energy balance equation is shown in Equation (21):

$$\begin{gathered}\dot{m}_{solar}\cdot c_{p,w}\cdot\left(T_{buffer,mid,t-1} - T_{buffer,bot,t-1}\right) + \dot{m}\prime_{DHW}\cdot c_{p,w}\cdot\left(T_{DHW,t-1} -\right.\\ \left.T_{buffer,bot,t-1}\right) + \dot{m}_{heat}\cdot c_{p,w}\cdot\left(T_{heat,out,t-1} - T_{buffer,bot,t-1}\right) -\\ \frac{\left(T_{buffer,bot,t-1} - T_{amb}\right)}{R_{buffer,bot}} = f_{bot}\cdot M_{buffer}\cdot c_{p,w}\cdot\frac{T_{buffer,bot,t} - T_{buffer,bot,t-1}}{\Delta t},\end{gathered} \tag{21}$$

where $f_{bot} = 0.4$ is the mass fraction of the bottom part of the buffer tank.

When the heating demand is satisfied by the buffer tank, water temperature returning from the building ($T_{heat,out,t-1}$, in °C) depends on the heating demand of the building ($\dot{Q}_{heat,demand}$, in kW) and it is calculated according to Equation (22):

$$T_{heat,out,t-1} = T_{buffer,mid,t-1} - \frac{\dot{Q}_{heat,demand}}{\dot{m}_{heat}\cdot c_{p,w}}, \tag{22}$$

Otherwise, when there is no heating demand from the building ($\dot{Q}_{heat,demand}$ = 0) or heat is provided by the heat pump working in heating mode, $\dot{m}_{heat}$ = 0 and $T_{heat,out,t-1} = T_{buffer,mid,t-1}$ (as an alternative to Equation (22)). In case that water temperature at the middle part of the buffer tank is below 45 °C ($T_{buffer,mid,t} < 45$ °C), heat cannot be delivered to the building from the buffer tank, therefore pump P4 switches off ($\dot{m}_{heat}$ = 0).

Heat losses from the bottom part of the buffer tank to the ambient air depend on the thermal resistance of this part of the tank ($R_{buffer,bot}$, in K/kW), which can be calculated using Equation (23):

$$R_{buffer,bot} = \frac{R_{buffer} \cdot \left(A_{edge} + 2 \cdot A_{base}\right)}{f_{bot} \cdot A_{edge} + A_{base}}, \quad (23)$$

The overall electricity consumption associated to the buffer tank only consists of the electricity consumption of pump P4 (34 W) when the heating demand of the building is higher than zero and this demand is met by the buffer tank ($\dot{m}_{heat}$ >0) and not by the heat pump working in heating mode.

2.2.8. DC-Bus

The heat pump is driven by DC through a connection to the DC-bus. Electricity can be taken either from the PV panels and/or from the battery, depending on the PV production and the state of charge of the battery. Furthermore, in case that the power supplied by the battery and the PV panels is not enough to feed the heat pump, electricity can also be provided by the power grid through an AC/DC converter (not shown in Figure 1). Conversely, when the PV production is high and the battery is fully charged, surplus electricity can be delivered to the grid. The power generated by the PV panels (PV, in kW) was assumed to be always less than the maximum charging power of the battery ($PV < MaxB$) and the maximum discharging power of the battery ($MaxB$) was assumed to be always higher than the power demanded by the HP ($MaxB > HP$).

Three different operating modes were considered for the DC-bus, focusing on the control strategy of the battery, as summarized in Table 2. Two thresholds, E_1 (in %) and E_2 (in %), were used to drive the DC-bus in one of the three possible operating modes.

Table 2. Operating modes of the DC-bus and associated thresholds values.

Mode	Name	E_1 (%)	E_2 (%)
1	Charging	75	90
2	Discharging	10	25
3	Buffer	10	90

Charging mode (mode 1) takes place if $E_{min} \leq B_S \leq E_1$, where $E_{min} = 10\%$ is the lower charging level allowed for the battery, B_S (in %) is the state of charge of the battery and E_1 (in %) is a threshold below which the battery is automatically charged (i.e., when $B_S \leq E_1$). In this mode, the battery charges at a constant maximum rate ($B = MaxB = 3$ kW) until $B_S = 75\%$, after which the battery switches to buffer mode.

In charging mode, the power required by the heat pump (HP) can only be taken from the PV and/or from the grid (G), but not from the battery (Figure 3a).

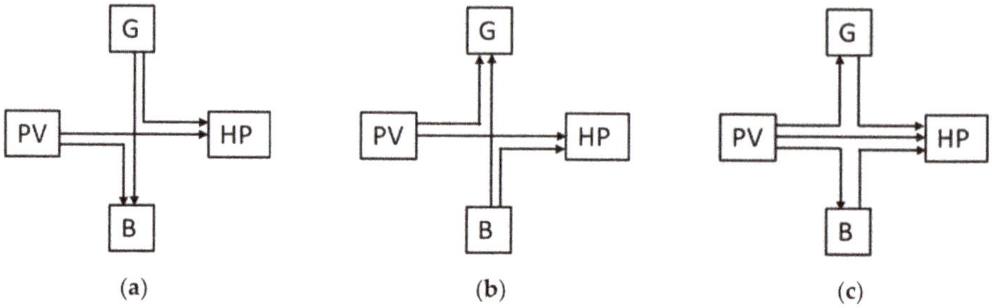

Figure 3. Schematic of the DC-bus operating in (**a**) charging, (**b**) discharging and (**c**) buffer modes.

The equations that describe the different energy streams in charging mode are shown in the set of Equation (24):

$$\begin{cases} HP_{PV} = \min(PV,\ HP) \\ B_{PV} = \max(0,\ PV - HP) \\ B_G = MaxB - B_{PV} \\ HP_G = \max(0, HP - PV) \\ G_{PV} = HP_B = 0 \end{cases}, \tag{24}$$

where HP_{PV} (in kW) is the power supplied to the HP from the PV panels, B_{PV} (in kW) is the power supplied to the battery from the PV panels, B_G (in kW) is the power required from the grid to charge the battery at maximum power, HP_G is the power required from the grid to feed the HP, G_{PV} is the power coming from the PV panels that is delivered to the grid and HP_B is the power supplied to the HP from the battery. The power exchanged between X and Y, where X and Y may refer to PV (PV panels), HP (heat pump), G (power grid) or B (battery), is assumed to be positive ($X_Y > 0$) if energy is incoming to X from Y ($X \leftarrow Y$).

Discharging mode (mode 2) takes place if $E_2 < B_S \leq E_{max}$, where E_2 (in %) is a threshold above which the battery automatically discharges (i.e., when $B_S > E_2$) and $E_{max} = 90\%$ is the upper threshold allowed for the charging level of the battery. In this mode, the battery is discharging at the maximum rate ($B = -MaxB = -3$ kW) towards both the HP and the grid (Figure 3b) until $B_S = 25\%$, after which the battery switches to buffer mode.

The equations that describe the different energy streams in discharging mode are shown in the set of Equation (25):

$$\begin{cases} HP_{PV} = \min(PV,\ HP) \\ HP_B = \max(0, HP - PV) \\ B_G = -MaxB + HP_B \\ G_{PV} = \max(0, PV - HP) \\ HP_G = B_{PV} = 0 \end{cases}, \tag{25}$$

Buffer mode (mode 3) takes place if $E_1 < B_S \leq E_2$. Whenever the optimizer decides to switch to the buffer mode, the following values are assigned: $E_1 = 10\%$ and $E_2 = 90\%$. In this way, the battery will be forced to switch to buffer mode whatever the value of B_S is.

In mode 3, the battery acts as a buffer, meaning that it charges if there is a surplus of electricity production from the PV panels or it discharges if the HP requires more power than is produced by the PV panels (Figure 3c). In this mode, there is no interaction between the battery and the grid, i.e., the battery cannot charge from the grid, neither can it deliver electricity to the grid.

The equations that describe the different energy streams in buffer mode are shown in the set of Equation (26):

$$\begin{gathered} HP_{PV} = \min(PV,\ HP) \\ B_G = 0 \\ \text{if } (-400\text{ W} < PV - HP < 400\text{ W}) \text{ then} \\ \begin{cases} G_{PV} = \max(0,\ PV - HP) \\ HP_G = \max(0, HP - PV) \\ B_{PV} = HP_B = 0 \end{cases} \\ \text{else} \\ \begin{cases} B_{PV} = \max(0,\ PV - HP) \\ HP_B = \max(0, HP - PV) \\ G_{PV} = HP_G = 0 \end{cases}, \end{gathered} \tag{26}$$

In this mode, the total power supplied to the grid (G) can be negative, positive or zero, depending on the relation between PV production and HP consumption. The total power supplied to the battery (B) can be positive if there is a surplus of PV generation, negative when the power demand of the HP cannot be met only from the PV panels, or zero, when the absolute difference between PV production and HP consumption is less than 400 W ($|PV - HP| < 400$ W).

For all operating modes, the state of charge of the battery at time instant t ($B_{S,t}$) is given by Equation (27):

$$B_{S,t} = B_{S,t-1} + \frac{\eta_B \cdot B \cdot \Delta t / 3600}{C_{B,max}}, \tag{27}$$

where $B_{S,t-1}$ is the state of charge of the battery at the previous time slot, η_B is the efficiency of battery charging/discharging process, $C_{B,max} = 7.3$ kWh is the maximum storage capacity of the battery and Δt (in seconds) is the time step of the simulation. For the sake of simplicity, the value of the efficiency of battery charging/discharging process (η_B) was assumed to depend on the sign of B: $\eta_B = 0.9$ if $B \geq 0$ (battery is charging) and $\eta_B = 1$ if $B < 0$ (battery is discharging).

2.2.9. Summary of the Main Model Parameters

The main model parameters considered for the training/testing scenarios that will be explained later are summarized below:

- Surface of the Fresnel solar collectors: 60 m^2.
- PV panels surface: 20.9 m^2.
- PV panels orientation: 0° (south).
- PV panels inclination: 30°.
- PCM tank storage capacity: ≈ 43,200 kJ (12 kWh).
- DHW tank capacity: 250 L.
- DHW electric heater power: 2 kW.
- Buffer tank capacity: 800 L.
- Battery energy storage capacity: 7.3 kWh.
- Maximum battery charging/discharging power: 3 kW.

2.3. *DRL Control Description*

2.3.1. General Description

Reinforcement learning is a class of solution methods that optimizes a numerical reward by interaction with the environment [9], in which a learning agent takes actions that drive the environment to new states, provoking some reward being observed by the agent. It is in this context that Markov decision processes (MDP) provide a useful mathematical framework to solve the problem of learning from interaction to optimize a given goal [35]. In a finite and discrete MDP, the environment is represented at each time step as a state. Based on this state, the agent, according to a given policy, decides to execute an action, obtaining a reward from the environment and moving it to the next state. Considering stochastic environments, one can think on state-transition probabilities that characterize the MDP. Furthermore, as each transition gives a reward, each state may be associated to a state-value function that represents all the expected MDP rewards given a state. These representations are the basis for the Bellman optimality equations [36], which must be solved to achieve an optimal solution for the problem.

Expressing the MDP abstraction more formally, and considering a discrete time MDP with time step $t = 0, 1, 2, \ldots$, the MDP consists of:

- A set of states S that represents the environment, being $S_t \in S$ the environment state at time t.
- A set of actions A that can be taken by the agent, being $A_t \in A(s)$ the action taken at time t from the subset of available actions at state s, $A(s)$.
- A numerical reward for the new visited state, $R_{t+1} \in \mathbb{R}$ that will depend on its trajectory: $S_0, A_0, R_0, S_1, A_1, R_1, \ldots, S_t, A_t, R_t$.
- Assuming that the system dynamics is Markovian, random variables S_t and R_t will only depend on its previous values, with a probability distribution, $p()$, which characterizes the system, defined as in Equation (28):

$$p(s', r | s, a) \doteq Pr\{S_t = s', R_t = r | S_{t-1} = s,\ A_{t-1} = a\}, \tag{28}$$

- An agent policy, π, which determines the chosen action at a given state. Defined as a probability, $\pi(a|s)$ results in the probability of choosing action a from state s.

Figure 4 shows a typical RL paradigm representation. In this case, policy π depends on a set of parameters θ that represents the neural network weights to be discussed later.

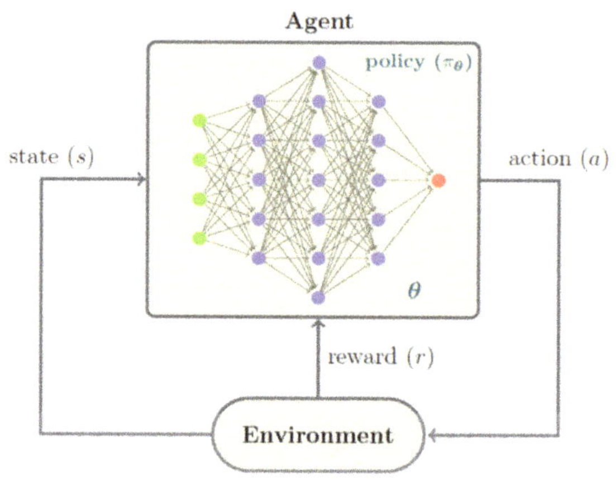

Figure 4. RL paradigm that represents the sequence: state, action, reward, of an MDP process.

The cumulative reward at a given time slot can be defined as in Equation (29):

$$G_t \doteq \sum_{i=0}^{T-t-1} \gamma^i R_{t+i+1}, \tag{29}$$

where T is the final time step and γ is a discount rate that determines the worthiness of future rewards. Equation (29) helps to define the concept of the value of being at a state for a given policy given in Equation (30):

$$v_\pi(s) \doteq \mathbb{E}_\pi[G_t | S_t = s], \tag{30}$$

and using Equation (28), Equation (30) becomes the Bellman equations for v_π, shown in Equation (31):

$$v_\pi(s) = \sum_a \pi(a|s) \sum_{s',r} p(s',r|s,a)[r + \gamma v_\pi(s\prime)], \tag{31}$$

for all $s \in S$.

Solving a RL problem implies to find an optimal policy (π^*) that solves the state-value function defined in Equation (32):

$$v^*(s) \doteq \max_\pi v_\pi(s), \tag{32}$$

and derives from Equations (28), (29) and (32), the Bellman optimality equations as in Equation (33):

$$v^*(s) \doteq \max_a \sum_{s',r} p(s',r|s,a)[r + \gamma v^*(s\prime)], \tag{33}$$

The solution of Equation (33) provides the best action (a), in terms of future rewards, from a given state (s). Once solved for every possible state, it gives the optimal policy, π^*, because the probabilities $\pi(a|s)$ are known.

It is at this point that the whole family of reinforcement learning algorithms is created, trying to solve these optimality equations by different means. Resolution techniques based on dynamic programming (DP) may solve the problem, i.e., find an optimal solution, by iteratively finding the state-values, $v(s)$, but these methods suffer from the so-called curse of dimensionality, because the number of states grows exponentially with the number of state variables. Such a curse is tackled by Monte-Carlo (MC) methods by sampling values of the state-values through experience, by interaction with the model. Even with a partial knowledge of those state-value functions, good MC algorithms converge to acceptable solutions. Even more, if those MC methods are combined with DP ideas, such as update regularly the estimated values, a new family of algorithms arises, called temporal-difference (TD) learning, such as Sarsa (λ), Q-Learning or TD (λ), proving to be highly efficient methods for a lot of optimal control problems.

Even the improvement of new RL methods, large and complex problems may require an enormous amount of computational power, particularly when the number of states is large, during the learning phase. Under this scenario, the ground-breaking concept of deep reinforcement learning (DRL) [13,37] appears to change the rules of the game, scaling up RL to space state dimensions previously intractable. DRL deals efficiently with the curse of dimensionality by using neural networks as substituting parts of traditional value tables, obtaining approximations of the optimal value functions trained by their corresponding neural network backpropagation mechanisms. The emergence of specialized libraries as TensorFlow [38] did the rest, by allowing parallelization across multiple CPUs or GPUs and permitting in this way to train huge neural networks able to cope the structure of complex systems in affordable running times.

2.3.2. Policy Gradient Algorithms

The above-mentioned RL algorithms based the resolution of the Bellman optimality equations on the learned value of the selected actions. Instead, policy gradient methods base their learning on a parameterized policy that selects the actions without the knowledge of a value function. Generically, one can consider a set of parameters $\theta \in \mathbb{R}^d$ that usualy correspond to the weights of a neural network. By doing so, one can rephrase the policy function as $\pi_\theta(a|s)$.

At this point, Equation (30) may work as an objective function, $J(\theta)$. Effectively, one can define the objective as in Equation (34):

$$J(\theta) \doteq v_{\pi\theta}(s_0) = \mathbb{E}_\theta[G_0|s_0], \tag{34}$$

i.e., the expected cumulative reward from $t = 0$. According to the policy gradient theorem [39], whenever the policy was differentiable with respect to θ, the gradient of the cost function obeys the proportionality shown in Equation (35):

$$\nabla_{\theta} J(\theta) = \mathbb{E}_{\pi}\left[G_t \frac{\nabla_{\theta}\pi(A_t|S_t,\theta)}{\pi(A_t|S_t,\theta)}\right], \tag{35}$$

Considering that Equation (35) can be instantiated at each time slot and that parameters θ are time-dependent, one can apply any gradient descent algorithm to compute θ as in Equation (36):

$$\theta_{t+1} \leftarrow \theta_t + \alpha G_t \frac{\nabla_{\theta}\pi(A_t|S_t,\theta)}{\pi(A_t|S_t,\theta)} = \theta_t + \alpha G_t \nabla_{\theta} \ln \pi(A_t|S_t,\theta), \tag{36}$$

being α a learning rate constant. Equation (36) is the fundamental idea that supports a new family of RL algorithms called REINFORCE [40]. As noted, REINFORCE is a MC-like algorithm because it can be implemented by sampling the environment, getting from it the cumulative reward and the logarithm of the policy gradient, presenting good convergence properties for small enough values of the learning parameter. The existence of gradient descent optimizers based on neural networks, as well as the existing softmax layers did the rest to allow efficient REINFORCE implementations.

2.3.3. HYBUILD Control Model

The HYBUILD control is based on a REINFORCE algorithm with no baseline. HYBUILD system model operates at two differently slotted time scales. First, a finer slot is considered in order to numerically compute the HYBUILD system behavior (3 min are typically considered). This smaller time slot is only considered for inner model operations and it is not relevant for control purposes. Second, a larger slot (T_s) is used to manage the control system (15 and 30 min were considered). Within T_s time slot, any action decided by the control system is invariant until reaching any subsystem limit. As an example, if during a given slot T_s one decides to charge the heat pump/PCM tank subsystem, the charging process will not stop unless the maximum state of charge was reached. Similarly, the input system variables for the control system are considered invariant in T_s.

HYBUILD control model for the Mediterranean system may be defined for cooling or heating purposes, but the heating model can be considered as a subset of the cooling model because heating operations for the Mediterranean system are much simpler. Actually, heating mode bypasses the PCM tank and sorption subsystems, resulting in only one operating mode for the heat pump subsystem.

The state vector (S_t) is an 8-dimensional vector in cooling mode (7-dimensional in heating mode) with the following components:

1. Thermal energy demand for cooling/heating in the current time slot (TE_t^{dem}).
2. Thermal energy demand for domestic hot water (DHW) in the current time slot (TE_t^{dhw}).
3. Ambient temperature ($T_{amb,t}$).
4. Energy cost for electric demand in the current time slot (C_t).
5. Direct normal irradiation, (DNI_t), as explained in Sections 2.2.1 and 2.2.2.
6. Charge level of the PCM tank subsystem, ($E_{PCM,t}$), as explained in Section 2.2.3. Not used in heating mode.
7. Buffer tank top temperature, ($T_{buffer,top,t}$), as explained in Section 2.2.7.
8. Battery state of charge in the DC-bus subsystem ($B_{S,t}$), as explained in Section 2.2.8. being t the corresponding time and all of them were standard normalized according to their ranges.

Choosing thermal energy demand as input, instead of temperature set-points, allow to decouple the model from building thermal mass dynamics, providing more consistency to the Markovian assumption. In this sense, considering the control process as an MDP results is a valid assumption as long as the heating/cooling subsystem models, detailed in Section 2.2, are time-depending on only previous time slots. As a counter effect, an on-site control implementation will require to model the building dynamics based on the temperature set-points in order to predict the thermal demand. In this sense, the models used in this study are based on reinforcement learning that accurately provide the thermal demand for a particular building under different weather conditions and set-points.

The set of actions ($\mathcal{A}$) that guide the control can be defined as $\mathcal{A} = \{\mathcal{C}, \mathcal{S}, \mathcal{B}\}$, where $\mathcal{C}$ is the set of cooling/heating operating modes, $\mathcal{S}$ is the set of activation modes for the sorption subsystem and $\mathcal{B}$ is the set of battery modes in the DC-bus subsystem. As only the set $\mathcal{C}$ differs for the cooling and the heating models, one can differentiate the set of actions accordingly: $\mathcal{A}_{cool} = \{\mathcal{C}_{cool}, \mathcal{S}, \mathcal{B}\}$ and $\mathcal{A}_{heat} = \{\mathcal{C}_{heat}, \mathcal{S}, \mathcal{B}\}$.

According to the operating modes defined in Table 1, $\mathcal{C}_{cool} = \{0,\ 1,\ 2,\ 3,\ 4\}$ and $\mathcal{S} = \{0,\ 1\}$ because the sorption subsystem may be on or off. For the heating modes, as sorption and heat pump/PCM tank subsystems are bypassed, only one operating mode is considered, being $\mathcal{C}_{heat} = \{0,\ 1\}$.

Concerning the actions related to the DC-bus subsystem, as detailed in Section 2.2.8, the high-level control may determine the E_1 and E_2 thresholds that define the area of DC-bus operation, as well as the maximum charging/discharging power when operating in charge/discharge areas. As the control model presented here only deals with discrete values, the DC-bus control operations were simplified according the following rules:

- Charging/discharging power is set to a fixed value, namely 3 kW.
- If from the high-level control the DC-bus is forced to operate in charging, buffer or discharging mode, the pair of values (E_1, E_2) is set to three fixed levels: (75, 90), (10, 90) and (10, 25), respectively, as a percentage of the battery state of charge, B_S.

Following these assumptions, $\mathcal{B} = \{0,\ 1,\ 2\}$, which corresponds to charging, buffer and discharging modes, respectively.

Finally, considering that during cooling mode 2 (all cooling energy is supplied by the PCM tank) the sorption chiller is in mode 0, the set of possible actions are:

$$\begin{aligned}\mathcal{A}_{cool} = \ & \{[1,\ 0,\ 0],\ [1,\ 0,\ 1],\ [1,\ 0,\ 2],\ [1,\ 1,\ 0],\ [1,\ 1,\ 1],\ [1,\ 1,\ 2],\\ & [2,\ 0,\ 0],\ [2,\ 0,\ 1],\ [2,\ 0,\ 2],\\ & [3,\ 0,\ 0],\ [3,\ 0,\ 1],\ [3,\ 0,\ 2],\ [3,\ 1,\ 0],\ [3,\ 1,\ 1],\ [3,\ 1,\ 2],\\ & [4,\ 0,\ 0],\ [4,\ 0,\ 1],\ [4,\ 0,\ 2],\ [4,\ 1,\ 0],\ [4,\ 1,\ 1],\ [4,\ 1,\ 2]\}\end{aligned}$$

and $|\mathcal{A}_{cool}| = 21$.

In heating mode, considering that the sorption chiller is always off, it follows that:

$$\mathcal{A}_{heat} = \{[1,\ 0,\ 0],\ [1,\ 0,\ 1],\ [1,\ 0,\ 2]\}$$

and $|\mathcal{A}_{heat}| = 3$.

It should be noted that all the cases where cooling/heating mode is 0 may be omitted because:

- If there is some energy demand, cooling/heating mode 0 is not an option.
- Otherwise, any cooling/heating mode will perform as mode 0 inside T_S.

In other words, mode 0 is adopted when energy demand is null.

For the purpose of this study, a policy gradient REINFORCE algorithm was implemented, with two three-layer fully-connected neural networks of sizes $N_{inp,heat} \times N_{hid,heat} \times N_{out,heat}$ and $N_{inp,cool} \times N_{hid,cool} \times N_{out,cool}$ for heating and cooling, respectively, with the following characteristics:

- $N_{inp,heat} = 7$ and $N_{inp,cool} = 8$ are the number of inputs, defined by the system state dimension. Their values are standard normalized with their corresponding ranges.
- $N_{hid,heat}$ and $N_{hid,cool}$ are the hidden layer sizes for heating and cooling modes, respectively. They use to be much larger than the size of inputs and outputs. Actually, the number of hidden layers, their size, the type activation functions, as well as other

parameters will be adjusted in a future study by hyper-parameter setting analysis, being out of the scope of this paper. The values $N_{hid,heat}$ = 100 and $N_{hid,cool}$ = 1,000 were adopted here, with exponential linear unit activation functions and a dropout rate of 0.8.

- $N_{out,heat}$ = 3 and $N_{out,cool}$ = 21 are the number of outputs corresponding to the cardinality of the actions set. Outputs represent softmax of logits and the corresponding action is taken as a multinomial of the logarithm of outputs.
- Learning rate, α = 0.0005.
- Discount rate, γ = 0.99.

The neural network was trained minimizing the cross entropy of the multinomial outputs using an Adam stochastic optimizer [41]. Under this scenario, one objective function was defined regarding an economic reward related to the cost associated to the system operation.

2.3.4. Minimum Cost Control Policy

In order to derive control policies focused on minimizing the cost of operation, the cumulative reward G_t used in Equation (36) and in Equation (29) should be calculated considering the reward function R_t defined in Equation (37):

$$R_t \doteq \left(EE_t^{fg} - 0.5 \cdot EE_t^{tg}\right) \cdot C_t + \left(TE_t^{dem} - TE_t^{hp} - TE_t^{pcm}\right) \cdot Penalty, \tag{37}$$

where:

- EE_t^{fg} is the electrical energy bought from the grid in slot t, either to feed the DC-bus or other equipment, such as the electric resistance of the DHW tank.
- EE_t^{tg} is the electrical energy sold to the grid in slot t. A discount factor of 0.5 was considered.
- TE_t^{hp} is the thermal energy provided by the heat pump subsystem for cooling/heating in slot t.
- TE_t^{pcm} is the thermal energy provided by the PCM tank for cooling/heating in slot t.
- $Penalty$ is the cost assumed for a non-covered demand. A value much higher than the energy cost is used.
- C_t and TE_t^{dem} as detailed in Section 2.3.3.

Note that TE_t^{dhw} is not part of the objective function because it is assumed that DHW requirements will always be fulfilled by the backup electric heater.

2.3.5. Rule-Based Control Policies

With the objective to evaluate the DRL control policy goodness, a simple rule-based control (RBC) policy for the cooling season was also implemented, which can be simplified for heating mode. The RBC policy is based on its own thresholds and can be described as follows:

- Battery mode—charging, buffer or discharging—is determined by two battery state of charge thresholds (B_{min}^{th} and B_{max}^{th}) and the grid cost (C_t).
- Cooling mode 1 (PCM tank charging) is set if there is no cooling demand. Otherwise, cooling mode 2 (PCM tank discharging) is set if PCM energy ($E_{PCM,t}$) is larger than a threshold factor (PCM_f^{th}) times the cooling demand (TE_t^{dem}). Otherwise, cooling mode 3 (simultaneous PCM tank charging and cooling supply to the building) or 4 (cooling supply using the standard HP evaporator) is set according to the energy stored in the PCM tank in relation to the PCM energy threshold (E_{PCM}^{th}).
- Sorption chiller mode is set depending on the buffer tank temperature threshold (BT^{th}) in comparison to the buffer tank temperature at the top region ($T_{buffer,top,t}$).

The details on both cooling and heating RBC policies are shown in Appendix A. In both RBC policies, a hyper-parameter optimization was applied in order to determine the optimal thresholds. Hyperopt python library [42] was used employing an adaptive Tree Parzen Estimator algorithm with 400 runs over the same training test set.

2.3.6. Implementation Aspects

Figure 5 shows the structure designed for the implementation of the HYBUILD control system. It is divided into three layers. Layer 3 is composed of the low-level controllers for each subsystem. It operates directly over the system components, including all sensors, actuators and low-level security protocols. Layer 2 is composed of the Supervisory Con-trol And Data Acquisition (SCADA) system. It monitors the system parameters, sends the state vector to layer 1 and executes the set of actions set by layer 1. Layer 1 is composed of the DRL control algorithm described in this paper. The communication between layers 1 and is done using MQTT(json) and the communication between layers 2 and 3 is performed using the OPC-UA communication protocol.

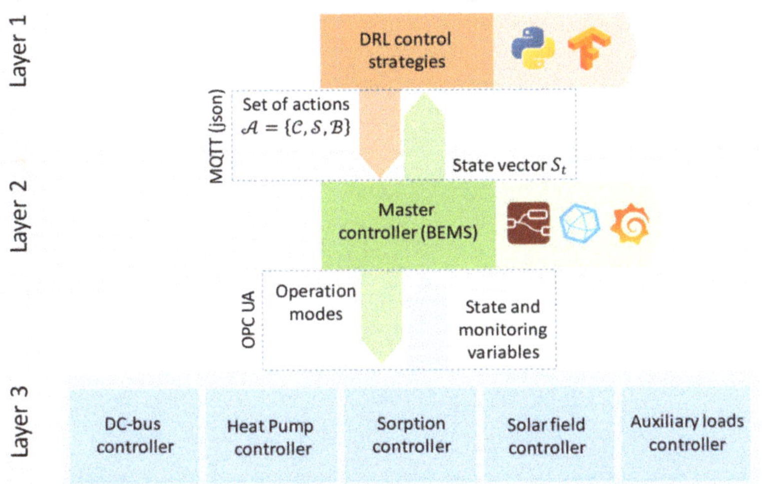

Figure 5. Diagram of smart control implementation in the HYBUILD system.

The HYBUILD control model was written in Python 3 [43]. Furthermore, Tensorflow libraries were used in control models [38]. The availability of a lite version of Tensorflow libraries makes suitable this implementation for light hardware or micro-controller environments that may be required for control scenarios in real time.

2.4. Network Trainizng

In this subsection, the data set used to train and test the network is described. The computations are performed with weather data for the reference building (assumed to be located in Athens), but it could be applied to any other location. The computing training time and its convergence issues are also presented in this subsection.

2.4.1. Training and Test Data

Cooling data set spans from day 120 to day 250 of the year, while heating data set spans from day 290 to day 365 and from day 1 to day 90. Such sets are shuffled and split into smaller subsets (batches). Each batch is composed of a fixed number of days ($\mathcal{T}$). Actually, its cardinality ($|\mathcal{T}|$) is a parameter. During the experimentation, batch sizes of 3 and 6 days were used, giving the last one better performance results. From the 130 days available for cooling, 18 days are taken for testing and the rest for training purposes. As mentioned in Section 2.3.3, control model inputs are: thermal demand for cooling/heating, thermal demand for DHW, ambient temperature, direct normal irradiation, cost of electricity, PCM state of charge (not used in heating mode), buffer tank top temperature and electric battery state of charge.

Ambient temperature and solar radiation are obtained from EnergyPlus weather data Europe WMO Region 6, Greece, Athens 167,160 (IWEC) [24]. Since the time slot for this data is one hour, data was linearly interpolated when T_s was smaller.

As already mentioned in Section 2.1, the energy demand profile for cooling, heating and DHW were obtained within the HYBUILD project [23] activities. For the grid electricity price, a two-period tariff was assumed:

- 0.2 €/kWh from 13:00 to 23:00 h.
- 0.1 €/kWh for the rest of the day.

2.4.2. Training Times

Before presenting the results of system performance, it is worth mentioning a few aspects of the training process. Inside a batch (3 or 6 days), a reward defined by Equation (37) is first computed, after which, gradients are computed and propagated. This process is repeated for all the sets in the training set, forming an iteration. After a small number of iterations, the trained model is applied to the test set in order to obtain the control system performance, always keeping the best model so far. Figure 6 shows the cumulative reward G_0 (or cost) for the test and training sets as a function of the number of iterations at two different scales, showing the learning process. During the first iterations, the network rapidly finds better strategies than the random one established at the beginning. It is a common behavior to get stuck at a local minimum during a large number of iterations. Even though the discovered strategies are quite good, they are still far from the best ones found beyond 2000 iterations. From this point, the strategies are slightly improved until reaching overfitting, where no improvement is observed.

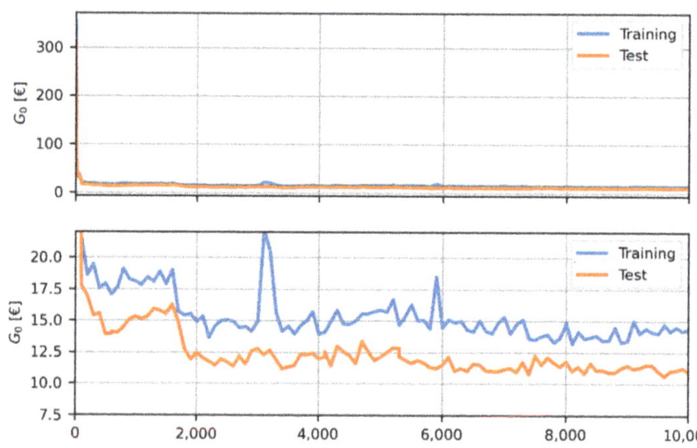

Figure 6. Cumulative reward for test and training set in cooling mode.

The following settings for the control system were considered:

- Time granularity for model computation: Δt = 3 min for cooling mode and Δt = 15 s for heating mode. Taking longer time slots for the period when the heat pump is switched on would surpass in excess the heating demand in that time slot, due to the fact that the heat pump has higher coefficient of performance in heating mode.

- Time slot between control decisions: T_s = 30 min.
- Batch size of 6 days. Test set consists of 3 batches (18 days or 864 control slots).

Note that, using an Intel i5-6600 4-cores at 3.3 GHz CPU, each iteration takes approximately 15 s (for T_s = 30 min and Δt = 3 min) and, consequently, the learning plot shown in Figure 6 took almost two days of CPU computation.

2.5. Robustness Analysis

A robustness analysis was also carried out to evaluate the effect that the uncertainties in some parameters of the mathematical models (not experimentally validated) of the main system components might have on the results. First, the network was trained as described in the previous subsection using the reference values of all the parameters of the component models. Second, the reference values of some of the model parameters were randomly altered following a uniform distribution within a certain error range around the reference value, as shown in Table 3. An "error multiplying" factor (n) was used to define different levels of errors affecting the parameters of the model.

Table 3. Set of parameters used to check the robustness of the model.

Variable	Symbol	Reference Value	Error Range	Units
Optical efficiency Fresnel	η_{opt}	Data from [22]	Ref$\cdot(1 \pm 0.2\cdot n)$	-
PV efficiency	η_{PV}	0.16	Ref$\cdot(1 \pm 0.25\cdot n)$	-
Maximum battery charging or discharging power	$MaxB$	3.0	Ref$\cdot(1 \pm 0.2\cdot n)$	kW
Battery charging efficiency	η_B	0.9	Ref$\cdot(1 \pm 0.11\cdot n)$	-
Sorption thermal efficiency	COP_{th}	0.55	Ref$\cdot(1 \pm 0.09\cdot n)$	-
Dry cooler electricity consumption	$\dot{W}_{dc}$	Equation (10)	Ref$\cdot(1 \pm 0.2\cdot n)$	kW
Heat pump cooling power	$\dot{Q}_{evap}$	Data from [30]	Ref$\cdot(1 \pm 0.2\cdot n)$	kW
Heat produced by the compressor	$\dot{Q}_{comp}$	Data from [30]	Ref$\cdot(1 \pm 0.2\cdot n)$	kW
Buffer tank thermal resistance	R_{buffer}	430.3	Ref$\cdot(1 \pm 0.13\cdot n)$	K/kW
RPW-HEX thermal resistance	R_{PCM}	424.5	Ref$\cdot(1 \pm 0.18\cdot n)$	K/kW
DHW tank thermal resistance	R_{DHW}	830.8	Ref$\cdot(1 \pm 0.19\cdot n)$	K/kW

Next, the performance of the model trained using the reference values was tested for ten different independent data sets obtained at each error level (defined by the value of the error multiplying factor n). To check the robustness of the DRL approach, the network was also trained using new training data sets generated for each of the ten deviated models at each error level. The average of the relative deviations in the results obtained using the network trained with the reference model and using the network trained with each of the deviated models was used to quantify robustness of the DRL approach. Finally, to compare the DRL and the RBC approaches, the RBC was also applied for each of the deviated data sets mentioned above, using the same thresholds obtained for the model without error.

3. Results and Discussion

This section details the results obtained with the trained system and abovementioned settings and model parameters in both cooling and heating scenarios. System performance results obtained using the smart control are compared against conventional RBC mechanisms.

Figure 7 shows the performance of the trained network for the test set. The plots, from top to bottom, show:

1. Cooling demand ('Demand') and global horizontal solar irradiation ('GHI tilted') on the tilted plane (PV surface). Green and orange areas show how the cooling demand was met: whether from the heat pump ('From HP') or from the PCM tank ('From PCM').
2. The state of charge of the PCM tank ('PCM SoC'), heat pump cooling mode ('Cool. mode') and mode of operation of the sorption chiller ('Sorption act.').

3. The values of E1 and E2 thresholds of the DC-bus subsystem as detailed in Section 2.2.8. The state of charge of the battery is also shown ('Battery SoC'), along with the cost of electricity ('Grid cost') as binary (0 corresponds to 0.1 €/kWh and 1 to 0.2 €/kWh).
4. Domestic hot water demand ('Demand DHW') and top region temperature of the buffer tank ('Buffer Tank top temp.'). Green and orange areas show how the DHW demand was met: whether from the heat pump ('From elect') or from the buffer tank ('From BT').
5. Cumulative cost associated to the energy delivered to and taken from the power grid during valley ('Ener. sold 0′ and 'Ener. bought 0′, respectively) and peak ('Ener. sold 1′ and 'Ener. bought 1′, respectively) electricity tariff, along with the total cost according to the cumulative cost ('Cost') defined as $\sum_{i=0}^{t} R_i$. The total amount of electricity consumption is also plotted ('Cumm. elec. energ.').

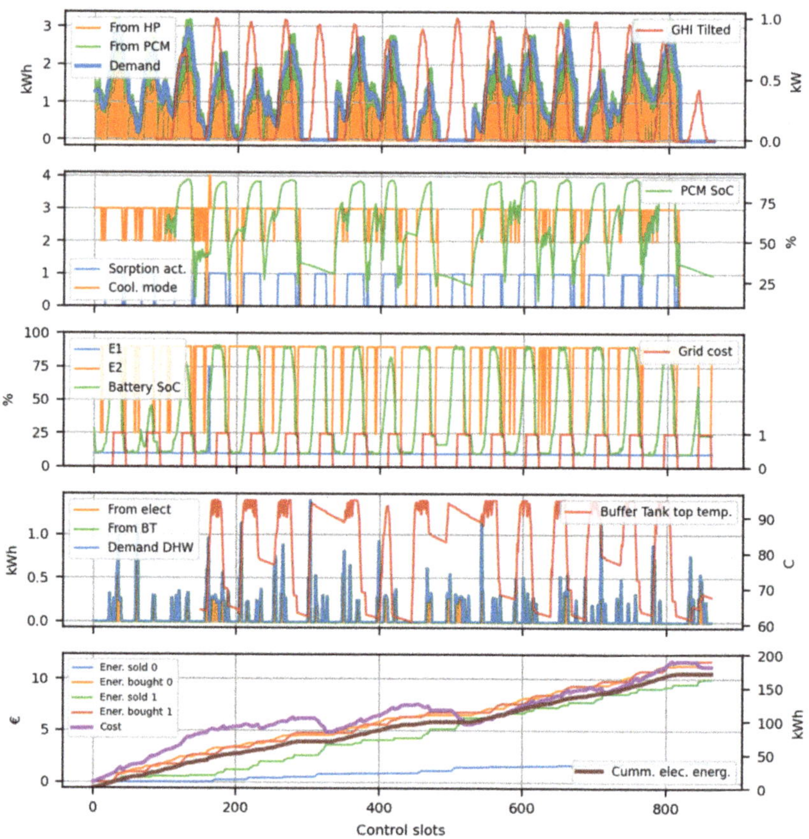

Figure 7. DRL performance results for the test set in cooling mode.

From Figure 7, some aspects of the DRL control policy can be highlighted:

- The operating cost for the 18 days of the test set is 11.1 €. As seen below, it is far less than the RBC policy tested under the same scenario, indicating that the deep learning control approach is highly efficient.
- Cooling demand is always covered, either from the HP or the PCM tank, in order to avoid penalties.
- Cooling modes 1 (PCM tank charging) and 4 (operation of the HP with the standard evaporator) are never (or rarely) used.

- All energy storage modules (PCM tank, buffer tank and electric battery) are fully exploited by charging and discharging them as much as possible on a daily basis within the allowed thresholds.
- The sorption chiller is also activated on a daily basis to assist the operation of the HP, which is beneficial for the overall system performance.

As seen from the bottom plot, the cost associated with the amount of energy sold in tariff period 0 (low cost) does not exceed the cost associated with the amount of energy bought during the same period. Depending on national regulations, an energy retailer may not reward consumers for the surplus of energy supplied to the grid during a certain period. No substantial differences were observed when running the control with a smaller time slot (T_s = 15 min).

For comparison purpose, Figure 8 shows the performance of the system for the same test set using an RBC control policy. The same variables as in Figure 7 are shown. The following optimal thresholds were used in the simulations based on an RBC policy:

- Minimum and maximum battery thresholds: B^{th}_{min} = 0.01 and B^{th}_{max} = 0.94, respectively.
- Threshold factor for PCM tank discharging: PCM^{th}_{f} = 1.98.
- Buffer tank temperature threshold: BT^{th} = 76.7 °C.
- Threshold of the (normalized) amount of energy stored in the PCM tank: E^{th}_{PCM} = 0.19.

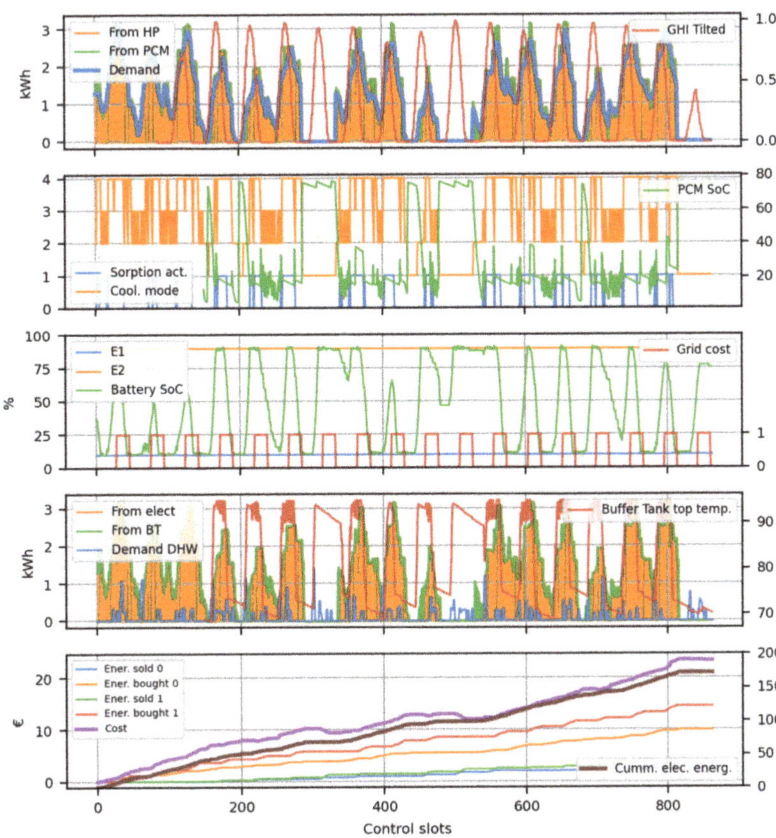

Figure 8. RBC performance results for the test set in cooling mode.

From Figure 8, the following aspects regarding the RBC policy can be highlighted:

- The operating cost for the 18 days of the test set is 23.5 €, which is more than double the cost obtained using an DRL policy.
- Cooling demand is always covered, either from the HP or the PCM tank.
- All cooling modes are used by the HP, with no clear predilection for a specific operating mode.
- Sorption chiller activation is much more irregular as compared with the DRL case.
- The full potential of the PCM tank is hardly exploited, while the buffer tank is charged and discharged as much as possible on a daily basis.
- Electric battery is reasonably well exploited, but the main difference with respect to the DRL policy is that it is not discharged when the electricity cost is high and electricity demand of the system is low.

Focusing on the DRL policy, a zoom view presented in Figure 9 shows that the battery is discharged at peak tariff periods by adjusting the E2 threshold, putting DC-bus in discharging mode. The control uses the PCM tank as a buffer and prevents its full discharge in order to ensure that the demand is met at all times and avoid penalties. Surprisingly, the energy required to meet the DHW demand is mostly supplied from the electric heater instead of the buffer tank. This could be explained by the fact that, from a cost point of view, it is better to use the heat stored in the buffer tank to drive the sorption module during periods of high cooling demand, which allows the heat pump to work with a higher efficiency leading to a lower electricity consumption and, therefore, to a lower operating cost.

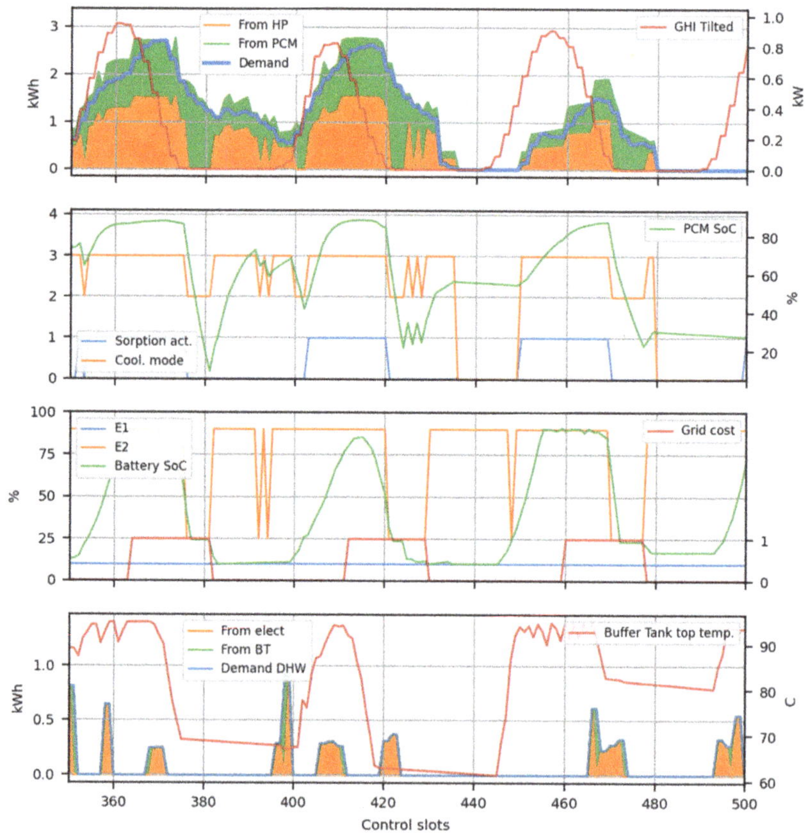

Figure 9. DRL performance results for the test set in cooling mode. Zoom view.

In relation to the heating mode, Figure 10 plots the results of system performance using the deep learning control strategy. The following aspects are worth noting:

- The upper plot (first) shows how the heating demand is covered, whether by the heat pump ('From HP') or the buffer tank ('From BT').
- It can be observed, in the third plot, how the buffer tank temperature in the middle layer drops when heat is provided to the building from the buffer tank.
- The cumulative cost results negative (bottom plot), meaning that economic benefit is obtained from selling energy to the grid. This is achieved by charging/discharging the battery during the corresponding valley/peak tariff periods, as observed in the second plot.
- Bottom plot shows that the amount of energy sold in valley/peak tariff periods is larger than the energy bought during the same periods. As mentioned previously, an energy retailer may not reward energy reinjection when the amount of sold energy surpasses the bought energy. If this is the case, the cumulative cost will be zero instead of negative.

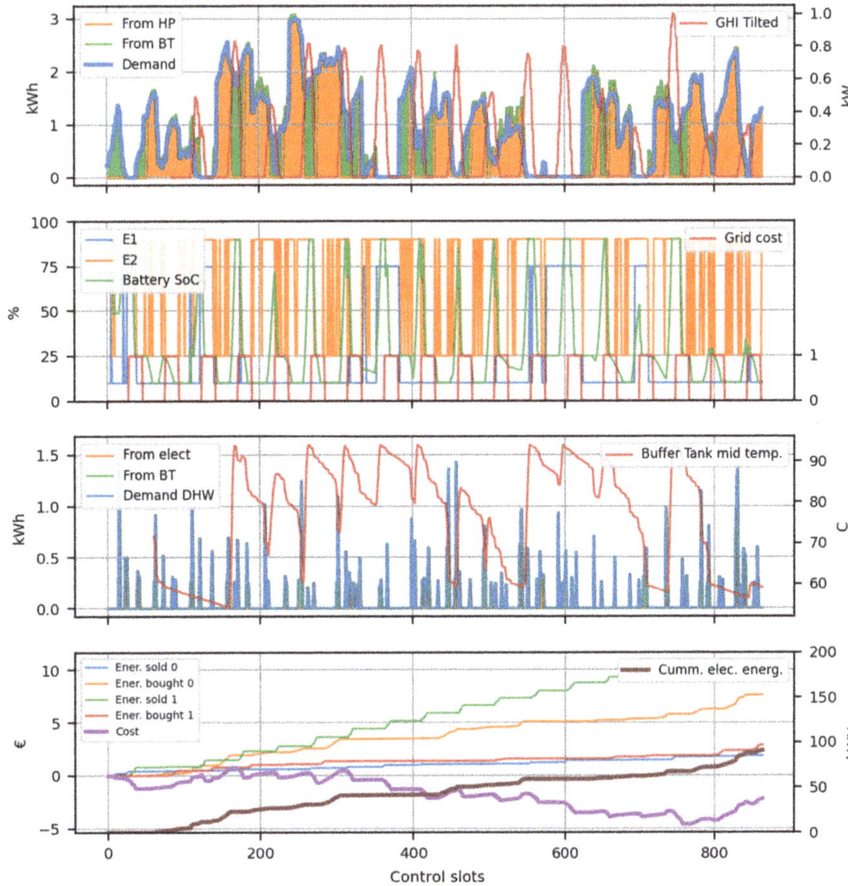

Figure 10. DRL performance results for the test set in heating mode.

Table 4 shows the results for DRL and RBC policies over the same test set. Clearly, the DRL policy outperforms the tested RBC policy.

Table 4. Operating cost (€) for DRL and RBC control policies.

Operating Mode	Policy	
	DRL	**RBC**
Cooling	11.1	23.5
Heating	−2.4	−0.1

However, the comparison between the DRL and RBC policies makes it clear that the DRL policy is able to achieve considerably better results in cooling mode, while in the heating mode it is only slightly better than the RBC policy. This is not surprising given the fact that the system investigated in this study was designed and sized mainly for use in Mediterranean climate regions, where the cooling demand is significant. In addition, the complexity of the system control is mainly associated to the subsystem that provides cooling, where there is a higher potential for improvement through an adequate control strategy. Indeed, the control of the subsystem that provides heating and DHW is relatively simple and it already includes some basic control rules at low (component) system level, which means there is not much room for improvement.

With regards to the robustness analysis, the two curves plotted in Figure 11 show the results obtained using the RBC and DRL approaches, both of them optimized for the model without errors (reference model), for values of the error multiplying factor from 1 to 4. As explained in Section 2.5, each point on the two lines is an average of the behavior of both controls on ten independent instances of the model with errors. The green dots denote the average over the same ten instances that correspond to the DRL approach in the case when the network was retrained with the deviated values of model parameters. As expected, the results obtained using the retrained network are better than the ones obtained using the network trained with the reference values of model parameters. Nevertheless, it can be seen that the original model (trained using the reference model) does not deviate too much from the optimum value for error multiplying factors lower than, or equal to, 2. It is only for value of the multiplying factor around 3 or higher that the deviation between the results becomes relevant. This would demonstrate the robustness of the solution over a wide range of model errors, since considerable deviations from the theoretical model (up to 40%) would have little impact (less than 3%) on the behavior of the control. Even if this were the case, the difference between the system performance using an RBC and a DRL approach would still be clearly in favor of the DRL strategy.

Figure 12 shows the behavior of the two DRL models (the one trained using the reference model and the one trained with the deviated models) for each individual instance and for four different values of the error multiplying factor. It can be seen that, for values of the error multiplying factor up to 2, the error in the cost obtained with the model trained with reference values are below 5%, even though the cost has large variations as a result of different model parameters. This confirms that the DRL model is able to adapt to different types of deviations in the actual components' behavior with respect to the mathematical model used in the simulations.

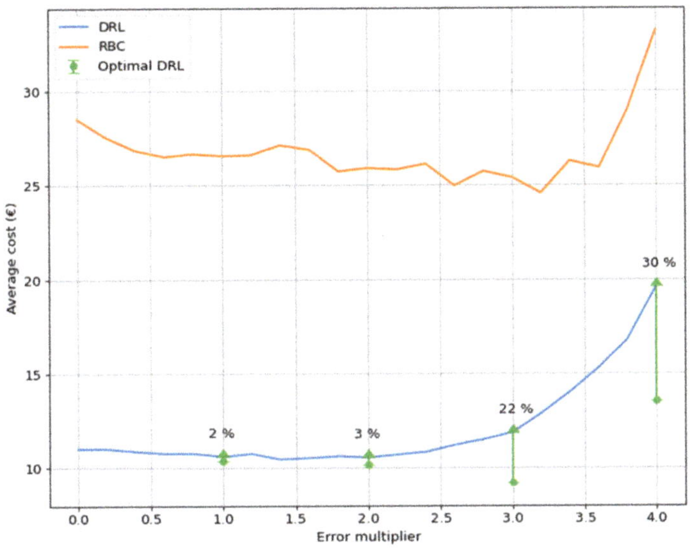

Figure 11. Results of the influence of the error level in the model parameters.

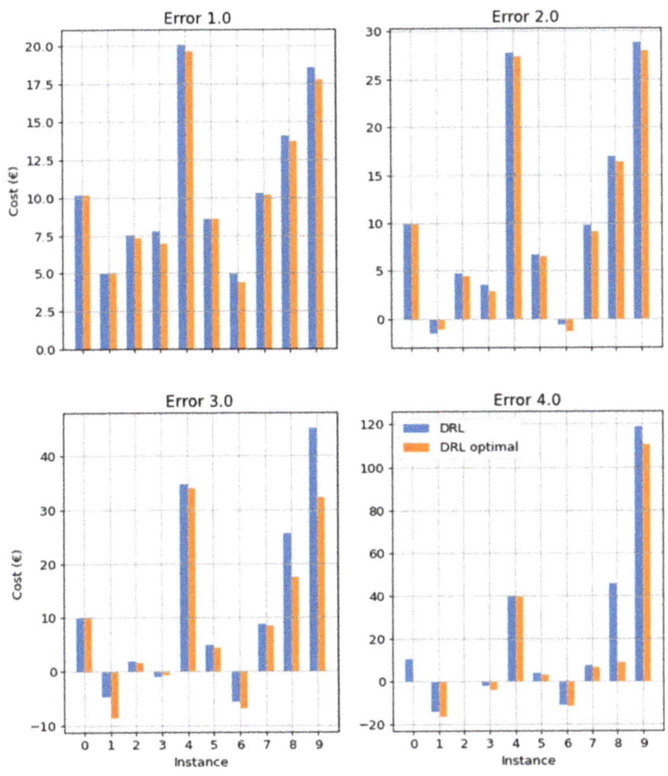

Figure 12. Results of the DRL models for individual instances at different error levels.

4. Conclusions and Future Work

This paper investigated a smart control based on a deep reinforcement learning control policy proposed for an innovative system developed within HYBUILD project. The main aim of the system is to reduce the energy demand for heating, cooling and domestic hot water of a standard single-family residential building by implementation of Fresnel collectors and PV panels combined with hybrid electrical and thermal storage components. The complexity of the system was a great challenge from the high-level control point of view, which was dealt with by applying deep learning techniques to optimize the operation of the overall system from a monetary point of view. To the best of the authors knowledge, this is the first study in which DRL has been applied to a complex TES system. The performance of the proposed control policy was compared with basic rule-based control policies for both cooling and heating modes. The results show that the deep learning control policy provides a proper system control that is able to efficiently manage the system and to obtain significant cost (and energy) savings with respect to a standard rule-based control. In addition, the results of the robustness analysis clearly showed that DRL model is able to adapt to any changes in the actual behavior of the system in a real test pilot plant, with deviations less than 3% in the average cost estimations for an error multiplying factor up to 2.

Immediate future work will consist of deploying the DRL control for a pilot plant in order to test its performance. Even the robustness analysis shows a good ability to deal with large mismatches between theoretical and real models, there are still big challenges before adopting this technology for the consumer market. Requiring accurate models for heating/cooling systems as well as for building thermal demand may be a time-consuming task, and more studies are required in order to determine the feasibility, in terms of time requirements, of self-training starting from a basic or general knowledge of the system.

Author Contributions: Conceptualization, C.F., G.Z., D.V. and L.F.C.; methodology, G.Z. and C.F.; software, C.F.; formal analysis, G.Z. and C.F.; investigation, C.F., G.Z. and D.V.; resources, C.F.; data curation, C.F.; writing—original draft preparation, G.Z. and C.F.; writing—review and editing, D.V. and L.F.C.; visualization, G.Z., C.F., L.F.C.; supervision, L.F.C.; project administration, L.F.C.; funding acquisition, L.F.C. All authors have read and agreed to the published version of the manuscript.

Funding: This project has received funding from the European Union's Horizon 2020 research and innova-tion programme under grant agreement No 768824 (HYBUILD). This work was partially funded by the Ministerio de Ciencia, Innovación y Universidades de España (RTI2018-093849-B-C31-MCIU/AEI/FEDER, UE) and by the Ministerio de Ciencia, Innovación y Universidades-Agencia Estatal de Investigación (AEI) (RED2018-102431-T). This work is partially supported by ICREA under the ICREA Academia programme.

Institutional Review Board Statement: Not applicable.

Informed Consent Statement: Not applicable.

Data Availability Statement: The data presented in this study are available on request from the corresponding author.

Acknowledgments: The authors would like to thank the Catalan Government for the quality accreditation given to their research group (2017 SGR 1537). GREiA is certified agent TECNIO in the category of technology developers from the Government of Catalonia.

Conflicts of Interest: The authors declare no conflict of interest.

Appendix A

The RBC used for the cooling mode is shown below (Figure A1):

Parameters: *Optimized thresholds:* B^{th}_{min}, B^{th}_{max}, BT^{th}, PCM^{th}, PCM^{th}_f
Inputs: $C_{b,t}$, $B_{S,t}$, TE^{dem}_t, $E_{PCM,t}$, $T_{buffer,top,t}$
Result: action $\in \mathcal{A}_{cool}$

```
if T_buffer,top,t > BT^th then
|  sorption_mode ← 1;                               /* Sorption On */
else
|  sorption_mode ← 0;
if C_b,t == 1 then
|  /* peak tariff                                                */
|  if B_S,t > B^th_max then
|  |  battery_mode ← 2;                             /* Discharging */
|  else
|  |  battery_mode ← 1;                                  /* Buffer */
else
|  /* valley tariff                                              */
|  if B_S,t < B^th_min then
|  |  battery_mode ← 0;                                /* Charging */
|  else
|  |  battery_mode ← 1;
if TE^dem_t > 0 then
|  /* Demand exists                                              */
|  if E_PCM,t > PCM^th_f · TE^dem_t then
|  |  cooling_mode ← 2; sorption_mode ← 0;
|  else
|  |  if E_PCM,t > PCM^th then
|  |  |  cooling_mode ← 4;
|  |  else
|  |  |  cooling_mode ← 3;
else
|  cooling_mode ← 1; sorption_mode ← 0;
return [cooling_mode, sorption_mode, battery_mode]
```

Figure A1. RBC for cooling.

The RBC used for the heating mode is shown below (Figure A2):

Parameters: *Optimized thresholds:* B^{th}_{min}, B^{th}_{max}
Inputs: $C_{b,t}$, $B_{S,t}$
Result: action $\in \mathcal{A}_{heat}$

```
if C_b,t == 1 then
|  /* peak tariff                                                */
|  if B_S,t > B^th_max then
|  |  battery_mode ← 2;                             /* Discharging */
|  else
|  |  battery_mode ← 1;                                  /* Buffer */
else
|  /* valley tariff                                              */
|  if B_S,t < B^th_min then
|  |  battery_mode ← 0;                                /* Charging */
|  else
|  |  battery_mode ← 1;
return [1, 0, battery_mode]
```

Figure A2. RBC for heating.

References

1. Afram, A.; Janabi-Sharifi, F. Theory and applications of HVAC control systems–A review of model predictive control (MPC). *Build. Environ.* **2014**, *72*, 343–355. [CrossRef]
2. Thieblemont, H.; Haghighat, F.; Ooka, R.; Moreau, A. Predictive control strategies based on weather forecast in buildings with energy storage system: A review of the state-of-the art. *Energy Build.* **2017**, *153*, 485–500. [CrossRef]
3. Cupelli, L.; Schumacher, M.; Monti, A.; Mueller, D.; De Tommasi, L.; Kouramas, K. Simulation Tools and Optimization Algorithms for Efficient Energy Management in Neighborhoods. In *Energy Positive Neighborhoods and Smart Energy Districts*; Elsevier BV: Amsterdam, The Netherlands, 2017; pp. 57–100.
4. Boudon, M.; L'Helguen, E.; De Tommasi, L.; Bynum, J.; Kouramas, K.; Ridouane, E.H. Real Life Experience—Demonstration Sites. In *Energy Positive Neighborhoods and Smart Energy Districts*; Monti, A., Pesch, D., Ellis, K.A., Mancarella, P., Eds.; Elsevier BV: Amsterdam, The Netherlands, 2017; pp. 227–250.
5. Tarragona, J.; Fernández, C.; de Gracia, A. Model predictive control applied to a heating system with PV panels and thermal energy storage. *Energy* **2020**, *197*, 117229. [CrossRef]
6. Gholamibozanjani, G.; Tarragona, J.; De Gracia, A.; Fernández, C.; Cabeza, L.F.; Farid, M.M. Model predictive control strategy applied to different types of building for space heating. *Appl. Energy* **2018**, *231*, 959–971. [CrossRef]
7. Achterberg, T. SCIP: Solving constraint integer programs. *Math. Program. Comput.* **2009**, *1*, 1–41. [CrossRef]
8. Vigerske, S.; Gleixner, A. SCIP: Global optimization of mixed-integer nonlinear programs in a branch-and-cut framework. *Optim. Methods Softw.* **2018**, *33*, 563–593. [CrossRef]
9. Sutton, R.S.; Barto, A.G. *Reinforcement Learning: An Introduction*; A Bradford Book; MIT Press: Cambridge, MA, USA, 2018; 427p.
10. Watkins, C.J.C.H.; Dayan, P. Technical Note: Q-Learning. *Mach. Learn.* **1992**, *8*, 279–292. [CrossRef]
11. Liu, S.; Henze, G.P. Experimental analysis of simulated reinforcement learning control for active and passive building thermal storage inventory: Part 1. Theoretical foundation. *Energy Build.* **2006**, *38*, 142–147. [CrossRef]
12. Liu, S.; Henze, G.P. Experimental analysis of simulated reinforcement learning control for active and passive building thermal storage inventory: Part 2: Results and analysis. *Energy Build.* **2006**, *38*, 148–161. [CrossRef]
13. Mnih, V.; Kavukcuoglu, K.; Silver, D.; Graves, A.; Antonoglou, I.; Wierstra, D.; Riedmiller, M. Playing Atari with Deep Reinforcement Learning. *arXiv* **2013**, arXiv:1312:5602. Available online: https://arxiv.org/abs/1312.5602 (accessed on 30 April 2021).
14. Wei, T.; Wang, Y.; Zhu, Q. Deep Reinforcement Learning for Building HVAC Control. In Proceedings of the 54th Annual Design Automation Conference, Austin, TX, USA, 18–22 June 2017.
15. Mason, K.; Grijalva, S. A review of reinforcement learning for autonomous building energy management. *Comput. Electr. Eng.* **2019**, *78*, 300–312. [CrossRef]
16. Yu, L.; Qin, S.; Zhang, M.; Shen, C.; Jiang, T.; Guan, X. Deep Reinforcement Learning for Smart Building Energy Management: A Survey. *arXiv* **2020**, arXiv:e2008.05074. Available online: https://arxiv.org/abs/2008.05074 (accessed on 30 April 2021).
17. Wang, Z.; Hong, T. Reinforcement learning for building controls: The opportunities and challenges. *Appl. Energy* **2020**, *269*, 115036. [CrossRef]
18. Cheng, C.-C.; Lee, D. Artificial Intelligence-Assisted Heating Ventilation and Air Conditioning Control and the Unmet Demand for Sensors: Part 1. Problem Formulation and the Hypothesis. *Sensors* **2019**, *19*, 1131. [CrossRef]
19. Liu, S.; Henze, G.P. Evaluation of Reinforcement Learning for Optimal Control of Building Active and Passive Thermal Storage Inventory. *J. Sol. Energy Eng.* **2006**, *129*, 215–225. [CrossRef]
20. De Gracia, A.; Fernández, C.; Castell, A.; Mateu, C.; Cabeza, L.F. Control of a PCM ventilated facade using reinforcement learning techniques. *Energy Build.* **2015**, *106*, 234–242. [CrossRef]
21. De Gracia, A.; Barzin, R.; Fernández, C.; Farid, M.M.; Cabeza, L.F. Control strategies comparison of a ventilated facade with PCM – energy savings, cost reduction and CO2 mitigation. *Energy Build.* **2016**, *130*, 821–828. [CrossRef]
22. HYBUILD. Available online: http://www.hybuild.eu/ (accessed on 4 December 2020).
23. Macciò, C.; Porta, M.; Dipasquale, C.; Trentin, F.; Mandilaras, Y.; Varvagiannis, S. Deliverable D1.1-Requirements: Context of Application, Building Classification and Dynamic Uses Consideration. 2018. Available online: http://www.hybuild.eu/2018/12/20/requirements-context-of-application-building-classification-and-dynamic-uses-consideration-deliverable-released/ (accessed on 30 April 2021).
24. Weather Data by Location. All Regions—Europe WMO Region 6—Greece. Available online: https://energyplus.net/weather-location/europe_wmo_region_6/GRC//GRC_Athens.167160_IWEC (accessed on 4 December 2020).
25. Solar PV Panel Module Aleo S79 Characteristics. Bosch Solar Services. Available online: https://bit.ly/2VQ91l1 (accessed on 16 September 2019).
26. Zebner, H.; Zambelli, P.; Taylor, S.; Obinna Nwaogaidu, S.; Michelsen, T.; Little, J. Pysolar. Available online: https://github.com/pingswept/pysolar (accessed on 15 December 2020).
27. Reindl, D.; Beckman, W.; Duffie, J. Diffuse fraction correlations. *Sol. Energy* **1990**, *45*, 1–7. [CrossRef]
28. Reindl, D.; Beckman, W.; Duffie, J. Evaluation of hourly tilted surface radiation models. *Sol. Energy* **1990**, *45*, 9–17. [CrossRef]
29. Loutzenhiser, P.; Manz, H.; Felsmann, C.; Strachan, P.; Frank, T.; Maxwell, G. Empirical validation of models to compute solar irradiance on inclined surfaces for building energy simulation. *Sol. Energy* **2007**, *81*, 254–267. [CrossRef]

30. Varvagiannis, E.; Charalampidis, A.; Zsembinszki, G.; Karellas, S.; Cabeza, L.F. Energy assessment based on semi-dynamic modelling of a photovoltaic driven vapour compression chiller using phase change materials for cold energy storage. *Renew. Energy* **2021**, *163*, 198–212. [CrossRef]
31. Palomba, V.; Vasta, S.; Freni, A.; Pan, Q.; Wang, R.; Zhai, X. Increasing the share of renewables through adsorption solar cooling: A validated case study. *Renew. Energy* **2017**, *110*, 126–140. [CrossRef]
32. Palomba, V.; Dino, G.E.; Frazzica, A. Coupling sorption and compression chillers in hybrid cascade layout for efficient exploitation of renewables: Sizing, design and optimization. *Renew. Energy* **2020**, *154*, 11–28. [CrossRef]
33. Chandra, Y.P.; Matuska, T. Stratification analysis of domestic hot water storage tanks: A comprehensive review. *Energy Build.* **2019**, *187*, 110–131. [CrossRef]
34. Duffie, J.A.; Beckman, W.A. *Solar Energy Thermal Processes*; John Wiley & Sons Inc.: Hoboken, NJ, USA, 1974; ISBN 9780471223719.
35. Bellman, R. A Markovian Decision Process. *J. Math. Mech.* **1957**, *6*, 679–684. [CrossRef]
36. Bellman, R. *Dynamic Programming*; Princeton University Press: Princeton, NJ, USA, 2010; 392p.
37. Silver, D.; Huang, A.; Maddison, C.J.; Guez, A.; Sifre, L.; Driessche, G.V.D.; Schrittwieser, J.; Antonoglou, I.; Panneershelvam, V.; Lanctot, M.; et al. Mastering the game of Go with deep neural networks and tree search. *Nat. Cell Biol.* **2016**, *529*, 484–489. [CrossRef]
38. Abadi, M.; Barham, P.B.; Chen, J.; Chen, Z.; Davis, A.; Dean, J.; Devin, M.; Ghemawat, S.; Irving, G.; Isard, M.; et al. TensorFlow: A system for large-scale machine learning. In Proceedings of the 12th USENIX Symposium on Operating Systems Design and Implementation, OSDI 2016, Savannah, GA, USA, 2–4 November 2016; pp. 265–283.
39. Sutton, R.S.; Mcallester, D.; Singh, S.; Mansour, Y. Policy gradient methods for reinforcement learning with function approximation. In Proceedings of the Advances in Neural Information Processing Systems, Denver, CO, USA, 27–30 November 2000; pp. 1057–1063.
40. Williams, R.J. Simple Statistical Gradient-Following Algorithms for Connectionist Reinforcement Learning. *Mach. Learn.* **1992**, *8*, 229–256. [CrossRef]
41. Kingma, D.P.; Ba, J. Adam: A Method for Stochastic Optimization. In Proceedings of the 3rd International Conference for Learning Representations, San Diego, CA, USA, 7–9 May 2015; Bengio, Y., LeCun, Y., Eds.; Scientific Research Publisher: Wuhan, China, 2015.
42. Bergstra, J.; Yamins, D.; Cox, D.D. Making a science of model search: Hyperparameter optimization in hundreds of dimensions for vision architectures. In Proceedings of the 30th International Conference on Machine Learning, ICML 2013, (PART 1), Atlanta, GA, USA, 16–21 June 2013; Dasgupta, S., McAllester, D., Eds.; PMLR: New York, NY, USA, 2013; pp. 115–123.
43. Van Rossum, G.; Drake Jr., F.L. *Python Tutorial*; 12th Media Services: Suwanee, GA, USA, 1995; pp. 1–156.

buildings

MDPI

Article

Energy Performances Assessment of Extruded and 3D Printed Polymers Integrated into Building Envelopes for a South Italian Case Study

Giovanni Ciampi *, Yorgos Spanodimitriou *, Michelangelo Scorpio, Antonio Rosato and Sergio Sibilio *

Department of Architecture and Industrial Design, University of Campania Luigi Vanvitelli, 81031 Aversa, Italy; michelangelo.scorpio@unicampania.it (M.S.); antonio.rosato@unicampania.it (A.R.)

* Correspondence: giovanni.ciampi@unicampania.it (G.C.); yorgos.spanodimitriou@unicampania.it (Y.S.); sergio.sibilio@unicampania.it (S.S.)

Abstract: Plastic materials are increasingly becoming used in the building envelope, despite a lack of investigation on their effects. In this work, an extruded Acrylonitrile-Butadiene-Styrene panel has been tested as a second-skin layer in a ventilated facade system using a full-scale facility. The experimental results show that it is possible to achieve performances very similar to conventional materials. A numerical model has then been developed and used to investigate the performances of plastic and composite polymer panels as second-skin layers. The experimental data has been used to verify the behavior of the numerical model, from a thermal point of view, showing good reliability, with a root mean square error lower than 0.40 °C. This model has then been applied in different refurbishment cases upon varying: the polymer and the manufacturing technology (extruded or 3D-printed panels). Eight refurbishment case studies have been carried out on a typical office building located in Napoli (Italy), by means of a dynamic simulation software. The simulation results show that the proposed actions allow the reduction of the thermal and cooling energy demand (up to 6.9% and 3.1%, respectively), as well as the non-renewable primary energy consumption (up to 2.6%), in comparison to the reference case study.

Keywords: ventilated facade; second-skin materials; 3D printed materials; additive manufacturing; TRNSYS; full-scale facility; retrofit action; energy saving

check for updates

Citation: Ciampi, G.; Spanodimitriou, Y.; Scorpio, M.; Rosato, A.; Sibilio, S. Energy Performances Assessment of Extruded and 3D Printed Polymers Integrated into Building Envelopes for a South Italian Case Study. *Buildings* **2021**, *11*, 141. https://doi.org/10.3390/buildings11040141

Academic Editors: Alessandro Cannavale, Francesco Martellotta and Francesco Fiorito

Received: 18 February 2021
Accepted: 25 March 2021
Published: 1 April 2021

Publisher's Note: MDPI stays neutral with regard to jurisdictional claims in published maps and institutional affiliations.

Copyright: © 2021 by the authors. Licensee MDPI, Basel, Switzerland. This article is an open access article distributed under the terms and conditions of the Creative Commons Attribution (CC BY) license (https://creativecommons.org/licenses/by/4.0/).

1. Introduction

Approximately 40% of the EU energy consumption can be directly attributed to the building sector, which is also responsible for about 36% of the greenhouse gas emissions [1,2]. In addition, in the EU-28, only 3% of the edifices have an efficient building envelope [3], mainly due to the fact that about 35% of the EU's buildings are over 50 years old and only around 1% of them are renovated each year [1]. Certainly, the constraints associated with the new buildings are fewer with respect to those associated with the refurbishment of existing constructions, so the new one allows for better-optimized design in terms of energy efficiency of the envelope [4]. However, in Italy, many buildings (about 4 million) were built in the early 1900s and about half of these have been classified as historical architectures and nowadays have been reused [5]. Therefore, in the Italian scenario, the improvement of the energetic performances of the existing building envelope represents a crucial aspect in the increasing of the building's energy efficiency and the indoor environmental quality on a large-scale [6]. In this context, different products and systems have been proposed to improve the buildings energy efficiency, visual and thermal comfort, as well as their sustainability [7–12] and, in recent years, the interest of the scientific community has seen an increase in the facade domain to improve the overall building energy efficiency [13]. In particular, the use of passive systems is raising more and more interest in the building sector [7,8,14]. A passive building is one in which the indoor environment is not regulated by using mechanical heating and cooling systems, but by means of a conscious structure and architectural design of the envelope and its components [7]. In recent years, as part of a shift towards more energy-efficient buildings, a lot of different new facade technologies and solutions have been proposed for the improvement of their energy performance by the introduction of better insulation, shading devices, as well as a second-skin layer (double-skin facades [15,16], building integrated photovoltaic [17,18] and opaque ventilated facades [19,20]). Among these, the double-skin facade (DSF) and opaque ventilated facades (OVF) have been suggested as one of the best solutions, thanks to their ability to ensure better thermal performance and indoor environmental quality, as well as to improve the aesthetic appeal of buildings [15,16,19,20]. The concept of DSF was introduced in the early 1900s, but little progress was made until the 1990s [7,21]; it consists of a standard facade, an air cavity, and an additional external skin. The material used as a second-skin is usually glass [7,21]; however, a shading device can also be installed within the cavity between the two layers of the facade to control the solar radiation [7,21]. The OVFs are passive systems that consist of multiple layer construction (external second-skin, an intermediate air cavity, and an internal wall). The OVFs have been, more and more frequently, chosen by contractors, designers as well as architects for different typologies (offices, schools, residential) of new and renovated buildings as well as in different climates [19,20]. Several papers [22–29] investigated the energy performances of OVFs through simulation software (EnergyPlus [30] and TRNSYS [31], the most widely used), highlighting the benefits achievable by these systems. In addition, the literature review [22–29] highlights that the materials usually used as second-skin layer are: glass, porcelain stoneware tile, natural stone, aluminum, OSB, and composite panels.

Nowadays, there are always more innovative materials [32–35] used in architecture, also as a second-skin layer, even if evaluating their impact on the envelope's energy performance is a complex task [6,11]. In this work, plastic and composite polymers have been investigated with the aim to evaluate their integration in the building facade and their potential benefits achievable in refurbishment case studies.

1.1. Plastic and Composite Polymers in Building Facades

In recent years, the use of polymers in building and engineering has increased substantially, thanks to their: (i) ease of production, (ii) ease of installation, (iii) durability, (iv) low maintenance requirements, (v) lightweight nature and (vi) ability to be formed into complex shapes [34,35]. Moreover, polymers form good thermal and electrical insulators that are not affected by chemical and biological risks [35]. They have not only been used to replace the traditional construction materials (cement, brick, concrete, wood, metal, and glass), but these materials have also been used in a complementary way to improve the building envelope performance to satisfy the modern demands of both new projects and refurbishment ones [32–35]. From an aesthetic point of view, these materials are available in several colors and texture alternatives. Several applications of plastic and composite polymer walls in buildings were reported in the literature [33]. In [33], 23 examples of architecture were reported, demonstrating that plastic and composite polymers can be used in dwelling domes, large-span volumes or envelop large facade surfaces, and transparent sky-lights on the roofs of industrial buildings. Another example can be found in [36,37], where the designers have realized a temporary pavilion with an envelope made of double-walled transparent corrugated sheets of PolyEthylene Terephthalate (PET) recycled plastic. Similarly, in [38], a massive pavilion, designed as an exhibition hall for the 2010 Taipei International Flora Expo, has been built using recycled PET for the building envelope, also proving strong resistance to fires and earthquakes. As reported in [39], polycarbonate multi-sheet systems are increasing their share of the glazing market since they provide good performance while weighing and costing significantly less than glass. For these reasons, several studies have been conducted to assess their characteristics upon varying geometries and installation typology. In [39], the authors assessed the thermal and optical characteristics of different polycarbonate sheets, highlighting a strong angular dependence in polycarbonate panels' optical properties, significantly different to the conventional multiple-layered glass, and a good overall thermal behavior. For these reasons, the polycarbonate could be used as a valid alternative, fulfilling the energy requirements and improving the visual comfort, reducing the glaring problem by diffusing the light, while providing more flexibility in the design and appearance of the buildings. In [40], a detailed experimental analysis of the thermal behavior of different polycarbonate multi-sheet systems has been carried out by varying the installation angle. The results highlighted a very low incidence of the angle of inclination on the equivalent value of thermal conductivity, thus allowing the material to be equally used in every part of the building envelope. Thus, polycarbonate has already found large usage, as in the Bavaria Brewery Tocancipá headquarters, by Construcciones Planificadas [41], where the plastic has been used to get an industrial appearance while providing for thermal and visual comfort, or the property registration offices, by Irisarri + Piñera, where polycarbonate has been used to complement and balance the appearance of the existing structure.

Across these materials already implemented in the traditional architectural paradigms, there is also a strong boost in the usage of plastics from the additive manufacturing (AM) sector, as several plastic products can also be utilized in AM processes, providing great freedom of form and enhancing designers', architects', and engineers' freedom in creating complex designs [42–46]. In addition, the global 3D printing filament material market volume was 1.8 billion US$ in 2019, growing at a compound annual growth rate (CAGR) of 27% [47]. The most popular filament materials are PolyLactic Acid (PLA) and Acrylonitrile-Butadiene-Styrene (ABS), holding about 47% and 29% of the market, respectively. In contrast, materials as PolyEthylene Terephthalate Glycol-modified (PETG) can be useful alternatives, despite not being as popular, providing similar mechanical properties while also offering excellent recyclability and scalability [47]. The AM in the facade industry presents new significant potential and requires relevant research to be conducted [46]. Indeed, more and more 3D printing materials are concurrently becoming utilized in contemporary architecture design [42–46], thanks to the lightness and effortless installation procedure, which results in a design solution useful for both new projects and refurbishment ones [44–46]. The 3D printing technology has been often used to create everything, from prototypes [42,43,46] and simple parts of facades, to give a distinctive signature to the constructions [44,45]. The 3D printing materials prove themselves to offer quite unique characteristics from an architectural and economic point of view [43–46]. Several scientific papers have been conducted into the loadbearing capacities and/or other essential qualities of AM products for the building industry, such as durability, water vapor diffusion resistance, thermal conductivity, or fire-resistance [48–51]. The authors emphasized the potential applications of additive manufacturing to build honeycomb panels that optimize mechanical properties and heat transfer [48–51]. However, scientific research related to 3D printing materials in building energy efficiency applications is limited due to its relatively new nature as a technology. Only Sarakinioti et al. [46] aimed their research at developing an integrated 3D printed for thermal insulation and building physics. In particular, they presented a 3D printed facade panel design for thermal insulation and movable liquid heat storage [46], providing an overview of the development process. The authors tested the prototype and, at the same time, simulated to verify the thermal effects of the proposed facade system on indoor spaces in different climates. The simulation results reported in [46] showed the potential of the proposed 3D printed facade panel for reducing heating and cooling energy demand. Therefore, the effects of adopting 3D printing materials as a second-skin layer on the indoor environment have been scarcely investigated.

Moreover, there is a lack of experimental testing and numerical model development of these materials in building simulation, even more, if considered in a second-skin in front of the building envelope, in a 3D printed composite facade arrangement. Indeed, in a facade arrangement realized with these innovative materials (ABS, PLA, PETG, etc.), the difficulties lie in predicting the behavior of the various facade sections, as the second-skin layer, the resulting air cavity, and finally, the effects on the indoor environment. Therefore, from an experimental point of view, standardization bodies, experts, and researchers are continually developing new methodologies or new procedures to correctly calculate the performances of these envelope components in simple and economical ways [6].

1.2. Research Aims

In this work, extruded ABS panels have been tested as second-skin materials in order to verify their performances in an OVF system. This novel material for building envelopes has been investigated through in-situ measurements by using two outdoor comparative test cells. The experimental data have been used to calibrate and validate a numerical model in TRNSYS 18 [31], also verifying the ability of the simulation software to effectively reproduce the behavior of a light material in an OVF system, which usually is made of materials as porcelain gres. Then, the validated numerical methodology has been used to implement different plastic materials in a set of refurbishment case studies, compared to a reference office building, in order to assess the potential benefits. The comparison has been performed in terms of (i) heating and (ii) cooling energy demands, as well as (iii) non-renewable primary energy consumption, upon varying the plastic material. Finally, additional refurbishment case studies have also been implemented considering 3D printed panels as a second-skin layer.

The aims of this research can be summarized as reported below:

- investigate the performances of extruded ABS panels as a second-skin layer for innovative building envelopes with experimental tests in-situ;
- calibrate and validate a simulation model to predict the energy performance of the plastic and composite polymer panels used as a second-skin layer in an OVF system;
- assess the potential energy saving achievable in office building refurbishment using the proposed materials (extruded and 3D printed polymers) through numerical simulation.

1.3. Structure of the Research

The research is structured as follows.

Section 2 describes the methodology used to carry out the research, showing in detail how (i) the experimental data have been acquired, (ii) the numerical model has been implemented in TRNSYS 18, and how (iii) the experimental data have been used to calibrate and validate this numerical model.

Section 3 reports the numerical results, in terms of the reduction of non-renewable primary energy consumption, achieved in an office building refurbishment through the installation of an OVF system, upon varying the material used as a second-skin layer, considering both extruded and 3D printed ones.

Finally, Section 4 discusses the integration capacity of the plastic and composite polymer panels in a second-skin layer of an OVF system, highlighting the advantages and limitations of such materials.

2. Methodology

This section describes in detail the measurement methodologies and the experimental results obtained during the in-situ test as well as the methods and results related to the validation of the implemented numerical model.

2.1. Description of the Gemini Facilities, Experimental Results, and Discussion

In this sub-section, a couple of experimental test cells and the experimental results are reported. Gemini facilities [11] are designed and built at the Ri.A.S.–Built Environment Control Laboratory [52] of the Department of Architecture and Industrial Design of the University of Campania Luigi Vanvitelli in Aversa (40°59′39.1″ N, 14°10′48.5″ E). These full-size outdoor test cells have been designed to experimentally evaluate double-skin facade module performances, in real outdoor weather conditions. The test cells are designed as an identical couple in order to carry out comparative measurements. The Gemini's internal dimensions are 2.20 m wide by 2.80 m deep and 2.40 m tall, oriented with the long side along the north-south axis. These dimensions correspond to the gross dimension of the main steel frame structure, on which the shell has been fixed externally and seamlessly in order to avoid thermal bridges. The shell has been realized in a single layer of 10 mm thick sandwich panels consisting of two galvanized steel sheets and a polyurethane rigid foam filling, with a thermal transmittance (U_{wall}) value of 0.23 W/m^{2}K [11]. Then, for the floor, a 0.10 m air gap and a wood flooring have been added above the structure, while, for the ceiling, a sheet metal roof has been placed 0.10 m above the outer panels, with a 2% slope, to allow a natural rainwater outflow.

The Gemini's facilities are designed to allow the in-situ characterization of innovative layers to be applied in double-skin facades with different geometries, layout, materials, and technologies. The acquired data can be used to evaluate the in-situ performances of the system under investigation and to realize, calibrate and validate simulation models.

The Gemini is well-instrumented to acquire different indoor and outdoor physical quantities. Table 1 shows the measurement range, the accuracy, and the response time of the sensors used for outdoor and indoor climate characterization. In particular, with the aim to evaluate the real weather conditions, sensors for wind direction, wind speed, air temperature, air relative humidity, air pressure, global horizontal radiation, and diffuse horizontal radiation were placed at about 6.50 m from the ground, in the best position to minimize the influence of external obstructions (i.e., the obstructions angles are less than 10°). In order to acquire diffuse horizontal radiation, one of the pyranometers is equipped with a shadow ring (diameter of 0.574 m and thickness equal to 0.052 m), and the data were corrected following the methodology proposed in [53], to take into account both the isotropic and anisotropic conditions. Figure 1 shows the weather station, with all the aforementioned sensors.

Table 1. Installed Gemini sensor measurement range and accuracy.

Number of Sensors	Measured Quantity	Type	Range	Accuracy
1	Wind speed	'Pro First Class' anemometer	0–50 m/s	±0.01 m/s
1	Wind direction	'Pro First Class' anemometer	0–356.9° ± 3°	±1°
3	Air Temperature and Relative humidity	Thermo-hygrometer with precision transducer	Temperature: −40–+60 °C Rel. Humidity: 0–100%	Temperature: ±0.2 °C Rel. Humidity: ±2%
1	Atmospheric pressure	Barometer with piezo-resistive transducer	800–1100 hPa	±0.3 hPa at 20 °C
3	Solar radiation	II class thermopile pyranometer	0–2000 W/m^2	±10 μV/(W/m^2)
2	Air cavity speed	Hot wire air speed transmitter	0.2–40.0 m/s	±0.2 m/s +3% f.s.
10	Temperature	T-Type thermocouple	−200–+350 °C	±1.5 °C

Figure 1. The weather station used to monitor the real outdoor conditions.

The air temperature inside the Gemini is monitored by a combined temperature-relative humidity sensor placed in the middle of the room. Also, when the test cells are configured to test a second-skin system, this is monitored through a set of ten thermocouples, placed on the significant interfaces and in the cavity.

The sensor layout follows the layout shown in Figure 2a. In particular, it can be noted that (i) six thermocouples are placed in the middle of both the inlet (T1, T2, and T3) and the outlet (T8, T9, and T10) sections of the air cavity, (ii) four thermocouples are placed in line at the center of the second-skin system (T4 on the back surface of the second-skin, T5 in the middle of the cavity, T6 on the external surface of the south wall of the test cell, T7 on the internal surface of the south wall of the test cell, respectively); this last set of thermocouples falls in line with the same thermo-hygrometer which monitors the temperature of the air inside the test cell, in order to have all the sensors aligned at the center of the system.

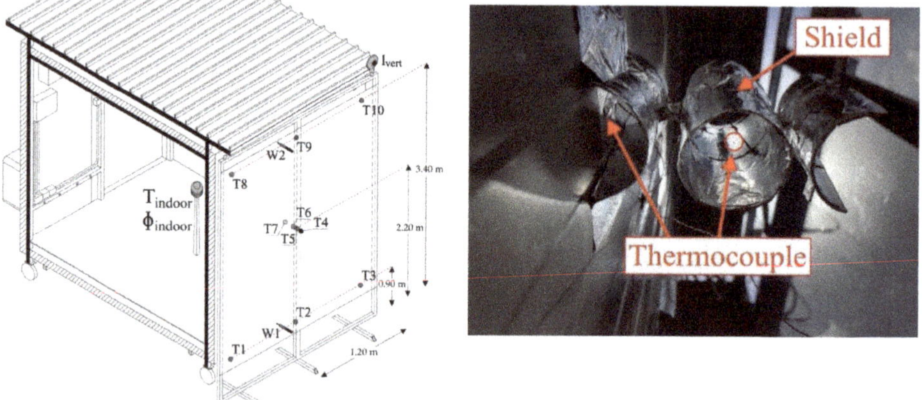

(a) (b)

Figure 2. (**a**) Axonometric views of the sensor layout; (**b**) shielding devices used in the experimental setup to avoid any direct solar radiation on the thermocouples.

In addition, all the thermocouples (Tx) have been shielded with high-reflective domes in order to avoid any direct solar radiation (Figure 2b).

The pyranometer (I_{vert}) has been installed to acquire the vertical solar radiation incident on the south surface and, finally, two hot-wire anemometers (W_{in}, and W_{out}) are placed in the air cavity, one in the inlet section and the other one in the outlet section, in order to monitor the airflow in the second-skin cavity.

In order to verify the measurement methodologies and characterize each test cell from the thermal point of view, preliminary data has been acquired in a standard configuration (both Gemini s without a second-skin system). These data are recorded with the aim of (i) verifying the operation of the different instruments and their correct positioning, (ii) comparing the thermal behavior of the two test cells, and (iii) defining a reference point for the following evaluation of the real performances of double-skin facades or smart windows. The preliminary experimental data were acquired and stored every 1 min on a period of 1 month (from 1 June to 30 June) and, later, averaged on an interval of 15 min. Figure 3 reports the indoor air temperature of Gemini 1 and Gemini 2, the external air temperature, and the global horizontal radiation for three typical days in June. This figure highlights that the difference between the indoor air temperature of Gemini 1 and the indoor air temperature of Gemini 2 ($T_{indoor,\ Gemini\ 1}-T_{indoor,\ Gemini\ 2}$) is negligible, varying within an interval with a maximum of 0.2 °C and a minimum of −0.2 °C.

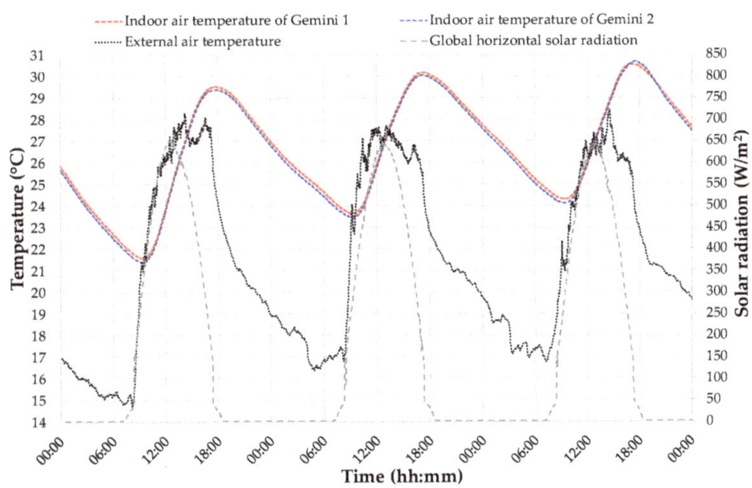

Figure 3. Gemini 1 and Gemini 2 preliminary experimental indoor air temperature trends for three typical June days.

After the preliminary experimental campaign, a second-skin system has been mounted and tested on a test cell (Gemini 1) with an air cavity gap equal to 0.10 m, while the other cell has been left unequipped and used as a reference (Gemini 2). In particular, the investigated second-skin system has been realized with extruded ABS panels [54]. The experimental data were acquired and stored every 1 min over a period of 1 month (from 8 December to 31 December) and, later, averaged on an interval of 15 min. This experimental campaign aims to verify the performances of plastic panels as a second-skin layer for innovative envelopes.

The extruded ABS panels have been selected with dimensions equal 600 mm × 1200 mm; such dimensions have been selected on the basis of the dimension of conventional panels in OVF systems, in order to guarantee an easy installation in a commercial OVF structure, as well as an easy substitution of conventional OVF system materials in a realistic scenario. In Figure 4a, a detailed view of the analyzed extruded ABS panel before the installation is displayed, while Figure 4b shows the Gemini 1 equipped with the second-skin realized with six extruded ABS panels.

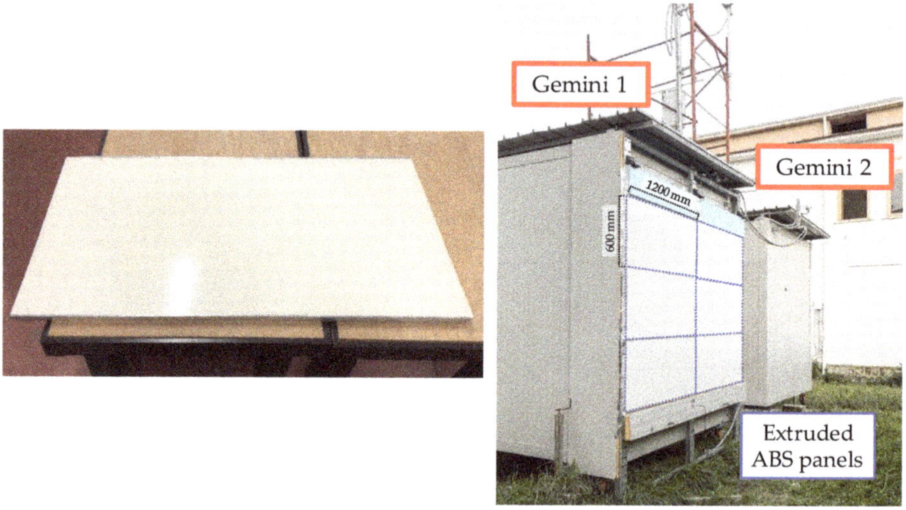

(a) (b)

Figure 4. (**a**) Extruded ABS panels [54] detail view; (**b**) Gemini 1 equipped with the OVF system.

The second-skin system equipped during the tests has been realized by mounting the six extruded ABS panels on a steel frame, then hanging the whole system to the brackets on the south facade of the test cell Gemini 1. Finally, the sides of the second-skin system were covered and sealed by means of panels, similar to those used for the test cells' envelope, in order to allow only for a vertical airflow in the cavity (Figure 4b). In this configuration, the acquisition period lasted for almost a month, in the wintertime. During the acquisition period, the temperatures were monitored following the layout reported in Figure 2a. Figure 5 shows an overview of the whole acquisition period, reporting the external air temperature, the global horizontal radiation, and the total vertical radiation on the south facade.

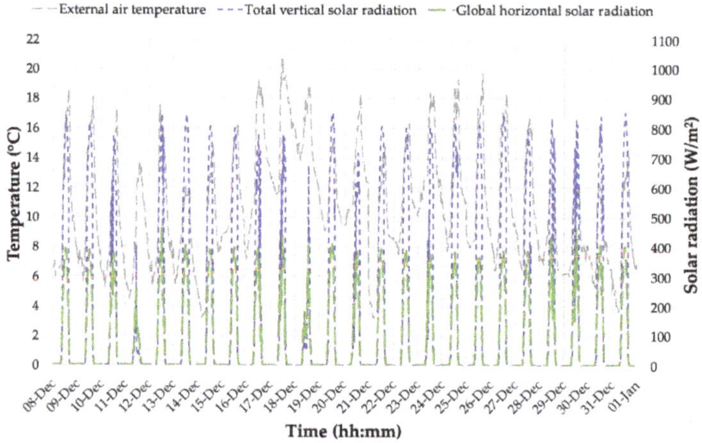

Figure 5. Overview of the whole winter acquisition period.

Figure 5 highlights that the external air temperature was quite warm, despite being the winter season. Also, the radiation values, both global horizontal and total vertical, show mostly high values, thus confirming good weather and clear sky across the whole acquisition period.

Figure 6a,b report a focus for four typical days on the weather conditions during the measurements with the Gemini 1 equipped with the second-skin system, and the Gemini 2 left uncovered as a reference. In particular, Figure 6a shows the temperature and solar radiation data (global horizontal solar radiation, diffuse horizontal solar radiation, and total vertical solar radiation on the south facade), while Figure 6b reports the wind characteristics acquired during the analyzed days; on the left axis the wind speed is reported, while on the right axis the wind direction is displayed, considering 0° as north direction, 90° as east direction, 180° as south direction, and 270° as west direction.

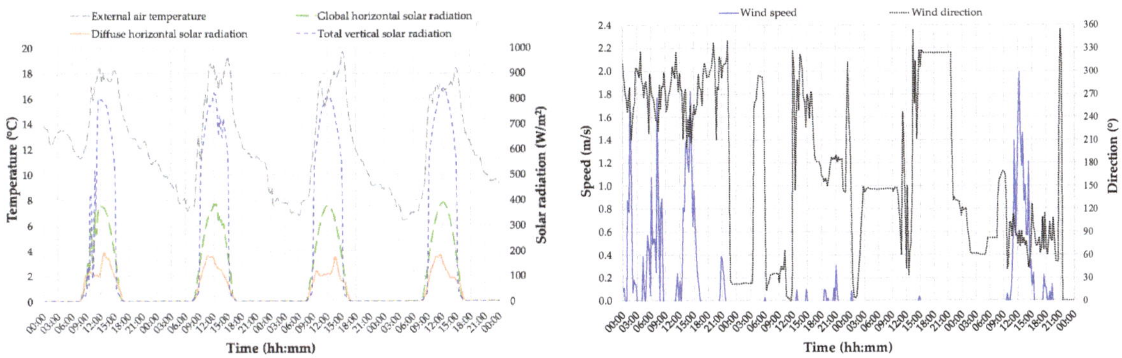

(**a**) (**b**)

Figure 6. Weather conditions during four typical acquisition days: (**a**) solar radiation and temperature; (**b**) wind speed and wind direction.

Figure 6a better highlights that, during the measurement period, sunshine days were acquired with atypical temperatures for the period, ranging between a minimum of about 5.9 °C and a maximum of about 19.4 °C.

Figure 6b shows a low wind speed in general during the measurement period and a slight wind predominance in the west/north-west direction. Also, the wind speed values acquired during the nighttime are equal to zero, due to a threshold value for the start/stop of the sensor equal to 0.15 m/s.

Figures 7 and 8 report the experimental data associated with the cavity with a 1-h timestep for a single acquisition day. In more detail, Figure 7 shows the trends of the daily values of the air temperature inside the cavity upon varying the height from the ground, while Figure 8 reports the airspeed at the inlet and the outlet of the air cavity, for the same day.

The data reported in Figure 7 corresponds to the measures of the three thermocouples positioned in the middle point of the cavity of the second-skin system, more specifically T2, in the middle of the air cavity inlet, T5, in the middle of the air cavity geometrical center, and T9, in the middle of the air cavity outlet, as shown in Figure 2a. As a first observation, the overall temperature distribution is directly related to solar radiation throughout the day, where the temperatures rise during the morning and drop during the afternoon. Also, during the day, the temperature trend seems to be substantially constant from the air cavity inlet to the middle of the facade, and then to increase to the air cavity outlet; this behavior is due to the chimney effect that is created thanks to the OVF system. Then, in the evening (starting from 16.00), the temperatures at the outlet of the cavity are only slightly higher than those at the center and the inlet; this is due to the reduction of solar radiation happening in the evening.

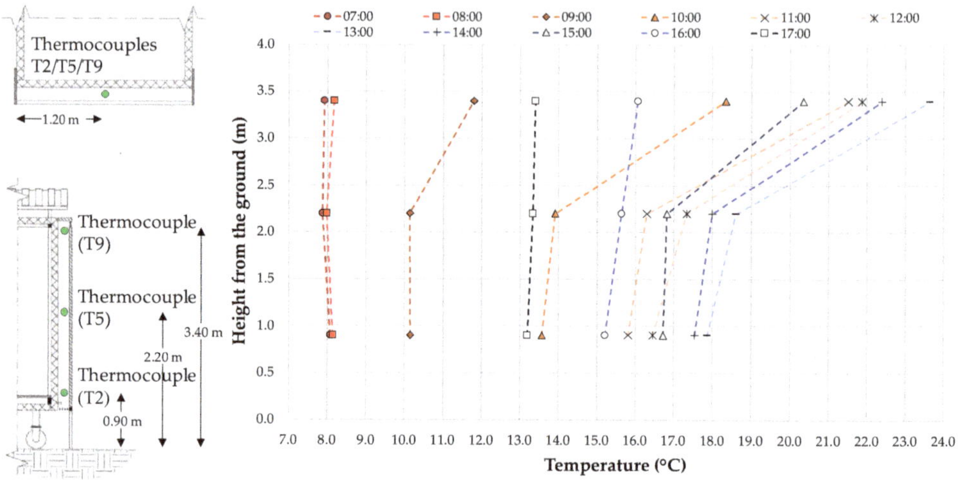

Figure 7. Daily values of the cavity air temperature.

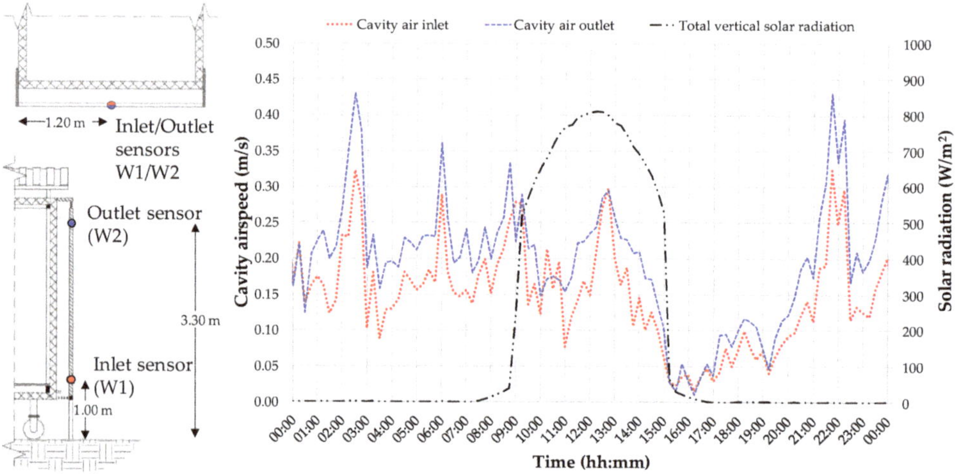

Figure 8. Daily values of the airspeed at the inlet and the outlet of the air cavity.

Figure 8 reports data acquired by the hot-wire anemometers placed based on the layout shown in Figure 2a, where W_{in} measured the airspeed value at the inlet of the cavity and W_{out} measured the value at the outlet of the cavity. This figure shows how, during the day, the values acquired by W_{out} is higher than those measured by W_{in}, referable to a direct effect of the solar radiation on the rising of the air temperatures along the air cavity, thus causing an increase in airspeed. After 15:00, the chimney effect in the air cavity is reduced because the temperatures are gradually decreasing over time due to the reduction of the solar radiation on the south facade; this causes a significant drop in the airspeed values measured in the cavity.

Therefore, the analysis of the experimental results shows that a plastic material (i.e., ABS) can be used as a second-skin layer in OVF systems.

2.2. Description of the Numerical Model

The software TRNSYS 18 [31] has been used to model the Gemini test cells and to develop the second-skin model. TRNSYS software adopts a modular approach by using Fortran subroutines. Each Fortran subroutine is called a 'Type' and contains the model for a single system component. Several studies have been carried out in order to validate the numerical models developed in TRNSYS from the Colorado State University experimental houses and other researchers around the world [55–58]. In this study, the following main TRNSYS Types [31,59–61] have been used:

- Type 56 to simulate the Gemini test cells [59];
- Type 1230 to model the second-skin system [61];
- Type 16c to estimate the solar radiation on the Gemini and second-skin system surfaces [59,60];
- Type 69b to determine the sky temperature [59,60];
- Type 33e to determine the moist air properties [59,60].

At first, the Type 56 subroutine has been used to model the thermal behavior of the Gemini test cells; in particular, two thermal zones have been modeled, one for each test cell. Also, Type 56 contains information about the test cells' surroundings (buildings, trees, bushes), which are described as 'shading objects'. The geometrical modeling occurred in the SketchUp software [62], where it was possible to model the shapes and the position of each element accurately. Then, by means of the Trnsys3D plug-in, the geometries were imported into the Type 56 subroutine. The physical properties of each test cell's external surface have been defined, on the basis of the data provided by the manufacturers. Lastly, the internal thermal gains have been set for each test cell, which was determined on the basis of the equipment installed inside the facilities (notebook, data acquisition systems, and uninterruptible power supply units).

The OVF system has been modeled using the Type 1230 subroutine, which effectively reproduces the behavior of an external second-skin layer with an air cavity behind it. Using this TRNSYS Type, the behavior of the OVF system has been correlated to that of the Gemini test cell modeled through the Type 56 subroutine. In particular, the last external layer of the Type 56 wall acts as an interface layer between the Type 1230 and the Type 56, by coupling its temperature and thermal resistance to model the wall heat transfer. Figure 9 shows a schematic of the boundaries of the two coupled Types (56 and 1230), highlighting the resistive interface layer. The Type 1230 parameters have been set following the data provided by the manufacturer of the extruded ABS panels [54], taking into account thickness, density, and thermal conductivity, specifically.

During the simulations, Type 1230 takes into account:

- the solar radiation, the longwave radiation, and the air convection on the external surface of the outside layer;
- the energy storage and the conduction in the outside layer;
- radiation exchange between the outside layer and the air cavity;
- the convective exchanges from all the surfaces facing in the air cavity;
- the conduction through the interface layer.

Type 16c has been implemented to model the solar radiation on all the external surfaces. This Type accepts global radiation, ambient temperature, and ambient relative humidity data as input, in order to output several quantities related to the position of the sun, as the diffuse radiation fraction on the horizontal, by estimating the cloudiness of the sky on the basis of the dry bulb temperature and the dew point temperature. Finally, the radiation on every external surface is computed, on the basis of their own orientation.

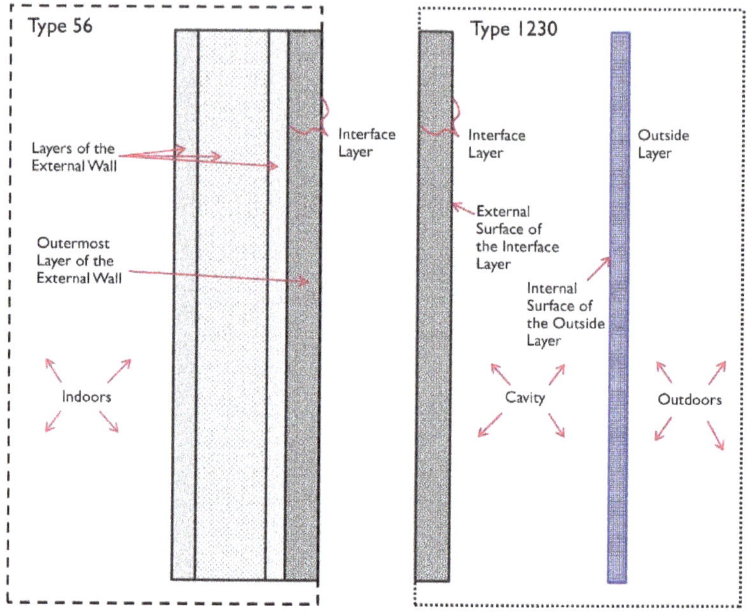

Figure 9. Coupling between Type 56 and Type 1230 [61].

The effective sky temperature is determined by means of Type 69b, which calculates the long-wave radiation exchanges between the external surfaces and the atmosphere. Type 69b calculates the cloudiness factor as well, on the basis of the dry bulb and the dew point temperatures.

Finally, Type 33e has been implemented in order to calculate the properties of the moist air, in particular, by taking the air temperature, the relative humidity, and the air pressure as input; it returns the density of the air mixture for every timestep, which is then used to calculate the airflow at the OVF inlet. In this way, the inlet airflow is not a fixed value, but it corresponds to the experimental data acquired through the hot wire anemometers, placed as reported in Figure 2a.

In this study, the experimental weather data acquired from 8 December to 31 December have been used as input data for the Type 16c, Type 69b, and Type 33e.

During the simulation, both the time base used to solve the differential equations and the simulation timestep has been set equal to 15 min in order to have a full correlation to the timestep of the experimental input data.

2.3. Validation of the Numerical Model

This sub-section reports the methods and results related to the validation of the numerical model. The model reliability has been verified in terms of indoor air temperature (T_{indoor}) and the average temperature of the air cavity (T_{cavity}) by comparing the experimental values with those obtained as an output of the simulation model above described, defining the following percentage differences ΔT_{indoor} and ΔT_{cavity}:

$$\Delta T_{indoor} = \left(T_{indoor,exp} - T_{indoor,sim}\right) / T_{indoor,exp} \tag{1}$$

$$\Delta T_{cavity} = \left(T_{cavity,exp} - T_{cavity,sim}\right) / T_{cavity,exp} \tag{2}$$

Figure 10a,b report the comparison between the simulation results and the experimental data acquired during the whole test period (from 8 December to 31 December) in terms of T_{indoor} and T_{cavity}, respectively.

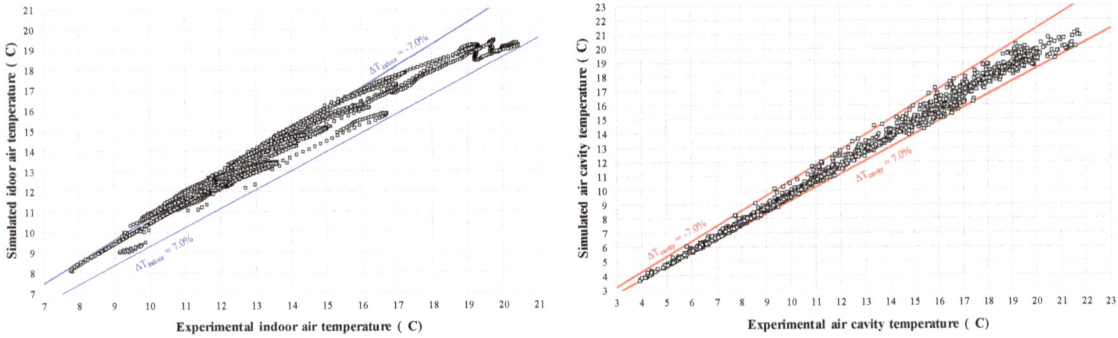

(a) (b)

Figure 10. Comparison between the simulated values and the experimental values acquired during the whole test period in terms of (**a**) T_{indoor} and (**b**) T_{cavity}.

These figures highlight how the developed model is quite reliable, with values of ΔT_{indoor} ranging between a minimum of −9.62%, and a maximum of 6.22%, as well as the values of ΔT_{cavity} between a minimum of −8.51% and a maximum of 7.85%.

The accuracy of the model has also been validated by calculating the mean error (*ME*), the mean absolute error (*MAE*), and the root mean square error (*RMSE*) as reported below:

$$ME = \sum_{i=1}^{N} (T_{\exp,i} - T_{sim,i})/N \quad (3)$$

$$MAE = \sum_{i=1}^{N} |T_{\exp,i} - T_{sim,i}|/N \quad (4)$$

$$RMSE = \sqrt{\sum_{i=1}^{N} ((T_{\exp,i} - T_{sim,i}) - ME)^2/N} \quad (5)$$

where $T_{exp,i}$ is the experimental value at time step i, $T_{sim,i}$ is the simulated value at time step i, and N is the number of measurements. Table 2 reports the values of the *ME*, *MAE*, and *RMSE* for both T_{indoor} and T_{cavity}.

Table 2. Values of *ME*, *MAE*, and *RMSE* obtained by comparing the simulated values and the experimental data acquired during the whole test period.

T_{indoor} (°C)			T_{cavity} (°C)		
ME	*MAE*	*RMSE*	*ME*	*MAE*	*RMSE*
−0.3	0.5	0.4	0.3	0.3	0.2

The values reported in Table 2 highlight that there is a slight difference between the measured and predicted results, in particular: (i) the ME associated to the T_{indoor} is equal to −0.3 °C, which means that the simulation model slightly overestimates the indoor air temperature, while that associated to the T_{cavity} is equal to 0.3 °C; (ii) the values of RMSE are equal to 0.4 °C and 0.2 °C for T_{indoor} and T_{cavity}, respectively.

Therefore, the results show the ability of the Type 1230 to accurately predict the behavior of the extruded ABS panels in an OVF system. Thus, the same methodology is used to carry out a complete numerical campaign on a set of case studies upon varying the polymer (selecting the ones more used in architecture, as highlighted in the literature review) and manufacturing technology (extruded and 3D printed).

3. Materials and Numerical Modeling Implementation

The software TRNSYS 18 [24] is used to assess the potential energy saving achievable in an office building refurbishment using plastic and composite polymers as the second-skin layer material.

The office building modeled in this work is the same for all configurations and it consists of three identical floors. Each floor has a surface of 451 m^2 and a volume equal to 1503 m^3, with a total window area ($A_{w,\,total}$) of 112.3 m^2 ($A_{w,\,North}$ = 24.5 m^2, $A_{w,\,South}$ = 87.8 m^2). It is located in Napoli (latitude = 40°51′ N; longitude = 14°16′ E), and as such, in order to simulate the weather condition, the corresponding EnergyPlus weather data has been used [30]. The office is firstly modeled in the SketchUp 3D modeling software (Figure 11). Then, the 3D model geometries were exported by means of the Trnsys3D plug-in and successively imported into TRNSYS 18 in order to model the building envelope (stratigraphy of the opaque wall and window typology), to define the infiltration, the internal gains, the operation period of the heating and cooling systems as well as the operation of the electric equipment and lighting system. In particular, the same TRNSYS Types described in Section 4 have been used to simulate the case study.

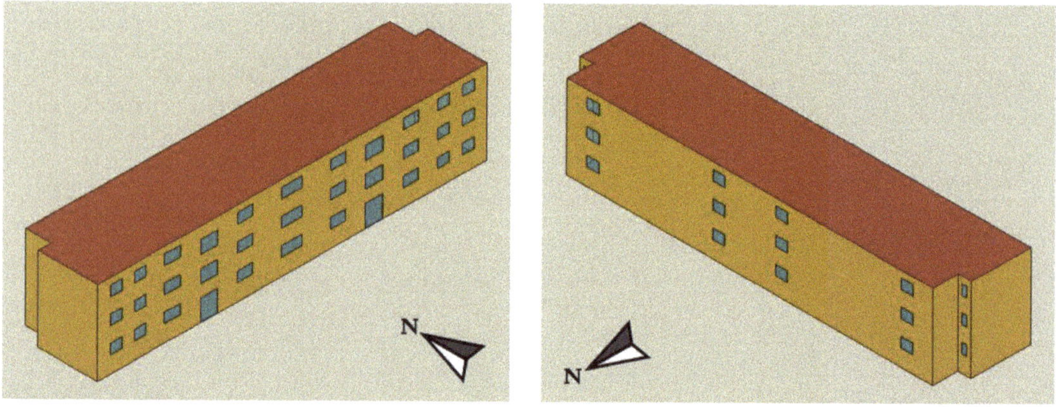

(a) (b)

Figure 11. Office building modeled in SketchUp 3D: (**a**) south view; (**b**) north view.

The same typical three-story office building is investigated upon, varying the insulation layer thickness on the south facade and the typology of second-skin material, for a total of eight case studies. Table 3 summarizes the eight simulation cases investigated in this work. In particular, this table reports the reference case (Case 0) without the second-skin (Figure 12a) and seven refurbishment case studies with the OVF system (Figure 12b) upon varying the second-skin material. These cases are:

- Case 1, with an OVF system made of a conventional second-skin material (Porcelain gres);
- Cases 2–5, where the OVF systems have been implemented by using the extrude plastic and polymer materials more used in architecture (polycarbonate multi-wall sheets, ABS, PETG, and PLA);
- Cases 3_3D–5_3D, where the second-skin materials used in the OVF are the most popular 3D printed polymers (ABS, PETG, and PLA);

Table 3. Summary of case studies investigated.

Case Study	Second-Skin Material	Insulation Thickness (m)	Air Gap (m)
Case 0	-	-	-
Case 1	Porcelain gres	0.072	0.10
Case 2	Polycarbonate multi-wall sheets	0.063	
Case 3	Extruded ABS panels	0.070	
Case 4	Extruded PETG panels	0.071	
Case 5	Extruded PLA panels	0.069	
Case 3_3D	3D printed ABS panels	0.065	
Case 4_3D	3D printed PETG panels	0.067	
Case 5_3D	3D printed PLA panels	0.063	

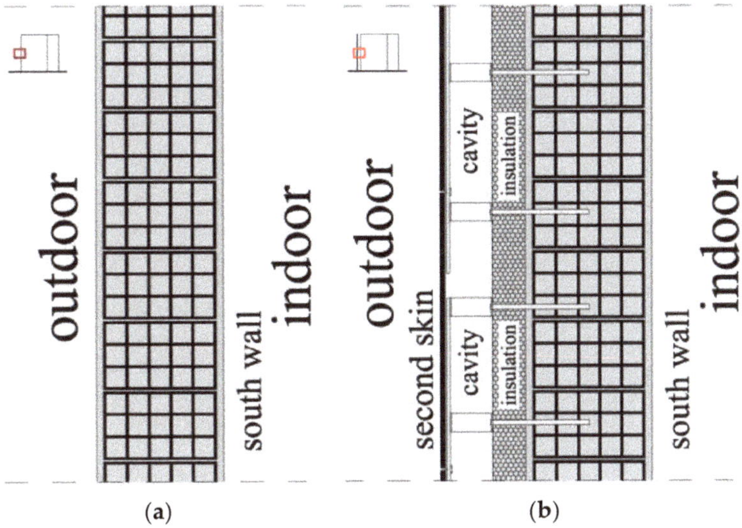

Figure 12. Section of the south wall of the office building: (**a**) reference case; (**b**) retrofit cases with the second-skin system.

In addition, an insulation layer has been added in each case study in order to reach the threshold values specified by the Italian Law [63] and equal to 0.36 W/m^{2}K for the climatic zone considered in this work. The different insulation thicknesses are also reported in Table 3, upon varying the simulation case.

In all the retrofit cases, the cavity inlet airspeed is directly related to the wind speed and direction, as only the wind coming from a similar orientation as the second-skin system (wind direction = 180° ± 45°) has been considered as input for the Type 1230. In addition, the second-skin system has a control logic for the air cavity shutters, which are considered open during the cooling period and closed during the heating period.

Table 4 shows the thermal-physical properties of the opaque walls of the envelope implemented in the case studies.

Table 4. Thermal-physical properties of the opaque walls implemented in the reference case study.

Surface	Material	Thickness (m)	Density (kg/m^3)	Thermal Conductivity (W/mK)	Thermal Capacity (kJ/kgK)
Vertical Walls	Plaster	0.015	1400	0.70	1.01
	Bricks	0.238	600	0.36	0.84
	Mortar	0.015	1800	0.90	0.91
Roof	Plaster	0.015	1400	0.70	1.01
	Lighter concrete	0.027	500	0.17	0.88
	Bricks	0.150	600	0.36	0.84
	Concrete	0.020	600	0.18	0.88
	Bitumen	0.005	1200	0.17	1.47
Floor	Tiles	0.020	2000	1.00	1.00
	Concrete	0.050	600	0.18	0.88
	Bricks	0.150	600	0.36	0.84
	Lighter concrete	0.030	500	0.17	0.88

Table 5 reports the general simulation parameters adopted in the eight different case studies. In particular, this table highlights: (i) the values of the thermal transmittance for both opaque walls and windows, (ii) the air infiltration rate, (iii) the target of the indoor air temperature, the operation period, and the characteristics of the heating and cooling system, (iv) the occupancy schedule and (v) the internal gains. As can be noticed from Table 5, in all the retrofit cases the values of thermal transmittance for the opaque surfaces are equal to those of the reference case (Case 0), with the exception of the south wall, where the OVFs have been implemented and the thermal transmittance has been set equal to the threshold values specified by Italian law [63], considering the second-skin materials, the air cavity and the insulation thicknesses reported in Table 3.

Two parallel-connected electric heat pump (EHP) devices, model CRA/K 91 [64], coupled with a multi-split type air conditioning system, have been used to cover both the heating and cooling demands. The national grid has been used to cover all the electrical energy demand.

The simulation timestep has been set to 30 min.

Finally, Type 1230 [61] has been used to model a second-skin layer in plastic and composite polymers. The parameters required by Type 1230 for each material (i.e., density, thermal capacity, and thermal conductivity), for Cases 1–5, have been derived on the basis of the manufacturers or literature data [40,54,65–67]. With respect to the 3D printed panels (Cases 3_3D–5_3D), several specimens have been printed (Figure 13), in order to measure the final dimensions and density.

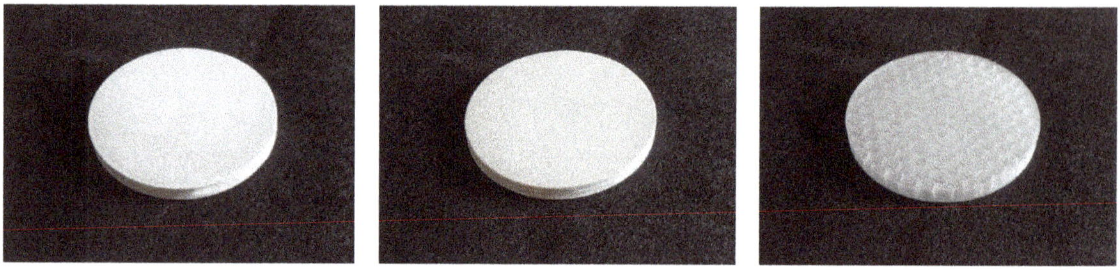

(**a**) (**b**) (**c**)

Figure 13. The specimens made through the 3D printing process: (**a**) ABS [68]; (**b**) PETG [69]; (**c**) PLA [70].

The thermal conductivity of the 3D printed materials (k_{3D}) has been calculated by means of the equation expressed by [48,51] and reported below:

$$k_{3D} = k_m \times \frac{2 \times \left(\frac{k_d}{k_m} - \frac{k_d}{a \times h_c} - 1\right) \times V_d + \frac{k_d}{k_m} + \frac{2 \times k_d}{a \times h_c} + 2}{\left(1 - \frac{k_d}{k_m} - \frac{k_d}{a \times h_c}\right) \times V_d + \frac{k_d}{k_m} + \frac{2 \times k_d}{a \times h_c} + 2} \quad (6)$$

where k_m is the thermal conductivity of the polymer as declared by their manufacturers [68–70], k_d is the thermal conductivity of the filler consisting of the air in the hexagonal cavities of the printed panels (equal to 0.026 W/mK [71]), a is the filler radius measured from the specimens (measured as 0.0045 m), h_c is the interfacial boundary conductance (considered equal to 12 W/m^{2}K [48]) and V_d is the volume fraction of the filler in the polymer matrix (calculated as 0.69 of the specimens' total volume). The internal geometries have been modeled as hexagons as suggested by [50], where hexagons specimens resulted as the most resilient to physical stress, thus more suitable for a building envelope integration.

Table 5. Simulation parameters used in this research.

Parameter	Detail	Value
Thermal Transmittance	Walls and South wall without insulation (Case 0)	U = 1.15 W/m^{2}K
	Roof	U = 1.10 W/m^{2}K
	Floor	U = 0.94 W/m^{2}K
	Windows (frame ratio of 15%)	U = 2.95 W/m^{2}K
	South wall with insulation (Cases 1, 2, 3, 4, 5, 3_3D, 4_3D and 5_3D)	U = 0.36 W/m^{2}K
Infiltration [72,73]	Air changes per hour	0.6 h^{-1}
Heating and Cooling systems	Heating system	Set point = 20 °C [74] Operation period = 16 November/30 March [74] COP = 2.67 [64]
	Cooling system	Set point = 26 °C [74] Operation period = 1 April/15 November [74] EER = 2.41 [64]
Occupancy schedule [75]	Workweek	Weekdays (8:00–18:00) Completely off on the weekends
Internal gains [76]	Lighting systems	Operation = Occupancy schedule Radiative = 11.13 W/m^2 Convective = 4.77 W/m^2
	Equipment	Operation = Occupancy schedule Radiative = 1.4 W/m^2 Convective = 5.6 W/m^2
	Occupants	Operation = Occupancy schedule Radiative/Convective = 2.5 W/m^2 Absolute humidity = 0.0055 kg/hm^2

Table 6 reports the parameters used to simulate the second-skin systems in the retrofit cases.

Table 6. Summary of the main simulation parameters for the Type 1230, upon varying the materials used as a second-skin layer.

Parameters	Case 1	Case 2	Case 3	Case 4	Case 5	Case 3_3D	Case 4_3D	Case 5_3D
Material	Porcelain gres	Polycarbonate multi-wall sheets	Extruded ABS	Extruded PETG	Extruded PLA	3D printed ABS	3D printed PETG	3D printed PLA
Thickness (m)	0.010							
Density (kg/m^3)	2000	300	1040	1300	1300	331	411	395
Thermal capacity (kJ/kgK)	0.840	1.05	1.40	1.20	1.80	1.21	1.07	1.25
Thermal conductivity (W/mK)	1.20	0.0453	0.17	0.29	0.13	0.0548	0.0818	0.0448
Resistance of interface layer (hm^{2}K/kJ)	0.486	0.427	0.472	0.479	0.467	0.438	0.455	0.427
Convective heat transfer coefficient of interface layer (kJ/hm^{2}K)	2.06	2.34	2.12	2.09	2.14	2.28	2.20	2.35

In conclusion, two additional parameters have been set for each case study, required by the Type 1230 as highlighted in Section 4: the resistance of interface layer, to be set in Type 1230 itself, and the convective heat transfer coefficient of the interface layer, to be set in the construction south wall in Type 56 instead. The convective heat transfer coefficients of the back of the south wall value have been set equal to the thermal transmittance value of the insulation layers, which act as interface layers between the two TRNSYS Types, while the thermal resistance values of the interface layer have been simply calculated as the inverse of the convective heat transfer coefficients.

These plastic and polymer materials do not differ only in terms of thermo-physical properties but also in terms of cost. In this work, the capital cost for each retrofit action has been neglected; however, Table 7 provides an overview of the costs per square meter associated with each plastic and polymer material implemented as a second-skin layer in the OVF system [54,77–79]. In general, the polycarbonate multi-wall sheets prove to be the cheaper material (10–25 €/m^2), being also the only one which is already used in the building sector; instead, the 3D printed panels are the more expensive ones (188–225 €/m^2). This cost limitation is typical for the 3D printing technology, especially compared to more traditional building materials. Despite this significant difference in cost, it must be noted that 3D printing is an emerging technology which usage is still not so widespread. However, the 3D printing technology is the only one that would allow for the obtaining of complex panels' shapes easily. Also, the price of the 3D printed panels reported in Table 7 are related to the brand-new spool, while it is possible to integrate also recycled filament spools in the production process [47]; indeed, the 3D printing manufacturing process is the only one where it's easy to fully integrate eco-compatible materials, like PLA.

Table 7. Costs per square meter of plastic and polymer materials used as a second-skin layer in this research [54,77–79].

	Polycarbonate Multi-Wall Sheets	Extruded ABS	Extruded PETG	Extruded PLA	3D Printed ABS *	3D Printed PETG *	3D Printed PLA *
Cost (€/m^2)	10–25	70–175	75–150	75–130	188–225	190–220	190–207

* Considering about 3.2 kg of material and including also the 3D printing cost (equal to 1.0 €/h, for 130 h of the whole printing process), for a panel of 1 m^2.

3.1. Energy Analyses: Methods

According to [12,75], the energy comparison between the proposed case (PC) and the reference case (RC) has been carried out considering the non-renewable primary energy consumption through the index PES (non-renewable primary energy saving):

$$PES = \left[\left(E_p^{RC} - E_p^{PC}\right)/E_p^{RC}\right] \times 100 \tag{7}$$

where E_p^{RC} is the non-renewable primary energy associated with the reference case (Case 0, see Table 3), while E_p^{PC} is the non-renewable primary energy associated with the eight proposed cases (Cases 1–5 and Cases 3_3D–5_3D, see Table 3).

The values of the E_p^{RC} and E_p^{PC} are calculated as reported below:

$$E_p^{RC} = \left(\frac{E_{th}^{RC}}{COP} + \frac{E_{cool}^{RC}}{EER} + E_{el,equipment} + E_{el,lighting}\right)/\eta_{PP} \tag{8}$$

$$E_p^{PC} = \left(\frac{E_{th}^{PC}}{COP} + \frac{E_{cool}^{PC}}{EER} + E_{el,equipment} + E_{el,lighting}\right)/\eta_{PP} \tag{9}$$

where η_{PP} is the Italian power plants' average efficiency, including the transmission losses, and it is assumed equal to 0.42 [74].

A positive value of the index PES means that the proposed refurbishment allows reducing the non-renewable primary energy consumption with respect to the reference case.

3.2. Energy Analyses: Results

In this section, the simulation results of the refurbishment case study are reported and commented on.

Figure 14 reports the values of PES as a function of the proposed case studies, while Figures 15 and 16 show the main energy flows of the building during the whole simulation period upon varying the simulation case. In particular, Figures 15 and 16 report the thermal energy flows in positive values, while the cooling energy flows in negative values.

These figures highlight that:

- all the proposed OVF systems return positive PES values in comparison to the reference case, which means a reduction of the non-renewable primary energy consumption ranging from 2.58% (Case 1) and 2.64% (Cases 2 and 5_3D); this is due to an average reduction of the thermal and cooling energy demands of about 6.9% and 3.0%, respectively;
- the retrofit actions where the plastic and composite polymers materials are used as a second-skin layer (Cases 2–5 and Cases 3_3D–5_5D, see Table 3) allow for a slight performance improvement with respect to those realized with a conventional second-skin material (Case 1), thanks to a reduction in the space cooling energy demand ranging from 31 kWh (Case 4) and 120 kWh (Case 5_3D);
- the results associated with the polycarbonate multi-wall sheets (Case 2) show a behavior similar to the Cases 3_3D–5_5D, mostly due to the fact that the polycarbonate panels have a structure assimilable to the 3D printing logic;

- considering the cases with the same polymers (Case 3 vs. Case 3_3D, Case 4 vs. Case 4_3D, and Case 5 vs. Case 5_3D), the 3D printed panels allow for a slight improvement in performances.

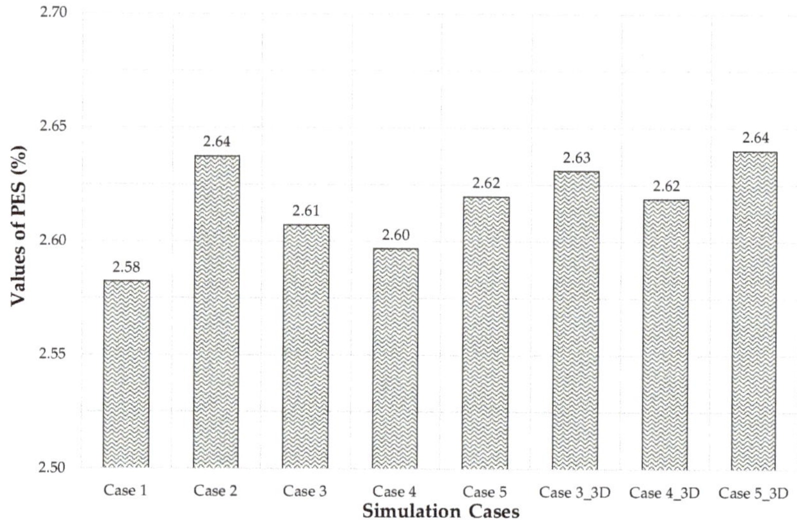

Figure 14. Values of PES upon varying the case studies.

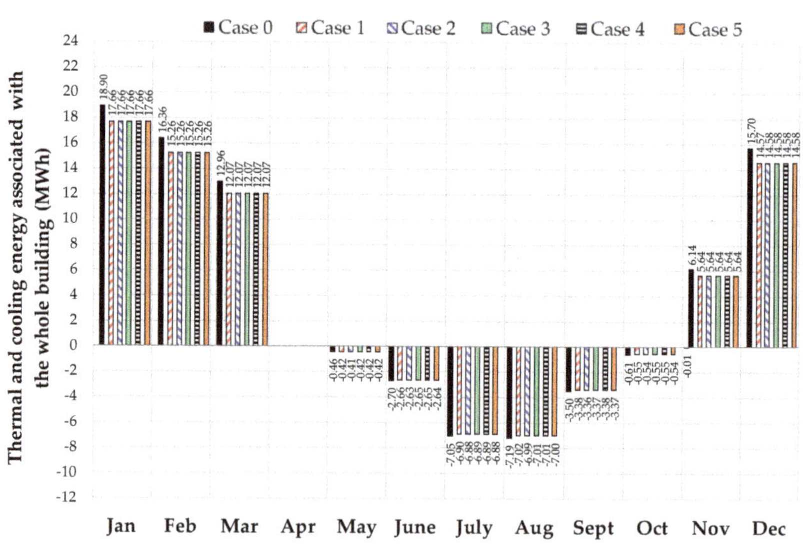

Figure 15. Main energy flows of the building during the whole simulation period associated with Case 0, Case 1, and retrofit Cases with extruded panels.

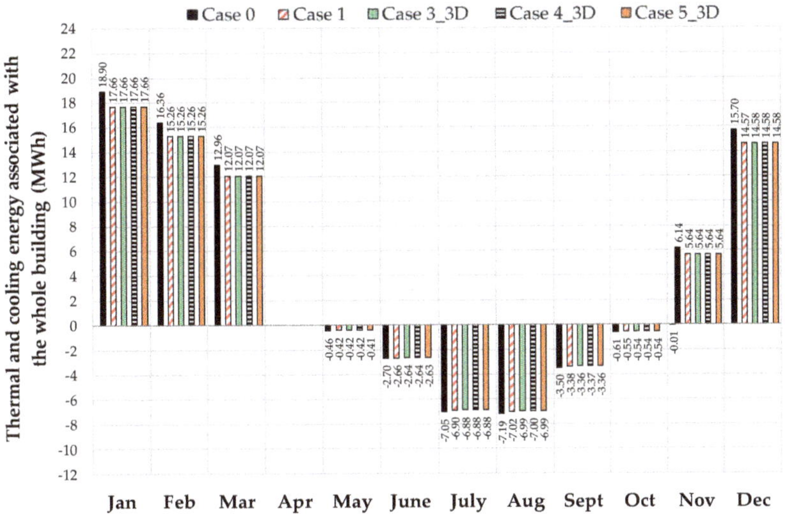

Figure 16. Main energy flows of the building during the whole simulation period associated with Case 0, Case 1, and retrofit Cases with 3D printed panels.

In order to better investigate the performance of the plastic and polymer materials used as a second-skin layer in the proposed OVF system, the trends of the values of the air temperature inside the cavity and the airspeed at the inlet and the outlet of the air cavity for the Case 5_3D in a typical summer day (2 August) is also reported in Figures 17 and 18 with a 1 h timestep. In general, similar trends in the values of the temperatures in the cavity as well as the airspeed at the cavity inlet and the outlet have been predicted for all the other cases.

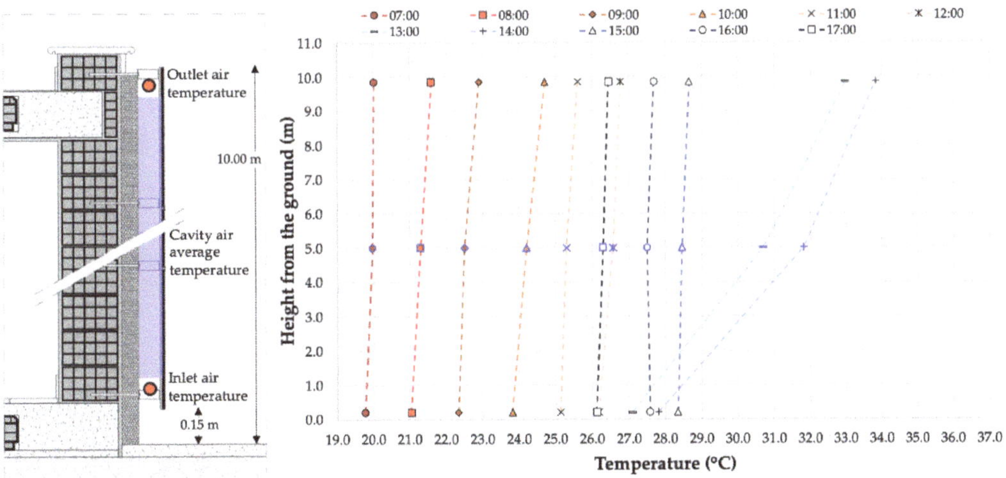

Figure 17. Simulated daily values of the cavity air temperature for a typical summer day.

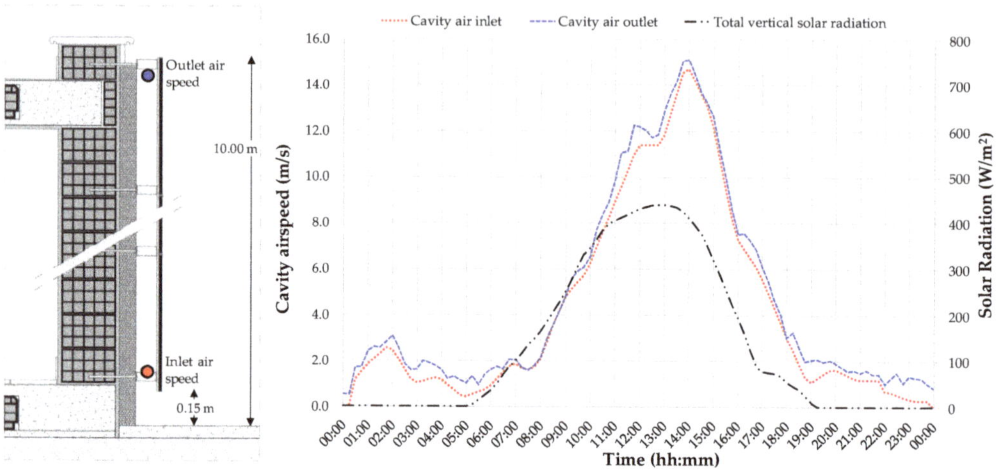

Figure 18. Simulated values of the airspeed at the inlet and the outlet of the air cavity for a typical summer day.

In more detail, Figure 17 shows the trends of the daily values of the air temperature inside the cavity upon varying the height from the ground, and, in particular reports, the temperature of the air at the cavity inlet (about 0.15 m from the ground), the average temperature of the air cavity (highlighted by the blue region in the building section) and the temperature of the air at the cavity outlet (about 10.00 m from the ground); Figure 18 reports the values of the airspeed at the cavity inlet (about 0.15 m from the ground) and the airspeed at the cavity outlet (about 10.00 m from the ground), as well as the total vertical radiation on the external surface of the second-skin layer.

The data reported in Figure 17 corresponds to the input and outputs of Type 1230. In particular, the inlet air temperature is an input for Type 1230, while the average air temperature in the whole cavity (T_{cavity}, the blue-edged markers in Figure 17) and the outlet air temperature are returned as output results by Type 1230 itself. These values represent the only two temperature values associated with the air cavity returned by Type 1230 [61]. As a first observation, the overall temperature distribution is directly related to solar radiation throughout the day, where the temperatures rise during the morning and drop during the afternoon. Also, during the day, the temperature trend seems to be constantly rising from the air cavity inlet to the air cavity outlet; this behavior is due to the chimney effect that is created thanks to the OVF system.

Figure 18 reports the simulation data corresponding to the cavity inlet airspeed (input of the Type 1230) and the cavity outlet airspeed (output of the Type 1230), as well as the total vertical solar radiation. This figure shows how, during the day, the values predicted at the cavity air outlet are always higher than those at the cavity air inlet (with a difference between outlet and inlet ranging from 0.01 m/s and 1.46 m/s), gradually rising to a maximum peak at around 14:00.

Finally, in order to verify the potential benefits coming from the best case, an additional simulation case, not reported in Table 3, has been carried out. In this last simulation, the OVF system has been implemented on the whole building, following the same installation methodology of the previous cases. The material selected as a second-skin layer is the 3D printed PLA, which proved to be one of the most effective in improving the non-renewable primary energy saving. The proposed OVF system returned a PES value equal to 8.10% if compared to the reference case.

4. Conclusions

The OVFs have been, more and more frequently, chosen for different building typologies (offices, schools, residential) and in different climates. Nowadays, there are always more innovative materials used in architecture and as a second-skin layer, even if the evaluation of their impact on the envelope's energy performance is a complex task. In particular, the use of polymers in building and engineering has increased substantially, thanks to their: (i) ease of production, (ii) ease of installation, (iii) durability, (iv) low maintenance requirements, (v) lightweight nature and (vi) ability to be formed into complex shapes. Several of these plastic products can also be utilized in additive manufacturing processes, providing excellent freedom of form, enhancing designers, architects, and engineers' freedom in creating complex designs.

In this work, the numerical model of extruded ABS panels in an OVF system has been developed and validated. Then, the simulation methods, suggested by the authors, have been applied in different refurbishment cases upon varying: the polymer and the manufacturing technology, extrusion (polycarbonate multi-wall sheets, ABS, PETG, and PLA), and 3D printing (ABS, PETG, and PLA).

The simulations have been carried out in order to assess the potential benefits achievable in terms of non-renewable primary energy saving, as well as thermal and cooling energy demand reduction. The simulation results highlight that: (i) all the proposed retrofit cases allow to achieve a benefit in terms of PES; (ii) the plastic and composite polymers materials allow for a slight performance improvement with respect to conventional second-skin material, such as porcelain gres; (iii) the best performances among the extruded polymers are returned by the polycarbonate multi-wall sheets (PES value equal to 2.64%); (iv) the best performances among the 3D printed polymers are achieved when using the PLA (PES value equal to 2.64%). However, the polymers' results show very similar performances, thus allowing building contractors, designers, and architects to select the material based on other project requirements, as mechanical strength, weather resistance, environmental impact, etc.

In this work, the thermal conductivity of the 3D printed materials has been calculated by means of an equation expressed in literature; therefore, in order to improve the accuracy of the model for the 3D printed material, in future works, the authors will carry out experimental investigations on full-scale 3D printed panels in OVF systems through the Gemini test cells. In addition, the capital cost for each retrofit action has been neglected. However, they represent an important parameter in the refurbishment typology choice; therefore, in future work, the authors will focus on a detailed economic analysis considering both the operating cost reduction and the simple payback period.

Author Contributions: Conceptualization, G.C., Y.S., M.S., A.R., and S.S.; methodology, G.C., Y.S., M.S., A.R., and S.S.; software, G.C. and Y.S.; validation, G.C., Y.S., M.S., A.R., and S.S.; formal analysis, G.C., Y.S., M.S., and A.R.; investigation, G.C. and Y.S.; resources, G.C. and S.S.; data curation, G.C. and Y.S.; writing—original draft preparation, G.C., Y.S., and S.S.; writing—review & editing, G.C., Y.S., M.S., and A.R.; visualization, G.C., Y.S., M.S., A.R., and S.S.; supervision, G.C., M.S., and S.S.; project administration, G.C., M.S., and S.S.; funding acquisition, S.S. All authors have read and agreed to the published version of the manuscript.

Funding: This research was co-funded by a collaborative research and development project, number F/050405/01-03/X32 "WALLED: Smart LED&OLED per Lighting e MediaBuilding"–Fondo per la crescita sostenibile–Call Horizon 2020 PON I&C and by the European Union-PON for Research and Innovation 2014–2020.

Data Availability Statement: The data presented in this study are available on request from the corresponding authors.

Acknowledgments: The authors would like to thank the Academic Editors for their invitation to the Special Issue "Novel Technologies to Enhance Energy Performance and Indoor Environmental Quality of Buildings" and the Assistant Editor for his support. The authors would also like to thank the three anonymous reviewers for their insightful suggestions and careful reading of the manuscript.

Conflicts of Interest: The authors declare no conflict of interest.

Nomenclature

Latin letters	
A	surface area (m^2)
a	filler radius (m)
ABS	acrylonitrile-butadiene-styrene
AM	additive manufacturing
CAGR	compound annual growth rate
COP	Coefficient of performance
DSF	double-skin facade
E	energy (kWh)
EER	energy efficiency ratio (-)
EHP	electric heat pump
h_c	interfacial boundary conductance (W/m^2K)
I_{vert}	vertical pyranometer on the south wall (W/m^2)
k_{3D}	thermal conductivity of the 3D printed materials (W/mK)
k_d	thermal conductivity of the filler (W/mK)
k_m	thermal conductivity of the selected 3D printable polymers (W/mK)
MAE	Mean Absolute Error (°C)
ME	Mean Error (°C)
N	number of measurements (-)
OVF	opaque ventilated facades
PC	proposed case
PES	non-renewable primary energy saving (%)
PET	polyethylene terephthalate
PETG	polyethylene terephthalate glycol-modified
PLA	polylactic acid
RC	reference case
RMSE	Root Mean Square Error (°C)
T	thermocouple/temperature (°C)
U	transmittance value (m^2K/W)
V_d	volume fraction of the filler
W	airspeed sensor
Greeks	
Δ	difference
η	efficiency (%)
Subscripts/Superscripts	
cavity	air cavity of the second-skin system
cool	cooling
el	electricity
exp,i	experimental value at time step i
indoor	indoor air
p	non-renewable primary energy
PC	proposed case
PP	power plant
RC	reference case
sim,i	simulated value at time step i
th	thermal
w	window

References

1. European Commission Energy Performance of Buildings Directive. Available online: https://ec.europa.eu/energy/topics/energy-efficiency/energy-efficient-buildings/energy-performance-buildings-directive_en (accessed on 4 August 2020).
2. Antonov, Y.I.; Heiselberg, P.; Flourentzou, F.; Pomianowski, M.Z. Methodology for evaluation and development of refurbishment scenarios for multi-story apartment buildings, applied to two buildings in Denmark and Switzerland. *Buildings* **2020**, *10*, 102. [CrossRef]

3. Cortiços, N.D. Improving residential building efficiency with membranes over facades: The Mediterranean context. *J. Build. Eng.* **2020**, *32*, 101421. [CrossRef]
4. Ascione, F.; Bianco, N.; Maria Mauro, G.; Napolitano, D.F. Building envelope design: Multi-objective optimization to minimize energy consumption, global cost and thermal discomfort. Application to different Italian climatic zones. *Energy* **2019**, *174*, 359–374. [CrossRef]
5. Scorpio, M.; Ciampi, G.; Rosato, A.; Maffei, L.; Masullo, M.; Almeida, M.; Sibilio, S. Electric-driven windows for historical buildings retrofit: Energy and visual sensitivity analysis for different control logics. *J. Build. Eng.* **2020**, *31*, 101398. [CrossRef]
6. Cattarin, G.; Causone, F.; Kindinis, A.; Pagliano, L. Outdoor test cells for building envelope experimental characterization—A literature review. *Renew. Sustain. Energy Rev.* **2016**, *54*, 606–625. [CrossRef]
7. Barbosa, S.; Ip, K. Perspectives of double skin facades for naturally ventilated buildings: A review. *Renew. Sustain. Energy Rev.* **2014**, *40*, 1019–1029. [CrossRef]
8. Ballarini, I.; De Luca, G.; Paragamyan, A.; Pellegrino, A.; Corrado, V. Transformation of an office building into a nearly zero energy building (NZEB): Implications for thermal and visual comfort and energy performance. *Energies* **2019**, *12*, 895. [CrossRef]
9. Sibilio, S.; Rosato, A.; Scorpio, M.; Iuliano, G.; Ciampi, G.; Vanoli, G.P.; de Rossi, F. A Review of Electrochromic Windows for Residential Applications. *Int. J. Heat Technol.* **2016**, *34*, S481–S488. [CrossRef]
10. Ciampi, G.; Rosato, A.; Scorpio, M.; Sibilio, S. Energy and economic evaluation of retrofit actions on an existing historical building in the south of Italy by using a dynamic simulation software. *Energy Procedia* **2015**, *78*, 741–746. [CrossRef]
11. Scorpio, M.; Ciampi, G.; Spanodimitriou, Y.; Rosato, A.; Laffi, R.; Almeida, M.; Sibilio, S. GEMINI: Test Cells for the Acoustic, Visual and Thermal Performances Evaluation of Double-Skin Facades. In Proceedings of the 14th Conference on Sustainable Development of Energy, Water and Environment Systems, Dubrovnik, Croatia, 1–6 October 2019; pp. 1–13.
12. Ciampi, G.; Scorpio, M.; Spanodimitriou, Y.; Rosato, A.; Sibilio, S. Thermal model validation of an electric-driven smart window through experimental data and evaluation of the impact on a case study. *Build. Environ.* **2020**, *181*, 107134. [CrossRef]
13. Krstić-Furundžić, A.; Vujošević, M.; Petrovski, A. Energy and environmental performance of the office building facade scenarios. *Energy* **2019**, *183*, 437–447. [CrossRef]
14. Mejía, K.J.; Barbero-Barrera, M.D.M.; Pérez, M.R. Evaluation of the impact of the envelope system on thermal energy demand in hospital buildings. *Buildings* **2020**, *10*, 250. [CrossRef]
15. Saroglou, T.; Theodosiou, T.; Givoni, B.; Meir, I.A. Studies on the optimum double-skin curtain wall design for high-rise buildings in the Mediterranean climate. *Energy Build.* **2020**, *208*, 109641. [CrossRef]
16. Ghaffarianhoseini, A.; Ghaffarianhoseini, A.; Berardi, U.; Tookey, J.; Li, D.H.W.; Kariminia, S. Exploring the advantages and challenges of double-skin façades (DSFs). *Renew. Sustain. Energy Rev.* **2016**, *60*, 1052–1065. [CrossRef]
17. Chen, X.; Yang, H.; Peng, J. Energy optimization of high-rise commercial buildings integrated with photovoltaic facades in urban context. *Energy* **2019**, *172*, 1–17. [CrossRef]
18. Cerón, I.; Caamaño-Martín, E.; Neila, F.J. "State-of-the-art" of building integrated photovoltaic products. *Renew. Energy* **2013**, *58*, 127–133. [CrossRef]
19. Ibañez-Puy, M.; Vidaurre-Arbizu, M.; Sacristán-Fernández, J.A.; Martín-Gómez, C. Opaque Ventilated Façades: Thermal and energy performance review. *Renew. Sustain. Energy Rev.* **2017**, *79*, 180–191. [CrossRef]
20. Stazi, F.; Ulpiani, G.; Pergolini, M.; Di Perna, C.; D'Orazio, M. The role of wall layers properties on the thermal performance of ventilated facades: Experimental investigation on narrow-cavity design. *Energy Build.* **2020**, *209*, 109622. [CrossRef]
21. Poirazis, H. Division of Energy and Building Design. In *Double Skin Façades for Office Buildings*; Department of Construction and Architecture, Lund University: Lund, Sweden, 2004; ISBN 9185147028.
22. Belleri, A.; Tarantino, S.; Lollini, R.; Arlati, E. Measurement and prediction of heat transfer and mass flow of a ventilated façade. In Proceedings of the AIRAH and IBPSA's Australasian Building Simulation 2017 Conference, Melbourne, Australia, 15–16 November 2017; pp. 1–15.
23. Alberto, A.; Ramos, N.M.M.; Almeida, R.M.S.F. Parametric study of double-skin facades performance in mild climate countries. *J. Build. Eng.* **2017**, *12*, 87–98. [CrossRef]
24. Souza, L.C.O.; Souza, H.A.; Rodrigues, E.F. Experimental and numerical analysis of a naturally ventilated double-skin façade. *Energy Build.* **2018**, *165*, 328–339. [CrossRef]
25. Kuznik, F.; Catalina, T.; Gauzere, L.; Woloszyn, M.; Roux, J. Numerical modelling of combined heat transfers in a double skin façade Full-scale laboratory experiment validation. *Appl. Therm. Eng.* **2011**, *31*, 3043–3054. [CrossRef]
26. Aparicio-Fernández, C.; Vivancos, J.L.; Ferrer-Gisbert, P.; Royo-Pastor, R. Energy performance of a ventilated façade by simulation with experimental validation. *Appl. Therm. Eng.* **2014**, *66*, 563–570. [CrossRef]
27. Diallo, T.M.O.; Zhao, X.; Dugue, A.; Bonnamy, P.; Javier Miguel, F.; Martinez, A.; Theodosiou, T.; Liu, J.S.; Brown, N. Numerical investigation of the energy performance of an Opaque Ventilated Façade system employing a smart modular heat recovery unit and a latent heat thermal energy system. *Appl. Energy* **2017**, *205*, 130–152. [CrossRef]
28. E2VENT Energy Efficient Ventilated Facades for Optimal Adaptability and Heat Exchange. Available online: http://www.e2vent.eu/ (accessed on 12 May 2020).
29. Pergolini, M.; Ulpiani, G.; Shehi, O.; Di Perna, C.; Stazi, F. Controlled inlet airflow in ventilated facades: A numerical analysis. *IOP Conf. Ser. Mater. Sci. Eng.* **2019**, *609*, 032009. [CrossRef]
30. EnergyPlus, Energy Simulation Software. Available online: https://energyplus.net/ (accessed on 29 March 2021).

31. Transient System Simulation Tool TRNSYS 18. Available online: http://www.trnsys.com/demo/ (accessed on 12 May 2020).
32. Trubiano, F. Performance based envelopes: A theory of spatialized skins and the emergence of the integrated design professional. *Buildings* **2013**, *3*, 689–712. [CrossRef]
33. Krivoshapko, S.N. The perspectives of application of thin-walled plastic and composite polymer shells in civil and industrial architecture. *J. Reinf. Plast. Compos.* **2018**, *37*, 217–229. [CrossRef]
34. Agarwal, S.; Gupta, R.K. *Plastics in Buildings and Construction*, 2nd ed.; Elsevier Inc.: Amsterdam, The Netherlands, 2017; ISBN 9780323390408.
35. Moradibistouni, M.; Vale, B.; Isaacs, N. Evaluating the use of polymers in residential buildings: Case study of a single storey detached house in New Zealand. *J. Build. Eng.* **2020**, *32*, 101517. [CrossRef]
36. Project DWG the Pet Pavilion, Public Space in a Changing Society. Available online: https://www.projectdwg.com/the-pet-pavilion-public-space-in-a-changing-society-in-enschede-by-projectdwg/ (accessed on 2 February 2021).
37. Loos.Fm the Pet Pavilion, Public Space in a Changing Society. Available online: http://loos.fm/en/project-pet-pavilion.php (accessed on 2 February 2021).
38. The Far Eastern Group EcoARK-World's First PET Bottle Green Building. Available online: https://www.expopark.taipei/en/cp.aspx?n=175 (accessed on 2 February 2021).
39. Moretti, E.; Zinzi, M.; Belloni, E. Polycarbonate panels for buildings: Experimental investigation of thermal and optical performance. *Energy Build.* **2014**, *70*, 23–35. [CrossRef]
40. Čekon, M.; Šikula, O. Experimental and numerical study on the thermal performance of polycarbonate panels. *J. Build. Eng.* **2020**, *32*, 101715. [CrossRef]
41. Construcciones Planificadas Hotel Grand Hyatt Bogotá. Available online: http://www.construccionesplanificadas.com/en/ (accessed on 2 February 2021).
42. Hager, I.; Golonka, A.; Putanowicz, R. 3D Printing of Buildings and Building Components as the Future of Sustainable Construction? *Procedia Eng.* **2016**, *151*, 292–299. [CrossRef]
43. Sakin, M.; Kiroglu, Y.C. 3D Printing of Buildings: Construction of the Sustainable Houses of the Future by BIM. *Energy Procedia* **2017**, *134*, 702–711. [CrossRef]
44. 3Dprint 3D Printed Building Facades. Available online: https://3dprint.com/187220/3d-printed-building-facades/%0A (accessed on 2 February 2021).
45. ARCHELLO 3D Printed Facade for Eu-Building. Available online: https://archello.com/en/project/3d-printed-facade-for-eu-building (accessed on 2 February 2021).
46. Sarakinioti, M.V.; Turrin, M.; Konstantinou, T.; Tenpierik, M.; Knaack, U. Developing an integrated 3D-printed façade with complex geometries for active temperature control. *Mater. Today Commun.* **2018**, *15*, 275–279. [CrossRef]
47. Vidakis, N.; Petousis, M.; Tzounis, L.; Maniadi, A.; Velidakis, E.; Mountakis, N.; Papageorgiou, D.; Liebscher, M.; Mechtcherine, V. Sustainable additive manufacturing: Mechanical response of polypropylene over multiple recycling processes. *Sustainability* **2021**, *13*, 159. [CrossRef]
48. Shemelya, C.; De La Rosa, A.; Torrado, A.R.; Yu, K.; Domanowski, J.; Bonacuse, P.J.; Martin, R.E.; Juhasz, M.; Hurwitz, F.; Wicker, R.B.; et al. Anisotropy of thermal conductivity in 3D printed polymer matrix composites for space based cube satellites. *Addit. Manuf.* **2017**, *16*, 186–196. [CrossRef]
49. Labonnote, N.; Rønnquist, A.; Manum, B.; Rüther, P. Additive construction: State-of-the-art, challenges and opportunities. *Autom. Constr.* **2016**, *72*, 347–366. [CrossRef]
50. Yazdani Sarvestani, H.; Akbarzadeh, A.H.; Niknam, H.; Hermenean, K. 3D printed architected polymeric sandwich panels: Energy absorption and structural performance. *Compos. Struct.* **2018**, *200*, 886–909. [CrossRef]
51. Hasselman, D.P.H.; Johnson, L.F. Effective Thermal Conductivity of Composites with Interfacial Thermal Barrier Resistance. *J. Compos. Mater.* **1987**, *21*, 508–515. [CrossRef]
52. University of Campania Luigi Vanvitelli; Department of Architecture and Industrial Design Ri.A.S.-Built Environment Control Laboratory. Available online: https://www.architettura.unicampania.it/images/ricerca/laboratori/EN/0-Lab_RIAS_ENG_30042020.pdf (accessed on 4 September 2020).
53. Batlles, F.J.; Olmo, F.J.; Alados-Arboledas, L. On shadowband correction methods for diffuse irradiance measurements. *Sol. Energy* **1995**, *54*, 105–114. [CrossRef]
54. Ensinger TECARAN ABS Grey. Available online: https://www.ensingerplastics.com/en/shapes/products/tecaran-abs-grey (accessed on 10 February 2021).
55. Dickinson, E.W. *Solar Energy Technology Handbook*; CRC Press: Boca Raton, FL, USA, 2017; ISBN 9781351076739.
56. Al-Saadi, S.N.; Zhai, Z. A new validated TRNSYS module for simulating latent heat storage walls. *Energy Build.* **2015**, *109*, 274–290. [CrossRef]
57. Vahidi Bidhendi, M.; Abbassi, Y. Exploring dynamic operation of a solar dish-stirling engine: Validation and implementation of a novel TRNSYS type. *Sustain. Energy Technol. Assess.* **2020**, *40*, 100765. [CrossRef]
58. Yasin, M.; Scheidemantel, E.; Klinker, F.; Weinläder, H.; Weismann, S. Generation of a simulation model for chilled PCM ceilings in TRNSYS and validation with real scale building data. *J. Build. Eng.* **2019**, *22*, 372–382. [CrossRef]
59. TRNSYS 18 TRNSYS Manual-Volume 3, Standard Component Library Overview. Available online: http://www.trnsys.com/assets/docs/03-ComponentLibraryOverview.pdf (accessed on 4 September 2020).

60. Solar Energy Laboratory, University of Wisconsin-Madison and Thermal Energy System Specialists. *Thermal Energy Systems Specialists TRNSYS Manual-Volume 4, Mathematical Reference*; Solar Energy Laboratory, University of Wisconsin-Madison: Madison, WI, USA, 2018; pp. 1–705.
61. TESS Component Library TESS Individual Components-TYPE 1230: Ventilated Air Cavity Wall (Ventilated Facade). Available online: http://www.trnsys.com/tess-libraries/individual-components.php.html (accessed on 14 March 2021).
62. Trimble Inc. SketchUp Pro 2014. Available online: https://3dwarehouse.sketchup.com/model/ua1d8251e-be08-4735-89c3-008b8e79ff3d/sketchUp-pro-2014?hl=it (accessed on 4 September 2020).
63. Italian Government DM 26 Giugno 2015. Available online: https://www.mise.gov.it/index.php/it/normativa/decreti-interministeriali/2032966-decreto-interministeriale-26-giugno-2015-applicazione-delle-metodologie-di-calcolo-delle-prestazioni-energetiche-e-definizione-delle-prescrizioni-e-dei-requisiti-minimi-degli-edifici (accessed on 4 September 2020).
64. G.I. Industrial Holding Technical Brochure-CRA/K 15÷131. Available online: https://clint.it/en/products/aircooled-liquid-chillers-and-heat-pumps-for-commercial-industrial-application-2/aircooled-liquid-chillers-and-heat-pumps/crak-15131/ (accessed on 4 September 2020).
65. García, E.; De Pablos, A.; Bengoechea, M.A.; Guaita, L.; Osendi, M.I.; Miranzo, P. Thermal conductivity studies on ceramic floor tiles. *Ceram. Int.* **2011**, *37*, 369–375. [CrossRef]
66. MakeItFrom.com. Glycol-Modified Polyethylene Terephthalate (PETG). Available online: https://www.makeitfrom.com/material-properties/Glycol-Modified-Polyethylene-Terephthalate-PETG-PET-G (accessed on 2 February 2020).
67. MakeItFrom.com. Polylactic Acid (PLA, Polylactide). Available online: https://www.makeitfrom.com/material-properties/Polylactic-Acid-PLA-Polylactide (accessed on 2 February 2020).
68. SD3D ABS Technical Data Sheet. Available online: https://www.sd3d.com/wp-content/uploads/2015/10/MaterialTDS-ABS-Web.pdf (accessed on 10 February 2021).
69. SD3D PETG Technical Data Sheet. Available online: https://www.sd3d.com/wp-content/uploads/2017/06/MaterialTDS-PETG_01.pdf (accessed on 10 February 2021).
70. SD3D PLA Technical Data Sheet. Available online: https://www.sd3d.com/wp-content/uploads/2017/06/MaterialTDS-PLA_01.pdf (accessed on 10 February 2021).
71. Stephan, K.; Laesecke, A. The Thermal Conductivity of Fluid Air. *J. Phys. Chem. Ref. Data* **1985**, *14*, 227–234. [CrossRef]
72. Ente Nazionale Italiano di Unificazione. *Energy Performance of Buildings Part 1: Evaluation of Energy Need for Space Heating and Cooling*; UNI/TS 11300-1; Ente Nazionale Italiano di Unificazione (UNI): Room, Italy, 2014.
73. Euporpean Committee for Standardzation. *Heating Systems in Buildings–Method for Calculation of the Design Heat Load*; EN 12831:2003 E; Euporpean Committee for Standardzation: Brussels, Belgium, 2003.
74. Rosato, A.; Ciervo, A.; Ciampi, G.; Scorpio, M.; Guarino, F.; Sibilio, S. Impact of solar field design and back-up technology on dynamic performance of a solar hybrid heating network integrated with a seasonal borehole thermal energy storage serving a small-scale residential district including plug-in electric vehicles. *Renew. Energy* **2020**, *154*, 684–703. [CrossRef]
75. Angrisani, G.; Canelli, M.; Roselli, C.; Russo, A.; Sasso, M.; Tariello, F. A small scale polygeneration system based on compression/absorption heat pump. *Appl. Therm. Eng.* **2017**, *114*, 1393–1402. [CrossRef]
76. SIA 2024:2015-Dati D'utilizzo di Locali per L'energia e L'impiantistica Degli Edifici. Available online: https://www.sia.ch/en/the-sia/ (accessed on 4 September 2020).
77. Veglas. Veglas s.r.l. Available online: http://www.veglas.it (accessed on 20 March 2021).
78. Plasting. Plasting s.r.l. Available online: https://www.plasting.biz/en/ (accessed on 20 March 2021).
79. SD3D. SD3D Printing. Available online: https://www.sd3d.com (accessed on 20 March 2021).

buildings

MDPI

Article

Influence of Ausforming Treatment on Super Elasticity of Cu-Zn-Al Shape Memory Alloy for Seismic Energy Dissipaters

Danko Ćorić and Irena Žmak *

Faculty of Mechanical Engineering and Naval Architecture, University of Zagreb, Ivana Lučića 5, 10000 Zagreb, Croatia; danko.coric@fsb.hr
* Correspondence: irena.zmak@fsb.hr

Abstract: In order to develop the application of the more cost-effective copper-based shape memory alloys (SMAs), rather than nickel–titanium as earthquake energy dissipaters, the influence of ausforming-induced plastic deformation on phase transformations, microstructure, super elasticity and mechanical properties of the shape memory alloy Cu-26Zn-4Al was examined. These specific SMA properties were targeted by applying appropriate parameters of the thermomechanical (the so-called ausforming) process: beta-phase homogenization at 800 °C for 20 min, one-step hot rolling at 800 °C and water quenching. The results showed significant microstructural changes, increased mechanical resistance and change in the phase transformation behavior. The SMA treated by ausforming retained the reversible austenitic–martensitic transformation ability, with the appearance of the super-elastic effect up to 6% of strain recovery. Although some strengthening occurred after hot rolling (an increase in true yield strength of 125 MPa was detected), all phase transformation temperatures were decreased. The smallest decrease was detected for the austenite finish temperature (32.8 °C) and the largest for the martensite finish temperature (42.0 °C), allowing both the expansion and the lowering of the temperature range of super elasticity, which is favorable for construction applications. It is concluded that it is possible to achieve an optimal combination of adequate strength and improved transformation behavior of Cu-Zn-Al alloy by applying the ausforming treatment.

Keywords: Cu-Zn-Al; shape memory; SMA; seismic; ausforming; transformation temperatures; super elasticity; microstructure

check for updates

Citation: Ćorić, D.; Žmak, I. Influence of Ausforming Treatment on Super Elasticity of Cu-Zn-Al Shape Memory Alloy for Seismic Energy Dissipaters. *Buildings* **2021**, *11*, 22. https://doi.org/10.3390/buildings11010022

Received: 1 December 2020
Accepted: 30 December 2020
Published: 6 January 2021

Publisher's Note: MDPI stays neutral with regard to jurisdictional claims in published maps and institutional affiliations.

Copyright: © 2021 by the authors. Licensee MDPI, Basel, Switzerland. This article is an open access article distributed under the terms and conditions of the Creative Commons Attribution (CC BY) license (https://creativecommons.org/licenses/by/4.0/).

1. Introduction

Shape memory alloys (SMAs) are novel materials which can recover the original shape after being heated to a determined temperature. They are able to "remember" their previous shape, i.e., return to the shape in which they were before a certain deformation. That is why they are called shape memory alloys, and the phenomenon is called shape memory effect (SME) [1].

The first SME was detected in gold–cadmium alloy in 1932, but the effect was not widely studied until 1971, when in Naval Ordnance Laboratories, USA, a nickel–titanium (mostly abbreviated as NiTi, but also Ni-Ti) alloy showed the possibility of considerable amounts of recoverable deformation. Since then, different alloy systems have been proposed as SMAs. Nevertheless, only NiTi, Cu-based and Fe-based SMAs have achieved a relevant commercial impact. Cu-based alloys include the Cu-Zn-Al and Cu-Al-Ni (also denoted as CuZnAl and CuAlNi) alloy systems [2].

Several shape memory alloys have been extensively studied and are available in many now almost everyday life applications, such as eyeglass frames, orthodontic wires and coronary stents [3]. These alloys are composed of nickel and titanium, which is why they have adequate biocompatibility as well. Besides medical applications, SMAs may nowadays be found in automotive applications, aeronautical applications and architecture [4].

Smart materials such as SMAs may be used in civil engineering for health monitoring of civil infrastructures, assessment of damage or integrity, structural control, maintenance and repair. Bedsides improving the reliability and durability, potential benefits include increased safety towards vibrations and natural hazards, such as earthquakes, wind blows, ocean waves, etc. Nowadays, the concept of self-sensing concrete is studied in the fields of smart concrete for self-healing, self-adjusting and self-heating. The stress or strain detection in concrete is studied by integrating concrete with optical fibers, piezoelectric ceramics and electrical resistance strain gauges. Carbon fiber-reinforced polymer composites have been studied as well, since they have very good electrical conductivity. SMAs may be used for monitoring deformations and estimation of cracks [5].

Since the 1990s, an increasing number of studies of application of SMAs as novel materials for seismic protection have been concluded. Currently, the construction industry is not commonly using SMAs because of the information gap between material science and practical civil engineering. Some of the recent successful applications of NiTi SMAs as seismic energy dissipaters include, for example, the retrofitting of the bell tower of the St. Giorgio Martire church in Trignano, RE, Italy, whose 18.5 m high tower was significantly damaged in the 1996 earthquake of 4.8 Richter magnitude. After retrofitting with SMA energy-dissipation devices, the bell tower successfully endured the next 4.5 Richter magnitude earthquake in 2000. Following this success, several other churches in Italy, as well as bridges in seismic-active areas of the USA, have been retrofitted in the same way [6].

Recent studies of possible application of nickel–titanium (NiTi) SMAs instead of steel to reinforce novel bridge columns offers long-term column durability and excellent seismic resistance [7]. One of the more economic SMAs which could be used instead of NiTi are Cu-Zn-Al SMAs, nowadays commercially available. For example, vibration damping of the steel frame engineering structures with Cu-Zn-Al SMA damping elements installed can be significantly reduced, up to 10 times, for both austenitic or martensitic structures [8]. SMAs improve the ability of concrete columns to withstand strong seismic excitations when used for confining concrete cylinders [9]. For the dissipation of earthquake energy, adequate curvature ductility is needed in critical areas of the construction where inelastic flexural deformations arise, so hysteretic behavior is necessary [10]. For preventing the collapse from strong earthquakes, buildings and bridge columns have to be designed with as high ductility as possible. Adequate transverse reinforcement of the plastic hinge regions in columns made from reinforced concrete may be provided either by hoops or spirals of a ductile material, typically steel, in order to avoid shear failure of the column [11]. Plastic hinges of reinforced concrete elements made out of hybrid SMA/steel have excellent effects against permanent deformation of the building after a seismic event, the so-called residual drift. Studies have shown that coupling of SMA with fiber-reinforced composite instead of steel additionally improves corrosion resistance [12]. Corrosion of the SMA itself may also occur, thus possibly lowering the shape memory effect. Chemical corrosion of the copper-based shape memory alloys is only recently being studied, for example, in solutions which simulate the marine environment, acid rain in the urban environment and acid rain in the industrial environment, where the shape memory effect was maintained, although corrosion products were made part of the porous corrosion layer [13].

The specific behavior of SMA's recovery to the original shape by heating or unloading is based on the presence of a reversible crystallographic austenite–martensite transformation. In general, there are two variants of martensite phase in alloys: twinned and detwinned martensite. Twinned or thermal martensite is formed when the alloy starts to cool below the temperature at which the martensitic transformation starts, i.e., below martensite start temperature (M_s), without any stress applied to the material. This martensite is called twinned because of twin domains in one martensite grain, which have the same crystallographic characteristics, except for the local orientation. On the other hand, there is the variant of the detwinned or stress martensite: this martensite is formed when stress, either tensile or compressive, is present. In the detwinned martensite, all domains are aligned according to the direction of the applied stress. The term detwinning originates from the crystallographic reorientation, which happens in the grain, leaving it without any twin variants. The applied stress has to be above a defined level for this phenomenon to occur, while the alloy shows an exemplary large deformation, which looks like typical plastic deformation observed in alloys in general [14].

The initial structure of shape memory alloys is the high-temperature austenite phase, usually denoted as beta (β), which transforms into the low-temperature martensitic phase, typically marked with alpha (α_M) [1,15,16]. β and α_M represent two different crystal structures, i.e., solid phase constituents of the Cu-Zn-Al alloy. β-phase is thermodynamically stable (i.e., will not change over a long period of time) at high temperatures, while α_M is stable at low temperatures. Both are solid solutions, where β has the body-centered cubic (BCC) unit cell or lattice and α_M has the cubic close packing (CCP) unit. During this transformation, the diffusion rate is negligible, as well as the atomic mobility, because the transformation takes place at relatively low temperatures. In the course of transformation, atoms move simultaneously and there is not any individual atom movement present. Rather, atoms are redistributed by a homogeneous shear, which occurs at levels lower than the parameter of the unit cell. This phenomenon is observed by a change in shape and dimensions of shape memory alloys, and is commonly known as the super elasticity (SE) or the shape memory effect (SME) [17–19]. The super elasticity effect may be induced by applying external mechanical stress, either with a temperature change or without it. In SE, the material returns to its previous shape after the load that has induced a large deformation is removed. On the other hand, the shape memory effect is connected to reversing to the original shape by applying heat, Figure 1 [20–23].

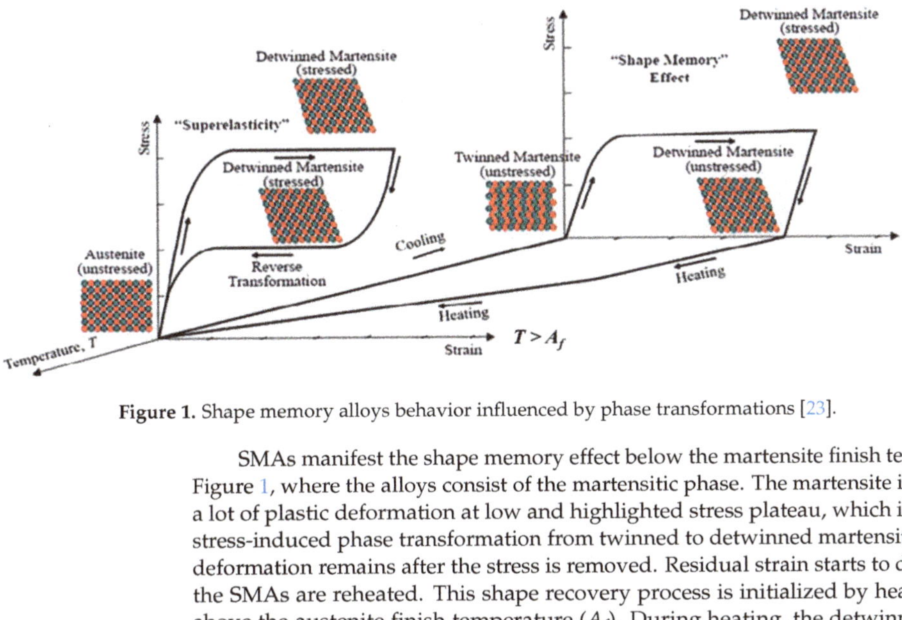

Figure 1. Shape memory alloys behavior influenced by phase transformations [23].

SMAs manifest the shape memory effect below the martensite finish temperature (M_f), Figure 1, where the alloys consist of the martensitic phase. The martensite is able to take in a lot of plastic deformation at low and highlighted stress plateau, which is a result of the stress-induced phase transformation from twinned to detwinned martensite. The induced deformation remains after the stress is removed. Residual strain starts to disappear when the SMAs are reheated. This shape recovery process is initialized by heating the SMAs above the austenite finish temperature (A_f). During heating, the detwinned martensitic phase changes back into the initial austenitic phase, i.e., it is a thermally induced reversible transformation. The twinned martensitic phase is reinstalled after cooling, by which the whole phase transformation cycle is completed, thanks to the shape memory effect [24].

The other phenomenon characteristic for SMAs is super elasticity, which appears at a temperature greater than the austenite finish temperature (A_f). Super elasticity is marked by the typical flag-shaped loop of hysteresis in the stress–strain diagram, Figure 1. Similar to the shape memory effect, the stress induces a considerable amount of reversible deformation by phase transformation, but now it is the transformation from austenite into detwinned martensite that occurs.

The austenitic phase of a SMA is able to fully recover to the original shape without any residual deformation. As indicated in Figure 1, austenitic SMAs have higher values of stress for the slip plateau than the martensitic SMAs.

The properties of SMAs are influenced by various factors, such as the temperature, chemical composition, crystal structure (polycrystal or monocrystal) and orientation, grain size, the presence of precipitate phase, heat treatment, etc. Sun et al. studied the super elasticity and the shape memory effect of polycrystalline SMAs and noticed that the thermomechanical response of SMAs shows a distinct grain size effect [25]. The polycrystal SMAs with smaller grain size shows lower capacity for shape memory and a narrower temperature interval for hysteresis, as well as higher hardening response for super elasticity. They also found that the crystal orientation has a clear influence on the stress-induced martensite transformation. Chen and Schuh studied the size of the super-elastic effect in CuAlNi microwires with a bamboo grain structure by conducting the tensile test and thermal cycling. The stress and temperature hysteresis were found to increase with the decrease in grain size [26]. Using the molecular dynamics method for the simulation of a single crystal NiTi alloy, thermomechanical response in adiabatic conditions, it was calculated that there is a significant temperature change of the SMA during both martensite and reverse transformation processes. For example, when the strain increased to 8%, the calculated temperature increased by 41.75 °C [27]. Similarly, the molecular dynamics method was used to reveal in detail the evolution of the microstructure of different NiTi compositions of SMAs. Several microstructural features obtained by simulation confirmed good agreement with literature reports and even new microstructural features were discovered [28].

Several alloying systems show the shape memory and super-elastic effects. For example, copper-based shape memory alloys were studied a lot over time [29]. Cu-based SMAs, such as Cu-Al, Cu-Al-Mn, Cu-Al-Ni and Cu-Zn-Al, have attracted a lot of scientific attention thanks to their satisfactory shape memory capacity. Compared to NiTi-based SMAs, Cu-based SMAs have a more limited transformation temperature interval, but are more easily produced, and hence have lower costs. On the other hand, polycrystalline copper-based SMAs are less tough, i.e., more brittle, which makes them harder to work than NiTi alloys. The brittleness comes as a result of a larger degree of order of the native austenitic β-phase, as well as a consequence of the elevated elastic anisotropy of the β-phase [30]. Copper-based SMAs are typically used for sensors and actuators, since they possess good mechanical properties at a comparably low price compared to NiTi SMAs [31,32]. Although NiTi SMAs possess higher super elasticity, better shape memory effect, higher tensile strength and better corrosion resistance than other SMAs, their main disadvantage is high cost. High costs arise from energy-intensive smelting of the alloying elements (primarily nickel and titanium) and a complex melting process which requires either vacuum melting or melting under an inert atmosphere. Therefore, where biocompatibility is not a primary requirement, the use of copper-based SMAs is preferred.

Copper-based SMAs may be heat-treated in order to modify yield and tensile strength and hardness, as specified by a distinct application. The desired properties are met through a change in microstructure or by introducing lattice defects to the crystal structure. The transformation interface propagation and the nucleation process may be altered using vacancies, dislocations, grain boundaries, solute chemical elements or precipitated particles [33,34]. For NiTi SMAs fabricated from powders by Spark Plasma Sintering (SPS), subsequent heat treatment at temperatures from 600 to 700 °C with slow cooling was found to be necessary to obtain good ductility [35].

The City of Zagreb, which is the Capital of Croatia, was hardly struck by the 5.5 Richter magnitude earthquake in March 2020. Currently, there is a need of an extensive retrofitting of several infrastructural facilities, as well as cultural and historic objects of national value. Although NiTi SMAs have been successfully applied in recent years in more developed countries for retrofitting of valuable civic building, their cost is very high. Research conducted by Si et al. indicated the possibility of using cheaper damping elements made of heat-treated (homogenized and quenched) Cu-Zn-Al SMAs with monophase austenitic or martensitic structure and super-elastic capacity around of 2% [8]. Therefore, the presented research is focused on using the thermomechanical treatment, which includes hot rolling at the homogenization temperature, for improving the properties (primarily, the super elasticity) of the Cu-Zn-Al SMA. This has already been proved successful on the example of the NiTi alloy, where Hornbogen et al., in their study, found that defects introduced by ausforming were related to strengthening of the austenite, super-elastic capability and the consequent shift to lower temperatures of the transformation cycle [36].

The primary objective of this study is to determine the influence of the ausforming processing on the SMA's microstructure, phase transformation temperatures, static mechanical properties and the recovery of deformation by the super elasticity effect. Thus, this study aims at improving the specific properties of copper-based SMAs by studying the capabilities and constraints of the ausforming processing.

2. Materials and Methods

The ternary Cu-Zn-Al shape memory alloy with chemical composition of Cu-26.38%Zn-3.72%Al (hereinafter Cu-26Zn-4Al) was prepared using high-purity copper (99.999 wt.%), zinc (99.99 wt.%) and aluminum (99.99 wt.%) by induction-melting furnace with argon atmosphere at Wieland-Werke A.G., Ulm, Germany, specialized in Cu-Zn-Al shape memory alloys. The studied alloy was cast into cylindrical bars of 50 mm diameter and the bars were cut in four pieces each of 18 mm thickness. Samples were cut in small pieces suitable for thermomechanical treatment. The ausforming heat treatment process, as described in [37], included β-phase homogenization, i.e., heating of the prepared alloy to form a β-phase homogeneous solid solution, at 800 °C for 20 min, followed by a one-step hot rolling process at 800 °C and finally quenching using water at room temperature, Figure 2.

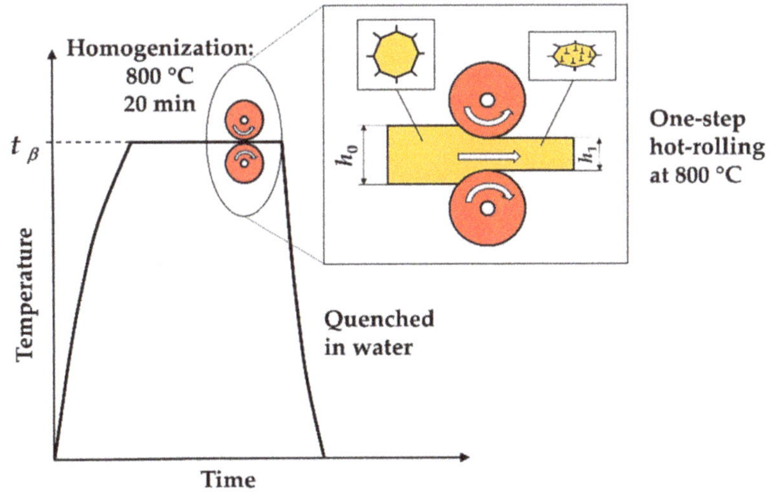

Figure 2. Applied ausforming process scheme: β-phase homogenization, hot rolling and quenching.

The β-phase in Cu-Zn-Al alloys is disordered at high temperatures and has a body-centered cubic (BCC) lattice [38]. According to Figure 3, it can be seen that the composition of the investigated alloy is situated in the monophase region, with only the BCC lattice (β-phase) being present at 800 °C, as indicated by the red dot. It may be noticed as well that the studied composition is also quite proximate to the dual phase, i.e., α_{Cu} + β area, where α_{Cu} (or simply α) is the solid solution of copper with zinc and aluminum, hence having the face-centered cubic lattice (FCC).

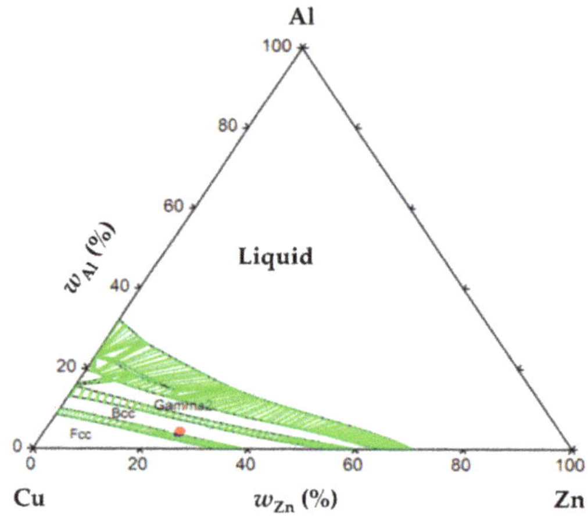

Figure 3. Isothermal section of Al-Cu-Zn calculated for 800 °C [38].

After the homogenization, the Cu-26Zn-4Al alloy was plastically deformed at 800 °C with various values of true plastic strain or true plastic deformation (φ_g):

$$\varphi_g = \ln \frac{h_0}{h_1} \times 100\ \% \quad (1)$$

where h_0 is the plate thickness before hot rolling and h_1 represents the thickness after the rolling, Figure 2. True strain (φ_g) is sometimes called logarithmic strain, Equation (1). Strain is a dimensionless quantity and is often expressed as a percentage, which will be used in this paper. Sometimes the derived unit mm/mm may be used as well. Out of four prepared samples, one was not deformed, while the other three samples were hot rolled to true plastic strain of 20%, 98% and 151%. After hot rolling, the samples were quenched in water in order to retain the β-phase and to remove segregations [39–41]. Moreover, the recrystallization and the recovery of structure were disabled by quenching. Quenching therefore restrained the removal of the lattice defects induced by hot deformation.

Samples used for optical microscopy were prepared using the standard metallographic procedures of grinding and polishing. After polishing, the samples were also electrochemically etched for 30 s in a solution containing 125 mL distilled water, 70 mL phosphoric acid, 70 mL ethanol, 12 mL propanol, 1.2 g urea and 0.5 mL dye. The applied voltage was 6 V and the current was 80 mA. Microstructure investigation of the prepared samples was carried out on the inverted metallurgical microscope type GX51F-5 (Olympus Corporation, Tokyo, Japan) with the integrated digital camera DP25.

The transformation temperatures were recorded by differential scanning calorimetry (DSC) using the DSC823e measuring module (Mettler Toledo International Inc., Greifensee, Switzerland). The DSC was performed five days after the ausforming process using the samples of approximately 60 mg, in the temperature range from 100 °C to −120 °C and using 5 °C/min cooling and heating rates. The typical martensitic and austenitic transformation temperatures were recorded: martensite start (M_s), martensite finish (M_f), austenite start (A_s) and austenite finish (A_f). The highest rate of martensitic and austenitic reaction can be detected in the DSC thermogram as they are marked by highest thermal flow peaks during cooling (M_m) and heating (A_m). The values of typical transformation temperatures (i.e., M_s, M_f, A_s and A_f) were detected from the DSC thermogram using the baseline tangents and the peak curve tangents.

Tensile tests and super elasticity measurements were carried out using the Mi 34 microtesting machine (Alfred J. Amsler & Co., Schaffhausen, Switzerland) at room temperature. The crosshead separation rate was set to 0.005 mm/s, i.e., 0.3 mm/min. All tensile test samples were oriented in the rolling direction. The shape and the dimensions of the microtesting sample are presented in Figure 4. Tensile tests and super elasticity measurements were performed on five samples for each thermomechanically treated state.

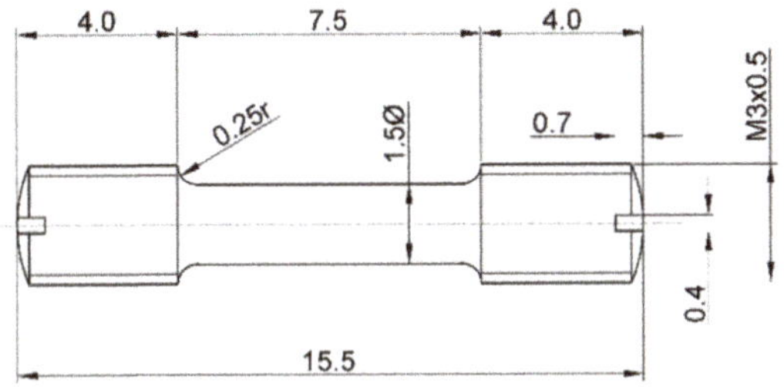

Figure 4. Shape and dimensions of sample for tensile microtesting.

Recorded mechanical properties of the studied Cu-26Zn-4Al SMA in conditions of tensile load were as follows: Transformation stress (R_{tr}), true yield strength (R_p) and tensile strength (R_m), Figure 5. Mechanically induced austenite to martensite transformation occurred at the transformation stress (R_{tr}) and at the corresponding super-elastic strain ($\varepsilon_{SE.}$).

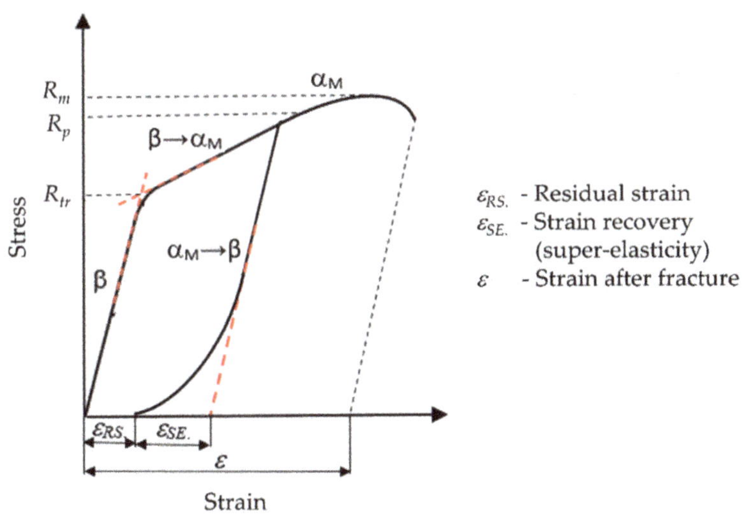

Figure 5. Stress–strain curve of polycrystalline Cu-Zn-Al shape memory alloy (SMA).

After the austenite was transformed into martensite, the martensite was deformed elastically in the beginning and then plastically only after the stress reached the value marked as R_p, Figure 5. When testing the super elasticity, the specimen began to unload before fracture, at the end of the strain range characteristic for the $\beta \rightarrow \alpha_M$ transformation [42].

3. Results

Based on the results of the optical microscopy testing, it was determined that the as-cast Cu-26Zn-4Al shape memory alloy has a dual phase microstructure, i.e., $\alpha + \beta$, Figure 6a. The microstructure of the as-cast state includes dendritic α-particles with FCC lattice, which are irregularly distributed in the β-phase matrix.

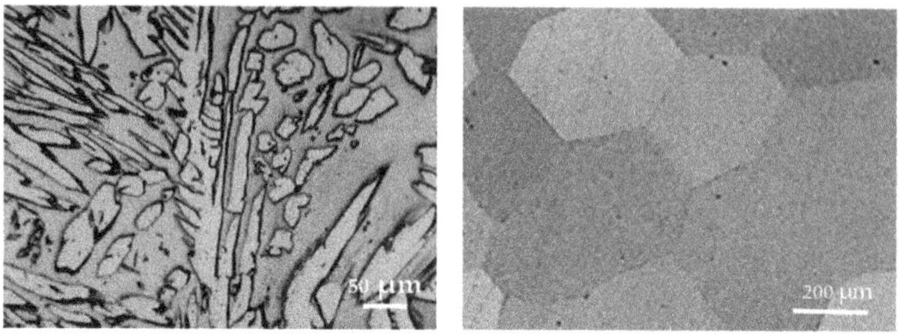

(a) (b)

Figure 6. Microstructure of SMA: (**a**) as-cast state; (**b**) homogenized and quenched.

Sample that was homogenized and quenched with room temperature water and that had not been hot rolled had a single phase austenite microstructure, as shown in Figure 6b. Based on the microscopic observations, it may be deduced that for the Cu-26Zn-4Al alloy, the martensite start temperature lies below room temperature.

A low level of true plastic strain introduced by the hot rolling process has produced the curling of the austenite grain boundaries, which is marked by a white arrow in Figure 7a. The microstructure of the plastically deformed sample (φ_g = 20%), besides the austenite phase, also presents very fine precipitates of the α-phase, which are located at the austenite grain boundaries. Larger plastic strain (φ_g = 98%) has resulted with changes in the crystal shape, Figure 7b,c. Effects of the recrystallization and the grain growth ware not observed. A greater amount of strain (φ_g = 151%) originates even more nucleation of the α-phase at the grain boundaries of the β-phase, as well as within the austenite (β-phase) grains, Figure 7b,c.

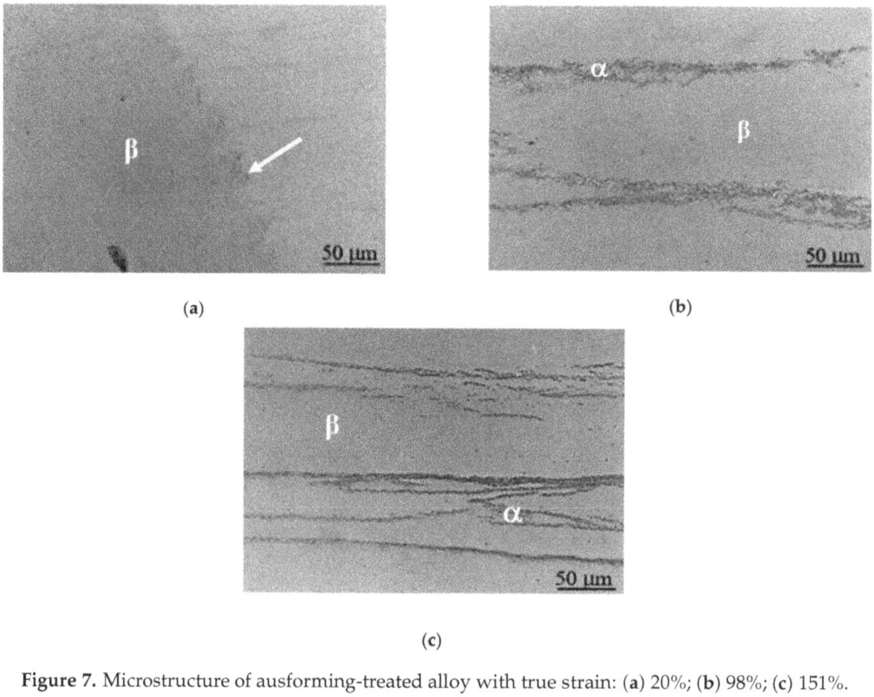

(a) (b) (c)

Figure 7. Microstructure of ausforming-treated alloy with true strain: (**a**) 20%; (**b**) 98%; (**c**) 151%.

Figure 8 shows the differential scanning calorimetry curves of heat flow for the Cu-26Zn-4Al alloy treated by the ausforming process to different plastic strains.

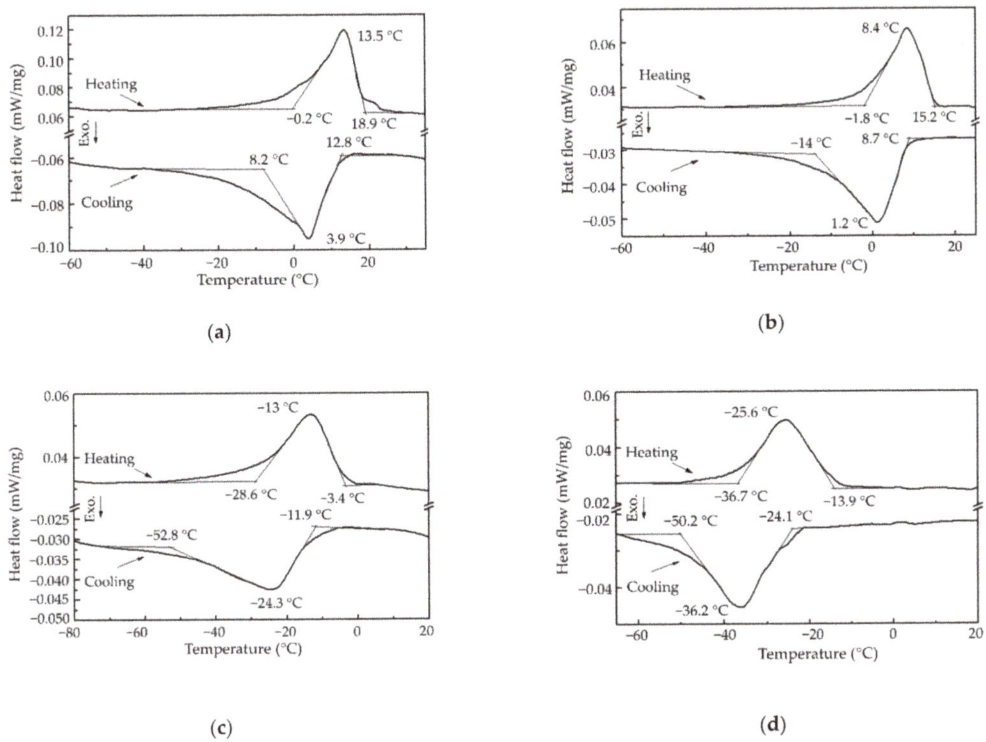

Figure 8. DSC curves of ausforming-treated alloy with true strain: (**a**) 0; (**b**) 20%; (**c**) 98%; (**d**) 151%.

Peaks of the heat flow of the martensitic and austenitic transformation can be clearly detected in Figure 8, both for cooling and heating. Based on the position of transformation peaks, the values of characteristic temperatures were determined and are presented in Table 1.

Table 1. Values of transformation temperatures of ausforming-treated alloy.

Transformation Temperature, °C	Amount of True Strain (φ_g, %)			
	0	20	98	151
M_s	12.8	8.7	−11.9	−24.1
M_m	3.9	1.2	−24.3	−36.2
M_f	−8.2	−14.0	−52.8	−50.2
A_s	−0.2	−1.8	−28.6	−36.7
A_m	13.5	8.4	−13.0	−25.6
A_f	18.9	15.2	−3.4	−13.9

When there is enough driving force for the diffusionless formation of α_M-crystals, the martensitic transformation starts and the set temperature is M_s. The martensitic transformation is an exothermic process. Therefore, in order not to interrupt the transformation process, the material must be cooled continuously through the martensitic transformation process. To achieve the complete transformation of martensite, the material must be cooled below the martensite finish temperature, M_f.

Conversely, the austenite transformation, which starts at the austenite start temperature (A_s), is an endothermic process. This means that it is needed to constantly heat up the material during the austenite transformation, in order not to interrupt this transformation process. When the material reaches a temperature higher than austenite finish (A_f), the whole of austenite transformation is completed.

It was observed that the austenitic transformation occurs within a narrower temperature interval in comparison to the martensitic transformation at all amounts of true plastic strain.

Both martensitic and austenitic transformation processes moved to lower temperature ranges when the true plastic strain of the alloy was increased. As shown in Table 1, all transformation temperatures were below room temperature, which satisfies the conditions for the super-elastic effect. The effect of the amount of true plastic strain on the Cu-26Zn-4Al shape memory alloy's transformation temperatures, M_s and M_f, as well as A_s and A_f, is shown in Figure 9. Equations in rectangles in Figure 9 represent linear regression equations between each transformation temperature and true plastic strain (φ_g).

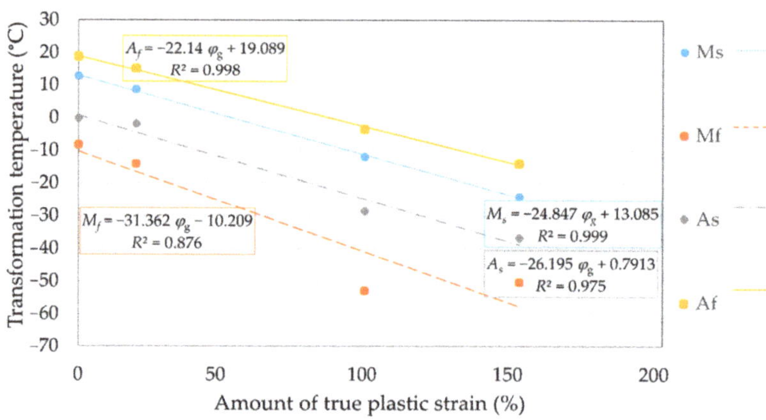

Figure 9. Transformation temperatures vs. true plastic strain of ausforming-treated alloy.

The data presented in Figure 9 confirm that the Cu-26Zn-4Al shape memory alloy is completely transformable for every studied amount of true plastic strain by the ausforming process. It may be deduced that both phase transformations, austenitic and martensitic reactions, start and finish at ever lower temperatures; the higher the strain, the lower the transformation temperatures. The highest influence of the amount of plastic strain on the change in phase transformation temperatures was observed for the M_f temperature (Δ 42 °C). The other phase transformation temperature changes were as follows: $\Delta\ M_s = 36.9$ °C, $\Delta\ A_f = 32.8$ °C and $\Delta\ A_s = 36.5$ °C.

The typical tensile stress–strain curves of the Cu-26Zn-4Al shape memory alloy treated by ausforming are presented with the red dashed line in Figure 10. Super-elastic behavior is clearly visible on all charts, presented with the blue line. All presented diagrams showing the super-elastic behavior were recorded during the first mechanical loading cycle.

The strain recovery of Cu-26Zn-4Al shape memory alloy by the super-elastic effect is evident from Figure 10. Nevertheless, a low plastic strain was also recorded, i.e., residual strain ($\varepsilon_{RS.}$), showing the presence of the stress-induced martensite, which is stable after removing the load. Mean values of the tensile testing results in terms of characteristic values: transformation stress (R_{tr}), true yield strength (R_p) and tensile strength (R_m), percentage elongation after fracture (ε), strain recovery ($\varepsilon_{SE.}$) and residual strain ($\varepsilon_{RS.}$) are presented in Figures 11 and 12, along with the appropriate standard deviation bars (error bars).

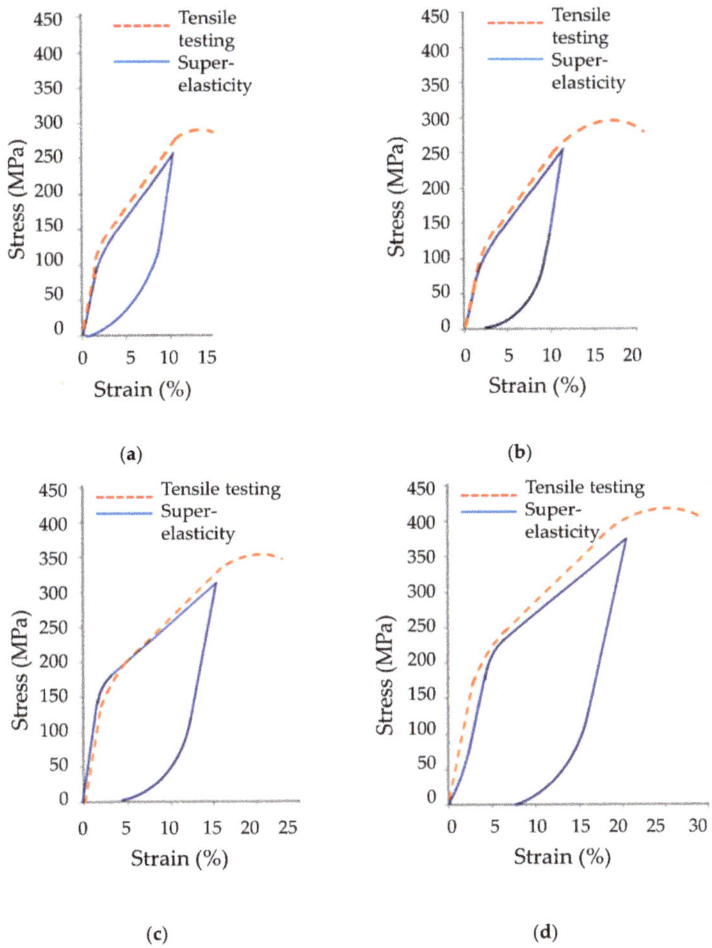

Figure 10. Stress–strain diagrams of tensile testing and super elasticity of ausforming- treated alloy with true plastic strain: (**a**) 0; (**b**) 20 %; (**c**) 98 %; (**d**) 151 %.

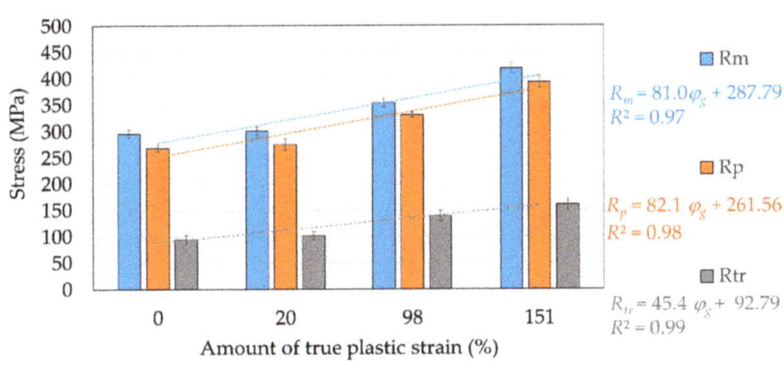

Figure 11. Transformation stress (R_{tr}), true yield strength (R_p) and tensile strength (R_m) as a function of true plastic strain (φ_g).

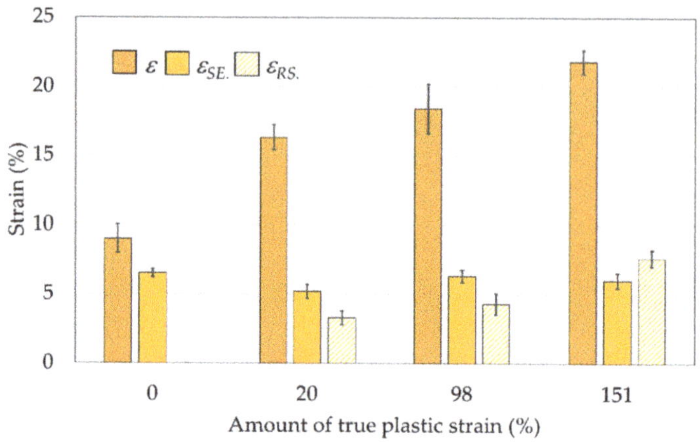

Figure 12. Elongation after fracture (ε), strain recovery ($\varepsilon_{SE.}$) and residual strain ($\varepsilon_{RS.}$) as a function of true plastic strain (φ_g).

4. Discussion

The transformation behavior of Cu-26Zn-4Al shape memory alloy is based on the reversible austenitic–martensitic transformation. After the austenitization process, i.e., homogenization in the austenite temperature range, the shape memory alloy is quenched in the aim of retaining the austenite (β-phase) at room temperature. Additionally, hot rolling may also be applied at the homogenization temperature to the austenitic microstructure. Hot rolling prolongs the β-crystals, and also promotes the deformation of grain boundaries. Furthermore, hot rolling intensifies the precipitation of α-phase at the β-phase grain boundaries as well as within the β-phase crystal grains. Moreover, hot rolling significantly increases the dislocation density in the β-phase.

DSC testing provided evidence of cooling-induced martensitic (α_M) reaction and reverse austenitic transformation induced by heating. The reversible $\beta \leftrightarrow \alpha_M$ phase transformation was detected in all ausforming-treated samples, despite the induced plastic deformation. As the amount of true plastic strain increased, all martensitic and austenitic transformation temperatures decreased to lower temperature ranges.

The observed relation between true plastic strain (φ_g) and M_s and other transformation temperatures is approximately linear. Martensite reaction temperature (M_m) without plastic deformation was 3.9 °C, while at a strain of φ_g = 151% it decreased to −36.2 °C; for the same true strains, austenite reaction temperature (A_m) decreased from 13.5 °C to −25.6 °C. The lowering of the transformation temperatures is a consequence of the increased defect density in β-crystals and precipitated α-grains. The α-phase has an adverse influence on the motion of the β/α_M transformation interface, thus making its mobility difficult. When all transformation temperatures were compared, it was observed that the M_f temperature decreases slightly more quickly, which can be ascribed to difficult forming of the final martensite portions in narrow corners of the deformed crystals.

Thermally induced reversible $\beta \leftrightarrow \alpha_M$ transformation is characterized by the existence of the transformation hysteresis. Austenitic transformation during heating will not start immediately above the M_f temperature, but a certain driving force in the form of heating, i.e., supplied heat, is necessary in order to start the process. A similar phenomenon is also found in the martensitic transformation, which will not start immediately at A_f temperature during cooling, but at some lower temperature M_s. The occurrence of hysteresis in the stress–strain chart is related to the friction between the crystal lattices on the β/α_M transformation interface.

The analysis of mechanical properties for different ausforming-treated states of Cu-26Zn-4Al alloy is fully consistent with the results of metallographic analysis and conducted

DSC tests. The results of the tensile test indicated the occurrence of the mechanically induced martensitic transformation. As the true plastic strain increased, so did the transformation stress as a consequence of the increasing share of non-transformable α-phase that occurs in the microstructure of hot rolled and quenched alloys, but also because of the consequent deformation of austenitic grains, Figure 7b,c. These microstructural changes affect the mobility of the β/α_M transformation interface, i.e., the transition of austenite into stress-induced martensite that requires additional driving force in the form of an external applied load. Besides increasing the transformation stress, microstructural alterations are equally responsible for lowering the transformation temperatures. Thermally induced $\beta \leftrightarrow \alpha_M$ transformation at higher degrees of deformation also requires additional energy, which is no longer mechanical but thermal energy, introduced by heating or removed by cooling. The increase in the transformation stress was followed as well by an increase in true yield strength (R_p) and tensile strength (R_m). It may be concluded that a larger true plastic strain results in a more strengthened Cu-26Zn-4Al shape memory alloy. R_p stresses vary within the range of 268 MPa (φ_g = 0) to 393 MPa (φ_g = 151%), and R_m from 295 MPa (φ_g = 0) to 419 MPa (φ_g = 151%). The increase in these stresses is the result of the strengthening processes caused by an increase in the density of dislocations introduced by hot rolling and retained by quenching (in the absence of recrystallization and recovery) and the appearance of deformed boundaries of austenitic grains.

Preliminary results of hardness measurements at temperatures lower than M_d (maximum temperature at which stress-induced $\beta \rightarrow \alpha_M$ reaction occurs) also support the hardening of the alloy, which indicate, for higher degrees of deformation, an increase in hardness of the stress-induced martensite. The higher slope outlining the trend of R_p and R_m changes in relation to the trend of R_{tr} change, Figure 11, is attributed to the $\beta \rightarrow \alpha_M$ transformation, which additionally introduces defects in the crystal structure that lead to subsequent hardening of the alloy.

All measured elongations increase with the increase in plastic deformation, Figure 12, due to the formation of finer grain structure, which is generally known for having higher strength, but also higher toughness and ductility. At the highest applied amount of true plastic strain (φ_g = 151%), the percentage elongation after fracture (ε) is almost three times higher than the percentage elongation in the undeformed state, while the super-elastic strain ($\varepsilon_{SE.}$) is approximately constant (around 6%, Figure 12) and independent of the degree of induced deformation. By increasing the true plastic strain, the residual strain after unloading residual strain ($\varepsilon_{RS.}$) also significantly increases due to the formation of highly deformed non-transformable martensite, which does not have the ability to reverse the transformation back into austenite.

5. Conclusions

The influence of the ausforming treatment on the microstructure, phase transformation temperatures, mechanical properties and super elasticity of Cu-26Zn-4Al alloy was examined in detail. These conclusions may be drawn based on the obtained results:

(i) After casting, the microstructure of the Cu-26Zn-4Al SMA is made up primarily of the β-phase with comparably brittle BCC lattice. A significant amount of α-particles was formed as irregular dendritic crystals, having the FCC lattice and being distributed in the β-matrix.

(ii) The ausforming treatment of hot rolling and quenching increases the strength of the Cu-26Zn-4Al shape memory alloy. The observed strengthening is proportional to the amount of true plastic strain, which is a result of the introduction of microstructural defects (dislocations) and observed deformed austenitic grain boundaries.

(iii) All phase transformation temperatures are lowered with the increase in plastic deformation level.

(iv) Greater amounts of plastic deformation have an adverse effect on both thermally and mechanically induced martensitic transformation, since a higher moving force is required to move the β/α_M transformation interface.

(v) Although the shape memory alloy was strengthened, the super elasticity effect was not lost. The strain recovery by super-elastic effect approximately amounts to 6%.
(vi) The applied ausforming thermomechanical treatment facilitates the achievement of enhanced strength and at the same time the reversible austenitic–martensitic transformation capability. This transformation is important for many mechanical and construction engineering demands where the super-elastic effect is or could be used, such as the earthquake energy dissipaters.

Author Contributions: Conceptualization, D.Ć.; methodology, D.Ć.; software, I.Ž.; validation, D.Ć. and I.Ž.; formal analysis, D.Ć. and I.Ž.; investigation, D.Ć.; resources, D.Ć.; data curation, D.Ć. and I.Ž.; writing—original draft preparation, D.Ć.; writing—review and editing, D.Ć. and I.Ž.; visualization, D.Ć. and I.Ž. Both authors have read and agreed to the published version of the manuscript.

Funding: This research received no external funding.

Institutional Review Board Statement: Not applicable.

Informed Consent Statement: Not applicable.

Data Availability Statement: The data presented in this study are available on request from the first author, D.Ć.

Conflicts of Interest: The authors declare no conflict of interest.

References

1. Yamauchi, K.; Ohkata, I.; Tsuchiya, K.; Miyazaki, S. (Eds.) *Shape Memory and Superelastic Alloys: Technologies and Applications*; Woodhead Publishing in Materials: Sawston, UK, 2011.
2. Huang, W.M.; Ding, Z.; Wang, C.C.; Wei, J.; Zhao, Y.; Purnawali, H. Shape memory materials. *Mater. Today* **2010**, *13*, 54–61. [CrossRef]
3. Simha, N.K.; Rama Sreekanth, P.S.; Venkata Siva, S.B. Shape-Memory Alloys. In *Reference Module in Materials Science and Materials Engineering*; Elsevier: Amsterdam, The Netherlands, 2017. [CrossRef]
4. Nematollahi, M.; Baghbaderani, K.S.; Amerinatanzi, A.; Zamanian, H.; Elahinia, M. Application of NiTi in Assistive and Rehabilitation Devices: A Review. *Bioengineering* **2019**, *6*, 37. [CrossRef] [PubMed]
5. Han, B.; Yu, X.; Ou, J. Challenges of Self-Sensing Concrete. In *Self-Sensing Concrete in Smart Structures*; Butterworth-Heinemann: Oxford, UK, 2014; pp. 361–376. [CrossRef]
6. Fang, C.; Wang, W. *Shape Memory Alloys for Seismic Resilience*; Springer: Singapore, 2020. [CrossRef]
7. Aryan, H. Seismic Resistant Bridge Columns with NiTi Shape Memory Alloy and Ultra-High-Performance Concrete. *Infrastructures* **2020**, *5*, 105. [CrossRef]
8. Si, N.-C.; Sun, K.-Q.; Sun, S.-C.; Liu, H.-X. Damping performance of Cu-Zn-Al shape memory alloys in engineering structures. *J. Cent. South Univ. Technol.* **2004**, *11*, 246–251. [CrossRef]
9. Hong, C.; Qian, H.; Song, G. Uniaxial Compressive Behavior of Concrete Columns Confined with Superelastic Shape Memory Alloy Wires. *Materials* **2020**, *13*, 1227. [CrossRef] [PubMed]
10. Laterza, M.; D'Amato, M.; Laksiri, P.; Thanthirige, L.; Braga, F.; Gigliotti, R. Comparisons of Codal Detailing Rules for Curvature Ductility and Numerical Investigations. *Open Constr. Build. Technol. J.* **2014**, *8*, 132–141. [CrossRef]
11. Mander, J.B.; Priestley, M.J.N.; Park, R. Theoretical Stress-Strain Model for Confined Concrete. *J. Struct. Eng.* **1988**, *114*, 1804–1826. [CrossRef]
12. Kabir, M.; Alam, M.; Said, A.; Ayad, A. Performance of Hybrid Reinforced Concrete Beam Column Joint: A Critical Review. *Fibers* **2016**, *4*, 13. [CrossRef]
13. De Filippo, B.; Brotzu, A.; Natali, S. Corrosion behavior of Cu-Zn-Al shape memory alloy in controlled environments. *AIP Conf. Proc.* **2020**, *2257*, 020013.
14. Marfia, S.; Vigliotti, A. 1D SMA Models. In *Shape Memory Alloy Engineering*; Butterworth-Heinemann: Boston, MA, USA, 2015; pp. 99–140. [CrossRef]
15. Lexcellent, C. *Shape-Memory Alloys Handbook*; Wiley-ISTE: Hoboken, NJ, USA, 2013.
16. Mohd Jani, J.; Leary, M.; Subic, A.; Gibson, M.A. A review of shape memory alloy research, applications and opportunities. *Mater. Des.* **2014**, *56*, 1078–1113. [CrossRef]
17. Sepulveda, A.; Muñoz, R.; Lovey, F.C.; Auguet, C.; Isalgue, A.; Torra, V. Metastable effects on martensitic transformation in SMA. *J. Therm. Anal. Calorim.* **2007**, *89*, 101–107. [CrossRef]
18. Pieczyska, E.A.; Gadaj, S.P.; Luckner, J.; Nowacki, W.K.; Tobushi, H. Martensite and Reverse Transformation during Complete Cycle of Simple Shear of NiTi Shape Memory Alloy. *Strain* **2009**, *45*, 93–100. [CrossRef]

19. Niendorf, T.; Krooß, P.; Somsen, C.; Eggeler, G.; Chumlyakov, Y.I.; Maier, H.J. Martensite aging—Avenue to new high temperature shape memory alloys. *Acta Mater.* **2015**, *89*, 298–304. [CrossRef]
20. Asanović, V.; Delijić, K.; Jauković, N. A study of transformations of β-phase in Cu-Zn-Al shape memory alloys. *Scripta Mater.* **2008**, *58*, 599–601. [CrossRef]
21. Pereira, E.S.; Peixoto, I.F.; Viana, A.C.; Oliveira, I.I.; Gonzalez, B.M.; Buono, V.T.; Bahia, M.G. Physical and mechanical properties of a thermomechanically treated NiTi wire used in the manufacture of rotary endodontic instruments. *Int. Endod. J.* **2012**, *45*, 469–474. [CrossRef]
22. Dasgupta, R. A look into Cu-based shape memory alloys: Present scenario and future prospects. *J. Mater. Res.* **2014**, *29*, 1681–1698. [CrossRef]
23. Seo, J.; Kim, Y.; Hu, J. Pilot Study for Investigating the Cyclic Behavior of Slit Damper Systems with Recentering Shape Memory Alloy (SMA) Bending Bars Used for Seismic Restrainers. *Appl. Sci.* **2015**, *5*, 187–208. [CrossRef]
24. Lobo, P.S.; Almeida, J.; Guerreiro, L. Shape Memory Alloys Behaviour: A Review. *Procedia Eng.* **2015**, *114*, 776–783. [CrossRef]
25. Sun, Y.; Luo, J.; Zhu, J.; Zhou, K. A non-isothermal phase field study of the shape memory effect and pseudoelasticity of polycrystalline shape memory alloys. *Comput. Mater. Sci.* **2019**, *167*, 65–76. [CrossRef]
26. Chen, Y.; Schuh, C.A. Size effects in shape memory alloy microwires. *Acta Mater.* **2011**, *59*, 537–553. [CrossRef]
27. Wang, B.; Kang, G.; Kan, Q.; Wu, W.; Zhou, K.; Yu, C. Atomistic study on the super-elasticity of single crystal bulk NiTi shape memory alloy under adiabatic condition. *Comput. Mater. Sci.* **2018**, *142*, 38–46. [CrossRef]
28. Chen, H.-Y.; Tsou, N.-T. The Analysis of Thermal-Induced Phase Transformation and Microstructural Evolution in Ni-Ti Based Shape Memory Alloys by Molecular Dynamics. *Comput. Model. Eng. Sci.* **2019**, *120*, 319–332. [CrossRef]
29. Sathish, S.; Mallik, U.S.; Raju, T.N. Microstructure and Shape Memory Effect of Cu-Zn-Ni Shape Memory Alloys. *J. Miner. Mater. Charact. Eng.* **2014**, *2*, 71–77. [CrossRef]
30. Ahlers, M. The martensitic transformation in the Cu-Zn based shape memory alloys as a tool for the evaluation of transformation mechanisms and phase stabilities. *Mater. Sci. Eng. A* **2008**, *481–482*, 500–503. [CrossRef]
31. Saud, S.N.; Hamzah, E.; Abubakar, T.; Farahany, S. Structure-Property Relationship of Cu-Al-Ni-Fe Shape Memory Alloys in Different Quenching Media. *J. Mater. Eng. Perform.* **2013**, *23*, 255–261. [CrossRef]
32. Ainul Haidar, M.; Saud, S.N.; Hamzah, E. Microstructure, Mechanical Properties, and Shape Memory Effect of Annealed Cu-Al-Ni-xCo Shape Memory Alloys. *Metallogr. Microstruct. Anal.* **2017**, *7*, 57–64. [CrossRef]
33. Aal, M.I.A.E. Influence of the pre-homogenization treatment on the microstructure evolution and the mechanical properties of Al-Cu alloys processed by ECAP. *Mater. Sci. Eng. A* **2011**, *528*, 6946–6957. [CrossRef]
34. Saud, S.N.; Hamzah, E.; Abubakar, T.; Bakhsheshi-Rad, H.R. Thermal aging behavior in Cu-Al-Ni-xCo shape memory alloys. *J. Therm. Anal. Calorim.* **2015**, *119*, 1273–1284. [CrossRef]
35. Salvetr, P.; Dlouhy, J.; Skolakova, A.; Prusa, F.; Novak, P.; Karlik, M.; Hausild, P. Influence of Heat Treatment on Microstructure and Properties of NiTi46 Alloy Consolidated by Spark Plasma Sintering. *Materials* **2019**, *12*, 75. [CrossRef]
36. Hornbogen, E.; Brückner, G.; Gottstein, G. Microstructure and Texture of Ausformed NiTi. *Zeitschrift für Metallkunde* **2002**, *93*, 3–6. [CrossRef]
37. Gurau, C.; Gurau, G.; Braz Fernandes, F. Effect of severe ausforming on the shape memory microstructure of a copper based alloy. In Proceedings of the 14th International Multidisciplinary Scientific GeoConference on Nano, Bio and Green Technologies for a Sustainable Future SGEM 2014, Albena, Bulgaria, 17–26 June 2014; pp. 61–68.
38. Stošić, Z.; Manasijević, D.; Balanović, L.; Holjevac-Grgurić, T.; Stamenković, U.; Premović, M.; Minić, D.; Gorgievski, M.; Todorović, R. Effects of Composition and Thermal Treatment of Cu-Al-Zn Alloys with Low Content of Al on Their Shape-Memory Properties. *Mater. Res.* **2017**, *20*, 1425–1431. [CrossRef]
39. Mussot-Hoinard, G.; Patoor, E.; Eberhardt, A. Influence of wire-drawing on the properties of a Cu-Al-Be polycrystalline shape memory alloy. *Mater. Sci. Eng. A* **2008**, *481–482*, 538–541. [CrossRef]
40. Yildiz, K.; Kok, M. Study of martensite transformation and microstructural evolution of Cu-Al-Ni-Fe shape memory alloys. *J. Therm. Anal. Calorim.* **2013**, *115*, 1509–1514. [CrossRef]
41. Saud, S.N.; Hamzah, E.; Bakhsheshi-Rad, H.R.; Abubakar, T. Effect of Ta Additions on the Microstructure, Damping, and Shape Memory Behaviour of Prealloyed Cu-Al-Ni Shape Memory Alloys. *Scanning* **2017**, *2017*, 1789454. [CrossRef]
42. Safdel, A.; Zarei-Hanzaki, A.; Shamsolhodaei, A.; Krooß, P.; Niendorf, T. Room temperature superelastic responses of NiTi alloy treated by two distinct thermomechanical processing schemes. *Mater. Sci. Eng. A* **2017**, *684*, 303–311. [CrossRef]

buildings

MDPI

Article

Definition of Optimal Ventilation Rates for Balancing Comfort and Energy Use in Indoor Spaces Using CO_2 Concentration Data

Alessandro Franco * and Eva Schito

Department of Energy, Systems, Territory and Constructions Engineering (DESTEC), University of Pisa, Largo Lucio Lazzarino, 56122 Pisa, Italy; eva.schito@for.unipi.it

* Correspondence: alessandro.franco@ing.unipi.it

Received: 12 June 2020; Accepted: 22 July 2020; Published: 24 July 2020

Abstract: Air ventilation rate plays a relevant role in maintaining adequate indoor air quality (IAQ) conditions in public buildings. In general, high ventilation rates ensure good indoor air quality but entail relevant energy consumption. Considering the necessity of balancing IAQ and energy consumption, a correlation between the number of occupants obtained from analysis of CO_2 concentration variation is presented as a general element for controlling the operation of heating ventilation and air cooling (HVAC) systems. The specific CO_2 exhalation rate is estimated using experimental data in some real conditions in university classrooms. A method for the definition of optimal values of air exchange rate is defined, highlighting that the obtained values are much lower than those defined in current technical standards with possibilities of relevant reduction of the total energy consumption.

Keywords: indoor air quality (IAQ); CO_2 production rate; CO_2 concentration; occupancy estimation; demand controlled ventilation (DCV); energy efficiency

1. Introduction

According to the current European Directives [1,2], many strategies have been developed to improve indoor air quality (IAQ) and thermal comfort of public buildings. Moreover, programs of energy efficiency for buildings to reduce their energy needs through either a functional design of new structures or a refurbishment of existing ones have been encouraged.

In this perspective, measures for energy efficiency improvement of heating, ventilation, and air cooling (HVAC) systems appears to be of relevant importance in order to achieve the objective of energy sustainability. The interest on such topic has recently been further exacerbated by the spreading of the well-known COVID-19 pandemic; a strict IAQ control is of relevant importance because it can prevent airborne virus transmission in occupied spaces.

As ventilation is fundamental for maintaining adequate indoor air quality (IAQ), including health and thermal comfort, among the various elements, the possibility of reducing energy needs associated with ventilation and air exchange rate is a topic of particular interest. The energy consumption for ventilation cannot be easily estimated. In the literature, several works report implementation of control strategies on specific case studies for the reduction of energy requirements but maintain quite high standards or thermal comfort and IAQ. Papers containing experimental analyses and methodologies considering both individual houses and public buildings are available.

Carbon dioxide can be considered a possible marker for a quite simple estimation of occupancy level of buildings such as schools and universities and consequently for estimating optimal ventilation rates. Belmonte et al. [3] provide a methodology to compare indoor air quality in residential buildings. The methodology is based on monitoring data of CO_2 concentration, and the authors also estimate the air flow rate profile necessary to maintain a selected CO_2 concentration in two different zones of the apartments.

Merema et al. [4] present monitoring campaigns in offices and school buildings, where IAQ is controlled through a demand controlled ventilation system. The system permits one to maintain good air quality, even in the presence of reduced air flow rates; furthermore, significant reductions of energy consumption are obtained.

In [5], the results of three different ventilation strategies for maintaining thermal comfort and good air quality level in Norwegian primary schools are presented. The authors compare a constant air volume ventilation system (in which the ventilation rate is constant and equal to the maximum value, depending on users) and two demand control ventilation strategies, where the choice of the air flow rate depends either on the CO_2 concentration or the presence/absence of users in the room. In the paper, the strategy based on CO_2 monitoring appears to be the most promising one, with energy savings of 62% with respect to constant air volume system. Similar results were found in [6], where the authors compare a constant air volume system and a demand control ventilation system based on CO_2 measurements. Simanic et al. [7] present the outcomes of seven monitoring activities in primary schools in Sweden, where variable air volume systems are usually employed. The results show that a low CO_2 concentration is achieved by the use of quite high values of ventilation rates (over $0.01 \div 0.02$ m³/s × occupant).

The reduction of energy consumption should not compromise the requirement for a correct IAQ level. A given amount of fresh air is required to determine comfortable indoor environments, maintaining recommended values of relative humidity and pollutants concentrations, which can hamper users' health and productivity. Several indicators can be linked to IAQ control, but one of the most used is the CO_2 concentration.

In engineering applications, humans' exhalations of CO_2 are often correlated to metabolic rate, and this is connected to the specific activity [8]. In medical and biological applications, the specific CO_2 production rate is an object of a detailed study, even through experimental measurements [9]; the values of those exhalation rates differ from the estimates of engineering applications.

In an IAQ analysis through CO_2 measurements, excessively high values of carbon dioxide in closed environments can be an indication of malfunctioning of mechanical ventilation systems and an excessive occupancy density, which can lead to undesired issues for users, such as headache and somnolence. Some studies have also reported a reduction of users' productivity in environments with higher CO_2 concentration [10,11]; another work [12] highlights the concept that decrease in users' performance at school and work is probably due to concurrent effects of both high CO_2 concentrations and other adverse conditions (e.g., bad thermal conditions, exposure to other pollutants).

Even if this latter aspect should be further studied, it is well established that a CO_2 concentration over 1000–1500 ppm should be avoided [13,14] to maintain good indoor air quality conditions. Strictly speaking, the objectives of maintaining good IAQ with reduced energy use for the operation of an HVAC system is difficult, because the second objective usually requires high ventilation rates, and this has a consequence in terms of increase of the energy necessary for air exchange and for air treatment. A compromise is not simple and would require detailed quantitative knowledge of the link between air quality parameters and operation of thermal and ventilation control systems in the specific building under analysis. This knowledge is not exactly evident from the analysis of the above mentioned references. The problem is particularly relevant for public buildings, where it is estimated that around a third of energy consumption related to thermal control is due to ventilation. In literature, in particular, it is found that the choice of correct ventilation rates is difficult, especially in educational buildings, as considered in two recent papers [15,16].

This is due to the fact that air exchange rate is often managed automatically without considering the real occupation profile of the building and of the various rooms but using maximum values imposed by the technical standards with a quite high occupation of indoor spaces (e.g., 80% of the maximum occupation). Finding an optimal value of air exchange rate based on the maintenance of maximum levels of CO_2 concentration in educational buildings is prescribed by various regulatory documents from EU levels down to national design guidelines for educational buildings, but the dynamic control of the air flow rate is not always considered.

Considering the aforementioned elements, in the present paper, the monitored data of CO_2 concentration obtained during examinations and lessons in several classrooms within the Engineering School of the University of Pisa, previously presented in [17], are here used to provide a possible correlation between CO_2 concentration increase and number of occupants. The experimental analysis was performed in various classrooms of university buildings. The rooms were characterized by different geometrical parameters, such as surface area, volume, and shape, and operating parameters, such as indoor occupation. Then, a multi-objective methodology was proposed to establish optimal profile of demanded controlled ventilation of indoor spaces based on the actual occupation's profile obtained by monitoring the increase of carbon dioxide concentration with time.

The multi-objective methodology was based on the possibility of obtaining a compromise between occupant's comfort and energy used for mechanical ventilation. The direct correlation of occupancy profile of the buildings permits one to obtain an adaptive solution based particularly on the simple measurement of CO_2 in order to define a dynamic and demanded controlled ventilation (DCV) strategy that permits one to maintain the required standards of CO_2 concentration with minimum energy use, thus pursuing the objective of energy saving while satisfying the occupants' comfort.

2. The Connection of Carbon Dioxide Concentration, Occupation of the Indoor Spaces, and Air Ventilation Rate in Indoor Spaces

The occupancy detection of buildings and indoor spaces in general is important both for maintaining adequate IAQ standards (in term of temperature, humidity, and concentration of specific pollutants such as CO_2) and for energy efficiency purposes. Due to the relevant energy consumption connected to HVAC system operation, for maintaining temperature and humidity standards and pollutant concentration, a quite accurate estimation of the actual number of users inside a building or of a single room can be useful in order to modulate the operation of the HVAC system, varying the supply temperature to terminal units and an appropriate value of the air exchange rate.

In particular, supplying a fixed value of design volume exchange of fresh air in any condition of occupancy can lead to relevant energy consumption, particularly in spaces where the number of users varies frequently, as in educational buildings, in which different groups of students occupy the same classroom in different part of the day [18].

Among the various methods for the determination of occupants' number, the method using CO_2 concentration and its rise appears to be a particularly interesting one. Furthermore, it also solves some of the major issues related to the adoption of the other methods, such as use of video-cameras [19].

The use of indoor CO_2 concentration data for estimating the occupants' number could be less accurate than other technologies (such as video-cameras or counters) for defining the exact number of persons present inside the indoor space; regardless, it can be considered quite useful for an approximate quantitative estimation of a percentage value of occupation without interferences due to other devices or privacy concerns.

In this work, CO_2 was considered as the only component to be monitored. This component, even if it is not considered a real indoor pollutant, is correlated to IAQ in several studies, and its generation is due to people breathing. As its monitoring is easier than other indoor pollutants (e.g., carbon monoxide, formaldehyde, volatile organic compounds VOCs) thanks to relatively "cheap" sensors, it can be monitored as a "pollutant" of an indoor environment, as reported in other studies in literature.

From a theoretical point of view, considering an indoor space, a mass balance for CO_2 can be written, considering valid the following assumptions:

- uniform concentration in the room, assuming a well-mixed model;
- constant value of outdoor CO_2 concentration;

- well-defined value of the generation rate of the occupants;

Assuming the previous hypotheses, the CO_2 indoor concentration rise can be assumed to be dependent on air ventilation rate, imposed with a mechanical device or dependent on infiltration losses through doors and windows frames, with the following equation:

$$V\frac{dC_{\{CO_2\}}(t)}{dt} = \dot{r}\, n_{occ} - \dot{m}\left(C_{\{CO_2\}}(t) - C_{ext}\right) \tag{1}$$

where $C_{\{CO2\}}$ is the indoor concentration of carbon dioxide, V the total volume of the room, n_{occ} the number of occupants, $\dot{r}$ the CO_2 generation rate per person, $\dot{m}$ the air flow rate due to ventilation (mechanical of natural due to infiltration) expressed in m^3/s, and C_{ext} is the outdoor CO_2 concentration. CO_2 concentration variation in the room can also be written in explicit form, thus, at time t, it is equal at:

$$C_{\{CO_2\}}(t) = C_{\{CO_2\}}(t = 0)\exp\left(-\frac{\dot{m}}{V}t\right) + \left(C_{ext} + \frac{\dot{r}\, n_{occ}}{\dot{m}}\right)\left(1 - \exp\left(-\frac{\dot{m}}{V}t\right)\right) \tag{2}$$

The two equations are simple mass balance for CO_2 concentration inside a definite volume. However, these formulas are reported in the literature, such as [15].

Considering Equation (2), it is easy to understand how the variation of CO_2 concentration can be easily connected to the number of occupants if the rate of generation and the air exchange rate are exactly known, while the use of such a kind of model is quite difficult if no control on ventilation rate is available.

In a recent paper [17], the correlation between the variation of CO_2 concentration and the occupancy profile of the room was analyzed, proposing a linear correlation between the CO_2 concentration rise and the specific volume per occupant. It was discussed how, for a specific activity and for a specific category of users (students and young people), the most relevant variables are the volume available for each person, the specific production, and the air exchange rate.

The correlation was also verified with other variables, considering room characteristics and occupation data as particularly relevant.

Equation (1) can be also reformulated in the case of "closed volume" neglecting infiltration and natural ventilation effects and in absence of mechanical ventilation, thus considering the production rate as the predominant factor for CO_2 concentration variation inside the room:

$$\frac{V}{n_{occ}} = \dot{r}\frac{1}{\frac{dC_{\{CO_2\}}(t)}{dt}} \tag{3}$$

Neglecting infiltration appears to be especially accurate in new or renovated buildings, where doors and windows are airtight. Using an accurate value of the production rate $\dot{r}$, knowing the volume of the room and the CO_2 concentration variation in a given time period, Equation (3) allows for the estimation of the number of occupants in the room, and the estimation of the CO_2 production rate per person is an essential element. Furthermore, one can observe a hyperbolic correlation between the CO_2 concentration rise and the specific volume per occupant.

The generation rate depends on different elements—such as age, weight, and type of activity—and typical values of exhalation per person for common metabolic activities found in literature should be used if validated through experimental results. Then, one can use Equation (1) to estimate the mechanical ventilation air flow rate to be injected in the room to maintain an optimal operating temperature as well as a value of CO_2; this air flow rate can be adjusted to reduce energy requirements, thus obtaining an advantageous value for both the objectives.

Figure 1 provides a correlation between the derivative of CO_2 concentration and the air volume per occupant based on the application of the model previously exposed in Equation (3) for three different values of CO_2 production rate: (i) 0.2 L/min, used as a typical lower value for people sitting or involved in light intensity activities, such as students during teaching lessons ($\dot{r}_1$); (ii) 0.8 L/min, used for people standing and operating a medium physical activity ($\dot{r}_2$); and (iii) 1.5 L/min, used for high intensity physical activities ($\dot{r}_3$). The validity of Equations (2) and (3) was tested using the results of a detailed experimental analysis in several classrooms of educational buildings with different shapes and occupation profiles [17].

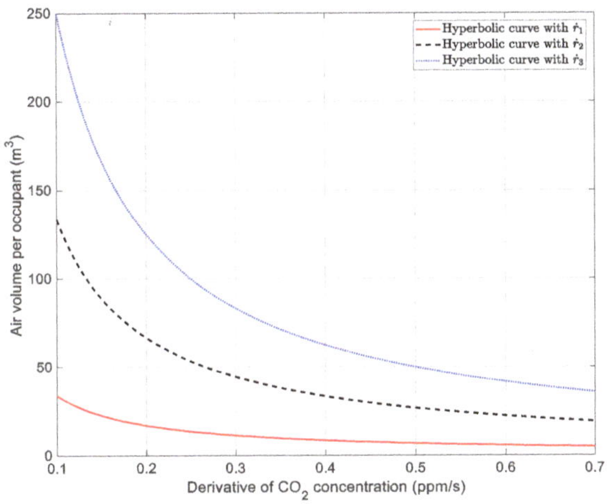

Figure 1. Correlation between derivative of CO_2 concentration and air volume per occupants for different values of CO_2 production rate per occupant.

3. Air Exchange Rate and Direct Correlation with Real Occupation: Data from a Monitoring Activity

In order to maintain the required condition of comfort concerning temperature, humidity, and CO_2 concentration, this last value being a typical pollutant marker in indoor spaces, the operation of HVAC in order to control the environmental parameters has to be correlated with the real occupation of the spaces. In particular, air exchange rate in closed rooms and indoor spaces generally causes relevant additional energy consumption. This is because the air that is circulated from the external environment, usually at lower (in winter) or higher (in summer) temperatures than the environment depending on the season, must be brought to the room temperature. The energy consumption is higher the greater the flow rate of air to exchange (air flow rate volume) and the difference in temperature between the external environment and the room are. The higher the required air exchange rate is, the better the air quality in the indoor space will be, but the ventilation power required and the overall energy consumption will also be higher.

In air conditioning systems, the air exchange can be total or partial. In the latter case, a percentage of the air taken from the room could be recirculated, and only a fraction of the same would be fresh air. After the COVID 19 health emergency, the legislation of many countries has imposed that, in newly designed systems and in public buildings in general, the percentage of air recovered from the external environment should be 100% and no recirculation of air from the room could be possible. This will certainly lead to greater energy consumption for the operation of the HVAC system and particularly of the ventilation rate. The energy consumption for the operation of an HVAC system is linked both to the transmission losses, proportional to the temperature difference between the room and the external environment, and to the thermal conductance of the structure (UA) and the mass transfer, connected to a percentage of volume of the room (V) that must be renewed in the unit time (usually one hour).

Under normal conditions (air exchange rate in the range 0.3–1 Vol/h), the energy consumption in connection with air ventilation rate can be estimated to be between 20% and 40% of the total energy consumption. However, if a more relevant air exchange rate is necessary, it would be possible that the energy consumption could be dependent on the air exchange rate, which could be higher than those connected to the heat transfer.

Usually, the fraction of the volume of the room that needs to be exchanged is established by technical regulations. This establishes a rather high value connected to a maximum use of the room, but no specific correlation with the real number of occupants is used. However, the fraction of air that must be renewed is certainly a function of the real occupation of the room. In the future, when there is reduction of social interaction and the rooms are less crowded than in the past, it will be very important to correlate the air mass exchange rate to the real number of occupants.

As previously mentioned, in this work, the results of a monitoring campaign in several classrooms within the Engineering School of the University of Pisa were used for defining an optimal value of the of the air exchange rate.

The main characteristics of the monitored classrooms in terms of volume, floor surface, and effective number of occupants during the various experiments are reported in Table 1. The monitoring campaign in each classroom lasted for a certain number of minutes, depending on the length of the exam or the lecture ongoing during the measurement. All the details of the experimental analysis are provided in [17]. The experimental analysis was carried on without action of mechanical ventilation and the air exchange rate. Thus, the air exchange rate was only possible due to infiltrations from window frames and doors.

Table 1. The main data of the 12 monitored classrooms.

Classroom	Building	Level	Max. Occupancy (N)	Floor Surface, S (m^2)	Volume V (m^3)	Ratio V/S (m^3/m^2)	Vol. for Student at Full Occupation (m^3)
0	LAB	1st	20	20	58	2.90	2.90
1	A	1st	108	88	482	5.48	4.46
CL1	B	1st	116	216	583	2.70	5.03
2	B	1st, 2nd	366	336	1426	4.24	3.90
3	C	4th	72	73	212	2.90	2.94
4	F	1st	309	286	1587	5.55	5.14
5	F	1st	208	216	1220	5.65	5.87
6	F	1st	109	130	721	5.55	6.61
7	F	1st	196	197	1094	5.55	5.58
8	F	1st	104	129	717	5.56	6.89
9	F	1st	109	128	711	5.55	6.52
10	F	1st	140	131	439	3.35	3.14

CL: computer lab.

One of the classrooms that particularly identified with label 0 was used as the benchmark room for experimentally defining typical values of CO_2 production rates for each student. The measurements of CO_2 concentration together with the measurements of indoor temperature and relative humidity were obtained using up to four different sensors, Chauvin Arnoux C.A 1510, [20]. Concerning the specific problem of CO_2 concentration measurement, the sensors permitted us to cover a measurement range 0–5000 ppm with an uncertainty of ± 3% of the reading ± 50 ppm at 25 °C at atmospheric pressure. Even in literature, CO_2 measurements are usually carried out through a single sensor placed next to the center of the room, and the uniformity of carbon dioxide concentration in the room must be proven. Thus, in some of the experiences, two or three sensors were placed in different zones of the room at breathing level to check if the assumption of well-mixed indoor volume could be realistic.

Two typical examples of CO_2 variation (increase) with time are shown in Figures 2 and 3. In Figure 2, the profile is related to an experimental analysis of room 5, where two sensors were used to measure the CO_2 concentration increase. Figure 3 provides the CO_2 concentration profile within the classroom identified as CL1 (computer lab).

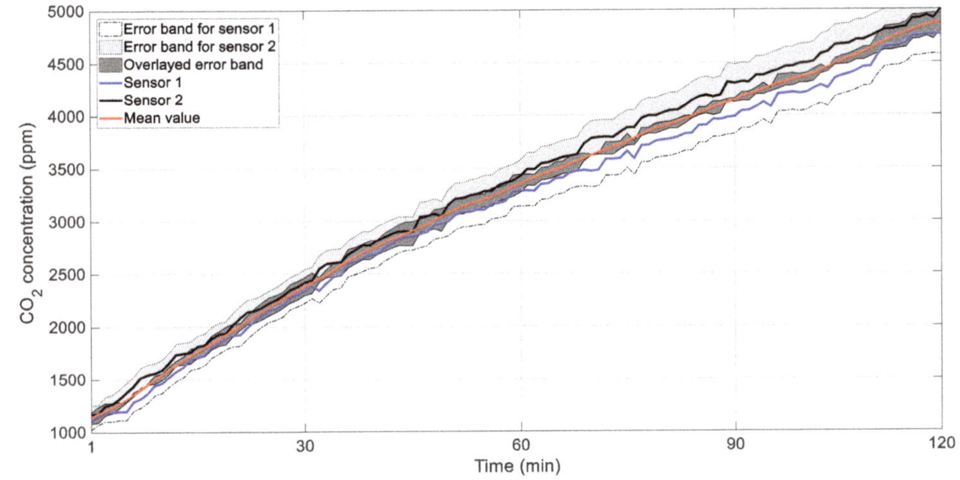

Figure 2. CO_2 concentration profiles in a particular experience in classroom 5.

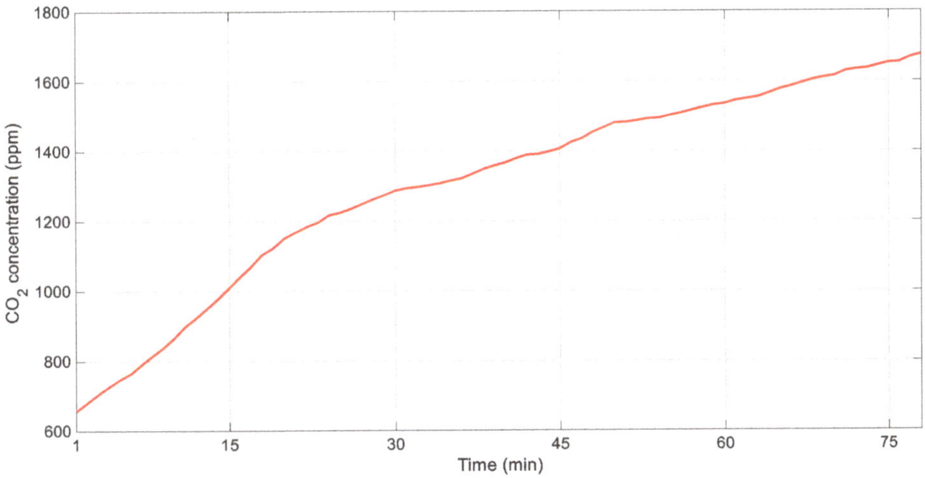

Figure 3. CO_2 concentration profiles in the computer lab (CL1) during typical didactic activity.

In all the cases where two or more sensor were used, similar profiles for the CO_2 measurements were found, similar to those provided in the previous Figure 2. Possible differences among the measurement of the sensors were within the error band of the sensors represented by the filled area, thus the hypothesis of well-mixed room was considered valid.

Table 2 provides the results obtained in all the critical cases when the threshold value of 1500 ppm for CO_2 was overcome. Results show the starting value and an average value of the CO_2 concentration increased in each classroom, which was evaluated in the first 10–15 min of monitoring, and an evaluation of the elapsed time between the start of the measurement and the time in with the CO_2 threshold value (1500 ppm) was reached. This time duration was used to evaluate the CO_2 concentration increase with time. Only in two cases was the threshold limit lower than the initial value, meaning that, during the previous didactic activity, the CO_2 concentration increased over the limit, and natural ventilation was not sufficient to have the concentration decreased; in those two cases, the estimation of the CO_2 concentration increase with time was done using the concentration at the starting value of the lesson.

Table 2. CO_2 concentration at the beginning and at the end of the monitoring experience.

Experiment Identification (Room/Number)	Number of Students	V/n_{occ} (m^3)	CO_2 Concentration at t = 0 (ppm)	Maximum CO_2 Concentration Measured	CO_2 Concentration Increase with t (ppm/s)	Time for Overcoming CO_2 Threshold (1500 ppm) (min)
0	20	2.90	704	1841	0.28	20
1	68	7.08	782	2566	0.17	72
CL1	38	15.08	659	1676	0.22	55
3	72	2.94	1264	2960	0.59	6
4#1	58	27.36	678	1741	0.14	95
4#3	93	17.06	596	2685	0.18	78
5#1	168	7.30	1138	4913	0.52	10
5#3	106	11.50	1100	3409	0.26	21
7#1	146	7.49	791	3297	0.40	24
8	54	13.27	1257	2383	0.37	8
9#1	54	13,16	1512	2334	0.43	0
9#2	59	12.05	1648	3369	0.26	0
10#2	50	8.78	695	2410	0.42	27

Even if all the mentioned variables seemed to be meaningful, the experimental results obtained in the particular conditions demonstrate that a hierarchical distribution of the variables could be identified; the most relevant one seemed to be, as expected, the volume available for each student.

The values of the rise of CO_2 concentration, expressed by the derivative $\frac{dC_{\{CO_2\}}}{dt}$ estimated in the experimental analysis referred in [17], ranged between 0.1 ppm/s up to 0.7 ppm/s, with maximum occurrence in the range between 0.2 and 0.5 ppm/s.

In order to establish a quantitative correlation between the number of persons inside a room and the evolution of CO_2 concentration with time, according to the model provided in Equations (1)–(3), accurate knowledge of the value of the CO_2 production rate per occupant was needed. This value can be obtained with reference to the available literature and technical standards, but it can also be measured. In the present paper, this second path was considered. In particular, the benchmark room was the one identified with label 0 in Table 1.

CO_2 was measured in two different point during the first phase of a short duration examination (a test of 30 min). A total number of 20 students were present inside the room at the initial time of the experimental analysis, and the windows were closed. The two sensors measured an almost uniform profile of CO_2 increase; the mean values are shown in Figure 4. In the first phase (five minutes), the sensors began to measure; CO_2 was almost constant as the classroom was unoccupied. Then, in a few minutes, twenty students entered and sat inside the room for about 30 min. Then, the door was opened, and the majority of the students left the room (phase 3), while only five of them remained inside.

Considering the experimental results obtained during this short time experimental analysis, using Equation (2), a mean value of CO_2 production rate equal to 0.20 l/min per person could be estimated. The value obtained agreed quite well with the specific activity when compared with CO_2 production rates available in the literature [8,9,21,22].

With reference to the available literature, represented by the four papers above, the values of this production rate in general cases can range between a minimum value of 0.2 l/min up to a maximum value of 0.45 l/min. In general, on technical engineering standards for a sedentary activity, a level of about 18 l/h that equals 0.3 l/min per adult person can be estimated. The values obtained considering the results of the experimental analysis during typical teaching activity were lower than those suggested by technical standards (e.g., [23]), but they were in accordance with some estimates evaluated by medical standards [21] and other monitoring campaigns [22].

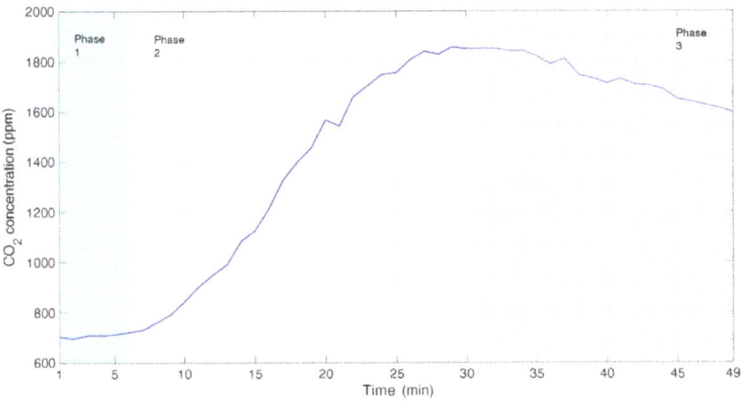

Figure 4. Evolution of CO_2 concentration in the room 0 for defining the CO_2 production rate.

4. Correlation of CO_2 Concentration Rise and Occupation for Defining Optimal Air Exchange Rate

As already discussed in the introduction, the idea of predicting the occupancy using physical and statistical CO_2 measurements has been diffusely expressed in the literature and has been object of recent analyses as well [24,25]. On the other hand, a demand controlled ventilation (DCV) strategy within the context of the American Society of Heating Refrigerating and Air-Conditioning Engineers (ASHRAE) Technical Standards is developed in [26] for public buildings using a digital control system.

In the present section, the methodology defined was used to define a correct air exchange rate for indoor volume with the objective of controlling the rise of CO_2 concentration. The analysis refers to some of the situations previously tested. The idea of defining a procedure for calculating a value of the air exchange rate, a demand controlled mechanical ventilation rate, has already been developed in the literature, and papers about this specific topic are already available [27,28].

A first outcome of the analysis was the validation of the correlation between the CO_2 increase and the specific volume per student in Equation (2) using the results of the specific experimental analysis. The problem in using Equation (3) is that it neglects air change in the room. Some of the results are shown in Figure 4, where the twelve pair of values $\left(\frac{\Delta C_{\{CO_2\}}}{\Delta t}, \frac{V}{n_{occ}}\right)$ obtained considering the data acquired in twelve specific experimental analyses are reported. Figure 5 also shows a hyperbolic curve, in red, representing the least-squares fitting curve for Equation (3). The value of CO_2 production rate per person as evaluated using the least-squares fit, $\dot{r}_{ls}$, was 0.20 l/min and was equal to the estimated value derived after the monitoring of room 0, defined as $\dot{r}_0$.

In particular, for classroom 5, a hyperbolic curve using the average value of CO_2 production among the monitored classrooms, $\dot{r}_m = 0.22$ l/min, fit exactly the experimental results obtained in the experimental analysis 5#1. The curve is represented in Figure 5 with a black dotted line.

Some of the monitoring campaign results slightly deviated from the hyperbole due to the quite high infiltration rate (classroom 3, which is hosted in an old building). Moreover, a hyperbolic correlation between the rate of increase of CO_2 concentration, $\frac{\Delta C_{\{CO_2\}}}{\Delta t}$, and the volume available for each student, $\frac{V}{n_{occ}}$, seemed to fit the results of the experimental analysis quite well. A hyperbolic function was used with model y = a/x, with "x" representing the derivative of CO_2 concentration (in ppm/s) and "y" the air volume per student (in m3). The fitting procedure provided the "a" value, generally in the range between three and four (average value equal to 3.33).

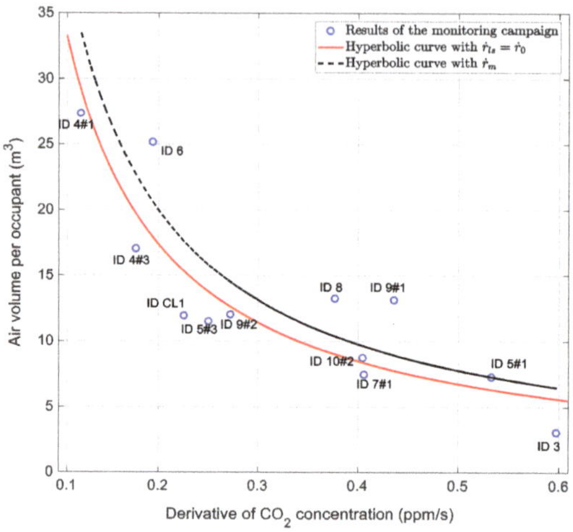

Figure 5. Correlation between CO_2 concentration and air volume per student, obtained in the different experiments provided in [17].

Thus, by evaluating the variation of CO_2 concentration in the room in a given time step (e.g., 10–15 min) in the absence of mechanical ventilation and knowing the volume of the room, it was possible to estimate, with quite good accuracy, the approximate number of occupants. This procedure can be repeated periodically for controlling the operation of the mechanical ventilation system in order to meet the desired target in terms of temperature relative humidity and IAQ.

Therefore, it is possible to seek the mechanical air flow rate necessary to maintain a suitable value of CO_2 concentration inside the room, $\dot{m}$, using Equation (2).

As an example, the case identified with the label ID 5#1 (classroom identified with label 5 and experience #1) is considered. The case was used as a reference because, according to the results of Table 2, it is considered as a representative one. In this case, the classroom had a volume of 1220 m^3, and several scenarios were assumed with different numbers of persons inside. An indoor steady-state CO_2 concentration between 800 ppm and 1500 ppm was considered acceptable for maintaining good IAQ conditions. Thus, for each threshold value, it was possible to estimate the air flow rate values that allowed for maintenance of the CO_2 concentration below the imposed value. The result of the application of the methodology is shown in Figure 6, where the thin colored curves represent the lines of equal air flow rate (iso $-$ $\dot{m}$). For example, for a number of occupants equal to 122 (corresponding to a reference value of the volume available for each occupant of 10 m^3) and a maximum allowable CO_2 concentration of 1000 ppm, an air exchange rate of 0.8 m^3/s was required.

The air exchange rate necessary for maintaining the desired CO_2 concentration can also be correlated with the energy requirements. This value depends on the mass flow rate and on indoor and external temperature; through the following equation (referred to an operating time, t), the energy amount required for ventilation can be calculated:

$$Q_v = \rho \, \dot{m} \, c_p (T_{int} - T_{ext}) t \tag{4}$$

where $\dot{m}$ is the air exchange rate, expesed in m^3/s. For users' comfort purposes, the indoor air temperature is limited within a given range (e.g., in winter, 18–22 °C and 24–27 °C in summer). Thus, it is possible to display ventilation loads such as in Figure 6, where a combination of curves are shown varying ventilation rate as well as external and indoor temperature. In the figure, the markers represent ventilation loads at the same indoor temperature (i.e., rhomboidal, plus, round, square, and triangular markers refer to five different values of the indoor temperature, $T_{in} = 18 - 22$ °C), while the colorbar on the right side represents the energy consumption for ventilation, expressed in kWh.

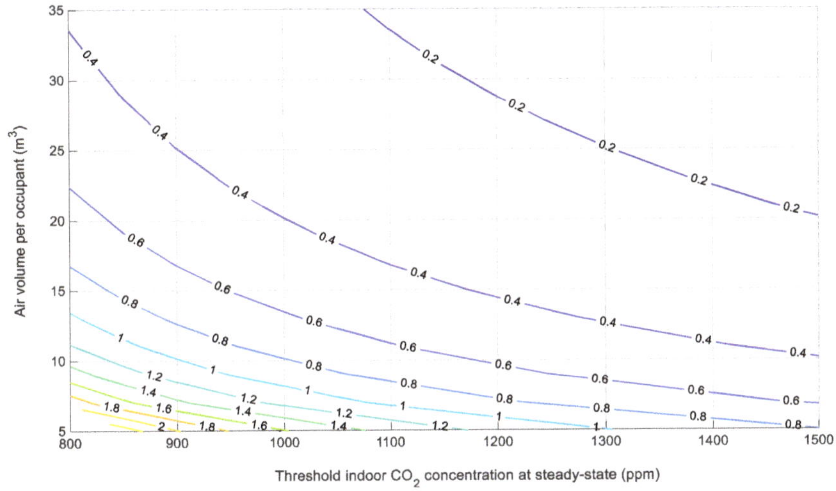

Figure 6. Air flow rates depending on desired CO_2 concentration and air volume per occupant.

Referring to the air exchange rate $\dot{m} = 0.8\ m^3/s$, considering an external temperature of 0 °C and an indoor set-point temperature of 20 °C, a "ventilation load" of about 20 kWh was required for each hour of operation in this specific classroom. In Figure 7, referring to the same case of classroom 5, it is possible to select a different value of the air exchange rate (e.g., $\dot{m} = 0.6\ m^3/s$ referred to a threshold CO_2 concentration, for example, 1200 ppm). In general, Figure 7 can be used to evaluate, for each couple of T_{in} and T_{ext}, the energy required for ventilation, (e.g., for a $T_{int} = 20$ °C and $T_{ext} = 0$ °C, $Q_v = 15$ kWh). If the air exchange rate necessary is at levels of 0.2 m^3/s, the energy used is reduced at 5–6 kWh.

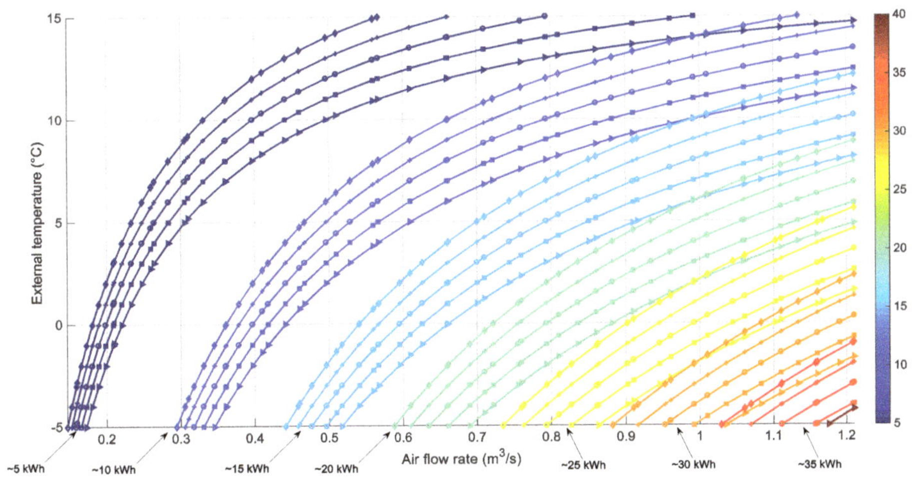

Figure 7. Energy consumption at different air exchange rates as well as external and indoor temperatures (classroom 5).

The definition of the air exchange rate imposed by a control on the upper value of CO_2 concentration obtained using the proposed methodology can be used as a dynamic method for ventilation rate control as an alternative to actual regulation solutions, as proposed in the European Union EN 16798 [23].

This technical standard proposes a ventilation rate to dilute bio effluents from the occupants and a ventilation rate to remove pollution produced inside the building. The former term depends on the occupancy, while the latter term depends on the type of building. In particular, the parameters are the maximum number of the persons in the room and consequently the ventilation rate for occupancy per person, expressed in l/(s person), and the floor area, m^2, corresponding to the ventilation rate for emissions from building, $l/(s{\cdot}m^2)$.

$$\dot{m} = n{\cdot}q + A_{floor}{\cdot}q_b \tag{5}$$

where n is the number of occupants, q is the ventilation rate for occupancy per person (expressed in l/s person), A_{floor} is the surface of the room, and q_b is the ventilation rate for emissions from building, expressed in $l/(s{\cdot}m^2)$. In the technical standards, four categories are presented depending on the level of required IAQ. Here, two methods were considered, conventionally labeled with TS1 and TS2 (Technical Standards 1 and 2), corresponding with the maximum and the minimum values contained in Technical Standard EN 16798 [21] for comparison with the method proposed based on an accurate estimation of the occupation. In particular, the values of q ranged from 2.5 l/s person (category IV) to 10 l/s (category I), while the values of q_b ranged from 0.3 $l/(s\ m^2)$ for category I up to 1.0 $l/(s\ m^2)$ in case of category I.

Method TS1. *The air exchange rate is calculated depending on the maximum number of occupants for the specific classroom under analysis,*$n_{occ,max}$ *, and on the building type, based on floor surface, using the lowest values of* q *and* q_b*, before considered: 2.5 l/s person and 0.3* $l/(s\ m^2)$*;*

Method TS2. *The air exchange rate is calculated depending on the maximum number of occupants and floor area using the upper values of the parameters* q *and* q_b*, in particular, those corresponding to category I of TS EN16798.*

It is worth noting that the two methods refer to the maximum number of occupants and not the actual number. The two methods reported in EN 16798 were compared with a ventilation rate defined with the proposed methodology, characterized by an assessed moderate IAQ (CO_2 threshold value equal to 1200 ppm). The test cases for which a comparison of the three different strategies were tested are those referred to in Table 2 and identified with labels 5#1 and 4#1. The characteristics of the rooms are provided in Table 1. The following relevant elements can be considered:

- actual number of users: 168 (5#1), 58 (4#1);
- initial CO_2 concentration: 1138 ppm (5#1), 678 (4#1);
- external CO_2 concentration: 500 ppm (value experimentally measured).

Table 3 reports the values obtained in term of air flow rates for maintaining acceptable values of CO_2 concentrations by using three strategies, the two derived from the European Directive based on the maximum occupancy of the classrooms and the one proposed in the paper based on the real estimation of occupation derived by CO_2 measurements. With respect to the two situations analyzed, in the first case (classroom 5#1), when the number of occupants of the room was about 80% of the maximum occupation (168 vs. 208), the air exchange rate obtained with the proposed strategy and based on the idea of maintaining an acceptable level of IAQ was similar to the one proposed with the lower values of the Technical Standard (TS1). However, if the room was only partially occupied (less than 20%, as in the case of classroom 4#1), the required air exchange rate was much lower than both the values obtained with the Technical Standards with important energy savings.

Table 3. Comparison among the air flow rate based on technical standards (TS) and with the proposed strategy for two of the classrooms under analysis.

	Classroom 5#1 Air Exchange Rate (m^3/s)	Classroom 4#1 Air Exchange Rate (m^3/s)
Strategy TS 1	0.58	0.85
Strategy TS 2	2.42	3.15
Proposed strategy (based on the actual number of occupants)	0.64	0.22

Table 4 shows that the strategy based on the control of the number of occupant (students), proposed in the present paper and available if a CO_2 monitoring is active, allowed for a reduced air exchange rate in both the analyzed situations with respect to the solutions recommended by the current Technical Standards. In particular, in three of the analyzed cases, for strategies TS1 and TS2, the values of the air exchange rate were higher than those based on the proposed strategy, indicating significant energy requirements. Those last values, estimated as a function of typical outdoor temperature, determines energy saving for mechanical ventilation; the values are reported in Table 4 for a reference indoor temperature of 20 °C and the actual value of the external temperature measured during the experimental analysis. The proposed strategy determined, in both cases under analysis, a reduction of value of the air exchange rate with respect to the one required according to the Technical Standards and consequently a reduction of the energy use.

Table 4. Energy consumption for air ventilation for the three strategies during a lessons of about 2 h.

	Classroom 5#1	Classroom 4#1
Monitoring day and starting hour	Jan 12	Jan 17
Average external temperature (°C)	8.0	11.9
Ventilation thermal load (Strategy TS 1) (kWh)	16.7	15.7
Ventilation thermal load (Strategy TS 2) (kWh)	69.7	60.5
Ventilation thermal load (Proposed strategy) (kWh)	18.4	4.2

Figure 8 shows the profiles of CO_2 concentration using the three different values of air exchange rates for the two classrooms analyzed using the same periods (minutes) of the monitoring campaign.

In both the figures, CO_2 concentration rise was observed in the basic experimental case, for which no mechanical air exchange rate was applied, and CO_2 concentration increased monotonically together with the trend observed with the three different values of air exchange rate. It was observed that a good IAQ could be obtained with each of the three strategies (instead, without mechanical ventilation, high values of indoor CO_2 were always measured).

Considering the action of a controlled ventilation, in all cases, the level of 1500 ppm was never reached, while the strategies based on the European Directive, TS1 and TS2, determined levels lower than those determined by the proposed strategy. In both cases, strategy TS2 maintained low values of CO_2 inside the classrooms but with high energy consumption.

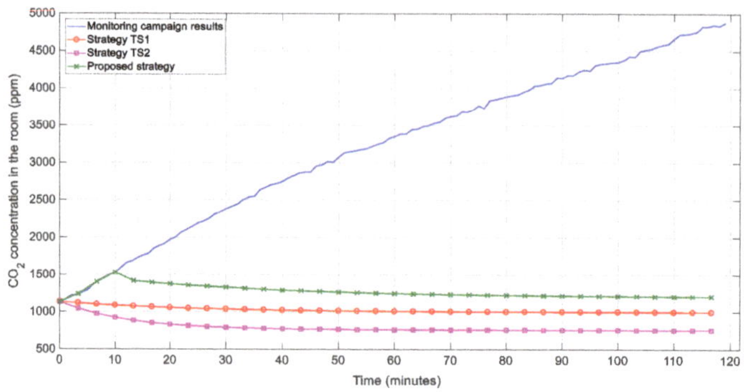

(**a**) Room 5#1

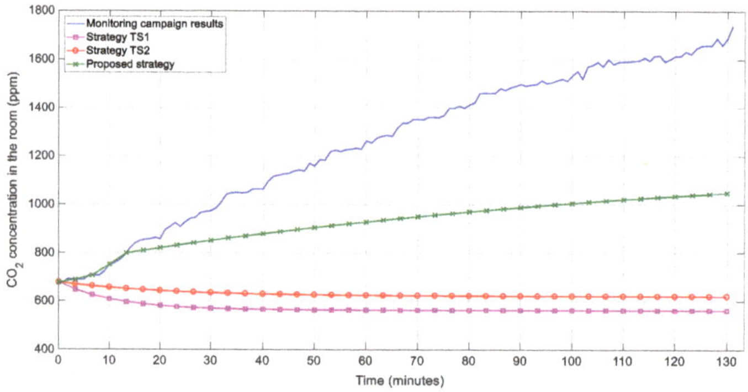

(**b**) Room 4#1

Figure 8. Comparison among monitoring results (without mechanical ventilation) and three strategies involving mechanical ventilation for a two hours experience.

5. Definition Optimal Ventilation Rate Based on a Multi-Objective Approach

The article should provide a methodology to choose the air change rate inside some classrooms, which are currently ventilated through infiltration and windows opening and where thresholds of CO_2 concentration are usually overcome, considering that, in these classrooms, the design ventilation rates are provided for the design value for the number of persons in the room. As university classrooms can accommodate a higher number of students than the actual number of occupants present during a standard lecture (as monitored), the choice of the design ventilation rate would lead to exaggerated ventilation rates (and, consequently, energy consumptions). Thus, the choice of modulating the ventilation rate on the basis of the actual number of occupants can reduce ventilation losses and energy consumptions of the fan and of the HVAC system.

The procedure analyzed in the previous section has a general value; the idea is to detect the presence of students inside the room and then to define a well-defined value of the air exchange rate, thus controlling the operation of HVAC system. This value is not fixed but depends on the actual number of occupants and, in particular, for a defined room, on the specific volume available for each student as well as the CO_2 production rate per person specific for the activity performed.

Once the number of occupants is estimated through a short time monitoring CO_2 concentration, then the air exchange rate can be calculated on the basis of the required IAQ and energy efficiency standards, thus obtaining maps similar to those reported in Figures 6 and 7.

The air flow rate can be chosen for a selected threshold of CO_2 concentration; all the values in the range between 800 and 1500 ppm can be considered valid and, for a given number of occupants, the minimum air exchange rate to be supplied in the room is provided together with the corresponding ventilation load. This last value functions on external and indoor temperatures.

The occupancy detection guarantees that optimal operation of an HVAC system could be performed and, in particular, the ventilation rate could be the lowest possible required for maintaining a well-defined value of CO_2 concentration. Moreover, solutions violating this limit (i.e., 1500 ppm) are not possible. Thus, in this way, an advanced demand controlled ventilation strategy as a result of the application of a multi-objective optimization strategy can be implemented and defined considering a compromise between maintaining high IAQ and controlled energy consumption.

In this case, it is possible to define an objective function for considering both the objectives using, for example, the "utility function method" available in the textbook by Rao [29] and already tested by Franco and Diaz Vazquez in a different context in [30]. This utility function *F* can be defined by aggregating in dimensionless form a series of functions $f_i(x)$, all expressed in dimensionless terms and representative of each objective (e.g., energy consumption, temperature difference with respect to the imposed value, pollutant concentration, relative humidity). It is also possible to consider different weights, ϕ_I, in order to give varying importance to them:

$$F = \sum_{i=1}^{n} \phi_i f_i(x) \tag{6}$$

In a simple case, it is possible to consider two functions, one representing the IAQ based on the control of the maximum permissible value of CO_2 concentration and the second one for controlling the energy consumption in connection with the imposed air exchange rate.

IAQ, for example, can be represented by the CO_2 concentration in the form of difference with respect to the available standards, while the second indicator accounts for the energy consumption required to maintain the level of CO_2 concentration imposed (e.g., 1500 ppm means upper acceptable value of IAQ), according to the European Standards [14].

As for the energy efficiency, the same air flow can be used as an indicator, as it is linearly correlated to the energy consumption due to the ventilation load (see Equation (5)). On one hand, high IAQ would require high ventilation rates to maintain the lowest CO_2 concentration inside the classroom. On the other hand, energy efficiency measures would encourage the reduction of air exchange rate in order to reduce energy consumption for ventilation. An optimal strategy choosing the most suitable air flow rate in the analyzed indoor space can be found by both considering the objectives and aggregating them using appropriate weights. If an equal weight is given to both the objectives, such as in this case, $\phi_1 = \phi_2 = 0.5$. In other cases, different values can be given to ϕ_i depending on the importance given to the two concurrent objectives. As example of application of the methodology is the case of the room identified with ID 5 with 122 occupants. The two objectives are, in this case, represented by the following dimensionless indicators:

$$\begin{cases} f_{IAQ} = \dfrac{C_{\{CO_2\},set} - C_{\{CO_2\},min}}{C_{\{CO_2\},max} - C_{\{CO_2\},min}} \\ f_{Energy} = \dfrac{\dot{m}_{set} - \dot{m}_{min}}{\dot{m}_{max} - \dot{m}_{min}} \end{cases} \tag{7}$$

For the first objective, it is considered that the maximum IAQ is obtained with $C_{\{CO_2\},min}$, set equal to 600 ppm, almost equal to the external concentration. The lowest IAQ is obtained with $C_{\{CO_2\},max}$, set equal to 1500 ppm, which is also the threshold limit for indoor spaces.

The $C_{\{CO_2\},set}$ is the effective concentration inside the indoor environment, depending on the air flow rate, which is the object of optimization. As for the second objective, $\dot{m}_{min}$ represents the air flow rate associated with minimum ventilation load, which is set equal to 0 m^3/s. Instead, $\dot{m}_{max}$ represents the air flow rate associated with the maximum IAQ, which is calculated equal to 4 m^3/s in this case;

$\dot{m}_{set}$ is the air flow rate, found by means of the utility function method. Both objective functions are singularly optimized when they are close to 0. In this case, the utility function is:

$$F = \phi_{IAQ} f_{IAQ} + \phi_{Energy} f_{Energy} \quad (8)$$

In the first case, the two objectives are considered equally important, thus ϕ_{IAQ} and ϕ_{Energy} are 0.5, and the utility function method gives an optimal value of air flow rate at 1.4 m^3/s and a CO_2 concentration equal to 800 ppm. If higher importance is given to energy efficiency, thus maintaining a lower IAQ level inside the room, different weights can be chosen ($\phi_{IAQ} = 0.3$ and $\phi_{Energy} = 0.7$). In this case, when CO_2 concentration is limited at 1000 ppm, an optimal air exchange rate of 0.8 m^3/s is found.

Figure 9 reports the results of the application of the utility function method for ϕ_{Energy} varying the two terms of weight from zero to one (with $\phi_{IAQ} = 1 - \phi_{Energy}$) in terms of CO_2 concentration at steady-state and optimal air flow rate. The results need to be verified and compared with those obtained in other recent papers about post-processing, such as [31,32].

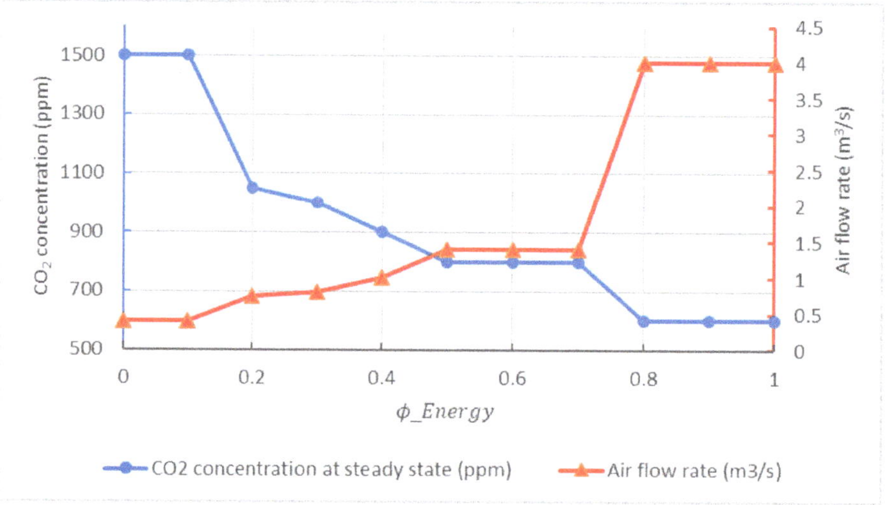

Figure 9. Optimal CO_2 concentration and air flow rates applying the utility function method.

6. Conclusions

The maintenance of indoor air quality (IAQ) in indoor spaces through mechanically assisted air exchange rate determines a higher value of the energy consumption. For defining an optimal operation of the HVAC system and, in particular, a suitable level of the ventilation rate, knowledge of the number of occupants in the indoor spaces is of fundamental importance.

Based on experimental data acquired in several university classrooms concerning the CO_2 concentration rise, the present paper proposes a methodology for correlating the value of air exchange rate required to maintain the maximum permissible CO_2 concentration level (1500 ppm) in connection with the real occupation of the same indoor space obtained through the measure of CO_2 concentration rise.

The method is based on the knowledge of a specific rate of CO_2 production for each person, in this case estimated as 0.2 L/min. The air flow exchange rates assessed through the proposed method were sensibly lower than those suggested by current Technical Standards and were sensitive to the occupation profile. Considering the two cases, an air exchange rate reduced by a factor 2–3 with respect to the values suggested by European Standards, mainly when the occupation of the room was reduced, was sufficient to maintain acceptable values of IAQ conditions with a relevant reduction of the energy consumption.

For the specific CO_2 production rate and a specific room tested (two cases were analyzed in particular), integrated maps were provided to find the most suitable air exchange rate, dependent on the number of occupants and on the imposed threshold CO_2 concentration value, that needed to be maintained by means of the demand controlled ventilation. In the two cases analyzed in detail, if the occupation was about 70% of the total, the energy consumption for ventilation was similar to the one obtained with the more conservative strategy suggested by Technical Standard (18.4 kWh vs. 16.7 kWh). If the occupation was of the order of 20% with respect to the maximum permitted, the energy consumption could be reduced in a relevant way (4.2 kWh vs. 15.7 kWh); the advantage in terms of energy saving appears to be relevant mainly when the occupation is reduced.

Finally, the paper presented a methodology for defining optimum values of the air exchange rates for balancing energy efficiency and user comfort in public buildings through the definition of a dimensionless utility function that considers both CO_2 concentration and energy consumption.

Considering that current solutions proposed for improving energy efficiency in public buildings require very accurate energy control, the adaptive solution proposed in the present paper, based on simple measurement of CO_2 concentration using commercial sensors, represents a considerable step forward with respect to traditional model-based systems, because it dynamically builds a "just accurate enough" model that could be sufficient to provide optimal decision strategies for operating ventilation rates and HVAC systems in general.

Author Contributions: Conceptualization, A.F.; methodology, A.F.; software, A.F., E.S.; data curation, A.F., E.S.; writing—original draft preparation, A.F. and E.S.; writing—review and editing, A.F. and E.S.; supervision, A.F.; funding acquisition, A.F. All authors have read and agreed to the published version of the manuscript.

Funding: "This research was funded by the University of Pisa (PRA 2018–19, project no. 2018_38)".

Acknowledgments: University of Pisa and School of Engineering are acknowledged for providing the classrooms in which the experimental analysis was carried out. Lorenzo Marchi, former student of the Master Degree in Energy Engineering, is acknowledged for the support given to data analysis.

Conflicts of Interest: The author declares no potential conflict of interest with respect to the research, authorship, and/or publication of this article.

References

1. European Parliament. *Directive 2012/27/EU of the European Parliament and of the Council on Energy Efficiency*; European Parliament: Brussels, Belgium, 2012.
2. European Parliament. *Directive 2018/844 of the European Parliament and of the Council of 30 May 2018 Amending Directive 2010/31/EU on the Energy Performance of Buildings and DIRECTIVE 2012/27/EU on Energy Efficiency*; European Parliament: Brussels, Belgium, 2018; p. 17.
3. Belmonte, J.F.; Barbosa, R.; Almeida, M.G. CO_2 concentrations in a multifamily building in Porto, Portugal: Occupants' exposure and differential performance of mechanical ventilation control strategies. *J. Build. Eng.* **2019**, *23*, 114–126. [CrossRef]
4. Merema, B.; Delwati, M.; Sourbron, M. Demand controlled ventilation (DCV) in school and office buildings: Lessons learnt from case studies. *Energy Build.* **2018**, *172*, 349–360. [CrossRef]
5. Mysen, M.; Berntsen, S.; Nafstad, P.; Schild, P. Occupancy density and benefits of demand-controlled ventilation in Norwegian primary schools. *Energy Build.* **2005**, *37*, 1234–1240. [CrossRef]
6. Wachenfeldt, B.J.; Mysen, M.; Schild, P.G. Air flow rates and energy saving potential in schools with demand-controlled displacement ventilation. *Energy Build.* **2007**, *39*, 1073–1079. [CrossRef]
7. Simanic, B.; Nordquist, B.; Bagge, H.; Johansson, D. Indoor air temperature, CO_2 concentration and ventilation rates: Long-term measurements in newly built low-energy schools in Sweden. *J. Build. Eng.* **2019**, *25*, 100827. [CrossRef]
8. Persily, A.K. Evaluating Building IAQ and Ventilation with Indoor Carbon Dioxide. *ASHRAE Trans.* **1997**, *103*, 193–204.

9. McLaughlin, J.E.; King, G.A.; Howley, E.T.; Bassett, D.R., Jr.; Ainsworth, B.E. Assessment of the Cosmet K4B2 portable metabolic system. *Med. Sci. Sport Exerc.* **2003**, *31*, S286. [CrossRef]
10. Satish, U.; Mendell, M.J.; Shekhar, K.; Hotchi, T.; Sullivan, D.; Streufert, S.; Fisk, W.J. Is CO_2 an indoor pollutant? Direct effects of Low-to-Moderate CO_2 Concentrations on Human Decision-Making Performance. *Environ. Health Perspect.* **2012**, *120*, 1671–1678. [CrossRef]
11. Myhrvold, A.N.; Olsen, E.; Lauridsen, O. Indoor environment in schools–pupils health and performance in regard to CO_2 concentrations. In Proceedings of the 7th International Conference on Indoor Air Quality and Climate, Nagoya, Japan, 21–26 July 1996; pp. 369–371.
12. Mendell, M.J.; Heath, G.A. Do indoor pollutants and thermal conditions in schools influence student performance? A critical review of the literature. *Indoor Air* **2005**, *15*, 27–52. [CrossRef]
13. American Society of Heating Refrigerating and Air-Conditioning Engineers (ASHRAE). *Ventilation for Acceptable Indoor air Quality*; ASHRAE Standard 62.1-2016; ASHRAE: Atlanta, GA, USA, 2016.
14. European Committee for Standardization (CEN). *EN 13779—Ventilation for Non-Residential Buildings—Performance* for Ventilation And Room Conditioning Systems; CEN: Brussels, Belgium, 2010.
15. Johnson, D.L.; Lynch, R.A.; Floyd, E.L.; Wang, J.; Bartels, J.N. Indoor air quality in classrooms: Environmental measures and effective ventilation rate modeling in urban elementary schools. *Build. Environ.* **2018**, *136*, 185–197. [CrossRef]
16. Chan, W.R.; Li, X.; Singer, B.C.; Pistochini, T.; Vernon, D.; Outcault, S.; Sanguinetti, A.; Modera, M. Ventilation rates in California classrooms: Why many recent HVAC retrofits are not delivering sufficient ventilation. *Build. Environ.* **2020**, *167*, 106426. [CrossRef]
17. Franco, A.; Leccese, F.; Marchi, L. Occupancy modelling of buildings based on CO_2 concentration measurements: An experimental analysis. *J. Phys. Conf. Ser.* **2019**, *1224*, 10. [CrossRef]
18. Chenari, B.; Dias Carrilho, J.; Gameiro Da Silva, M. Towards sustainable, energy-efficient and healthy ventilation strategies in buildings: A review. *Renew. Sustain. Energy Rev.* **2016**, *59*, 1426–1447. [CrossRef]
19. Szczurek, A.; Maciejewska, M.; Pietrucha, T. Occupancy determination based on time series of CO_2 concentration, temperature and relative humidity. *Energy Build.* **2017**, *147*, 142–154. [CrossRef]
20. Chauvin Arnoux Metrix. Chauvin Arnoux C.A 1510. Available online: https://catalog.chauvin-arnoux.it/it_en/c-a-1510.html?___from_store=it_it (accessed on 2 July 2020).
21. Hall, J.E.; Guyton, A.C. Diffusion of Oxygen and Carbon Dioxide through the Respiratory Membrane. In *Textbook of Medical Physiology*; Saunders Elsevier: Philadelphia, PA, UAS, 2011; pp. 485–494.
22. Leonard, W.R. Measuring human energy expenditure and metabolic function: Basic principles and methods. *J. Anthropol. Sci.* **2010**, *88*, 221–230.
23. European Committee for Standardization (CEN). EN 16798-1; *Energy Performance of Buildings. Ventilation for Buildings. Part 1: Indoor Environmental Input Parameters for Design and Assessment of Energy Performance of Building Addressing Indoor Air Quality, Thermal Environment, Lighting and Acoustics*; CEN: Brussels, Belgium, 2019; 82.
24. Zuraimi, M.S.; Pantazaras, A.; Chaturvedi, K.A.; Yang, H.J.J.; Tham, K.W.; Lee, S.E. Predicting occupancy counts using physical and statistical CO_2-based modeling methodologies. *Build. Environ.* **2017**, *123*, 517–528. [CrossRef]
25. Wolf, S.; Calì, D.; Krogstie, J.; Madsen, H. Carbon dioxide-based occupancy estimation using stochastic differential equations. *Appl. Energy* **2019**, *236*, 32–41. [CrossRef]
26. O'Neill, Z.D.; Li, Y.; Cheng, H.C.; Zhou, X.; Taylor, S.T. Energy savings and ventilation performance from CO_2-based demand controlled ventilation: Simulation results from ASHRAE RP-1747 (ASHRAE RP-1747). *Sci. Technol. Built Environ.* **2020**, *26*, 257–281. [CrossRef]
27. Nassif, N. A robust CO_2-based demand-controlled ventilation control strategy for multi-zone HVAC systems. *Energy Build.* **2012**, *45*, 72–81. [CrossRef]
28. Batterman, S. Review and extension of CO_2-based methods to determine ventilation rates with application to school classrooms. *Int. J. Environ. Res. Public Health* **2017**, *14*, 145. [CrossRef]
29. Rao, S.S. *Engineering Optimization: Theory and Practice*, 4th ed.; John Wiley & Sons: Hoboken, NJ, USA, 2009.
30. Franco, A.; Diaz Vazquez, A.R. A Thermodynamic Based Approach for the Multicriteria Assessment of Energy Conversion Systems. *J. Energy Resour. Technol.* **2006**, *128*, 346–351. [CrossRef]

31. Park, J.; Loftness, V.; Aziz, A. Post-Occupancy Evaluation and IEQ Measurements from 64 Office Buildings: Critical Factors and Thresholds for User Satisfaction on Thermal Quality. *Buildings* **2018**, *8*, 156. [CrossRef]
32. Baronti, P.; Barsocchi, P.; Chessa, S.; Mavilia, F.; Palumbo, F. Indoor Bluetooth Low Energy Dataset for Localization, Tracking, Occupancy, and Social Interaction. *Sensors* **2018**, *18*, 4462. [CrossRef] [PubMed]

© 2020 by the authors. Licensee MDPI, Basel, Switzerland. This article is an open access article distributed under the terms and conditions of the Creative Commons Attribution (CC BY) license (http://creativecommons.org/licenses/by/4.0/).

buildings

MDPI

Article

A Framework to Achieve Multifunctionality in Biomimetic Adaptive Building Skins

Aysu Kuru [1,*], Philip Oldfield [1], Stephen Bonser [2] and Francesco Fiorito [1,3]

1 Faculty of Built Environment, University of New South Wales, Sydney, NSW 2052, Australia; p.oldfield@unsw.edu.au (P.O.); f.fiorito@unsw.edu.au (F.F.)

2 School of Biological, Earth and Environmental Sciences, University of New South Wales, Sydney, NSW 2052, Australia; s.bonser@unsw.edu.au

3 Department of Civil, Environmental, Land, Building Engineering and Chemistry, Polytechnic University of Bari, 70126 Bari, Italy

* Correspondence: a.kuru@unsw.edu.au or aysuek@gmail.com

Received: 5 May 2020; Accepted: 23 June 2020; Published: 27 June 2020

Abstract: Building skins should host multiple functions for increased performance. Addressing this, their design can benefit by learning from nature to achieve multifunctionality, where multifunctional strategies have evolved over years. However, existing frameworks to develop biomimetic adaptive building skins (Bio-ABS) have limited capabilities transferring multifunctionality from nature into designs. This study shows that through investigating the principles of hierarchy and heterogeneity, multifunctionality in nature can be transferred into biomimetic strategies. We aim at mapping the existing knowledge in biological adaptations from the perspective of multifunctionality and developing a framework achieving multifunctionality in Bio-ABS. The framework is demonstrated through the case study of *Echinocactus grusonii* implemented as a Bio-ABS on a digital base-case building. The methods include the Bio-ABS case study demonstrating the framework and simulating the performance of the case study and base-case building to comparatively analyze the results. The outcomes are a framework to develop multifunctional Bio-ABS and simulation results on the performance improvement Bio-ABS offer. The performance comparison between the Bio-ABS and base-case building show that there is a decrease in the discomfort hours by a maximum of 23.18%. In conclusion, translating heterogeneity and hierarchy principles in nature into engineered designs is a key aspect to achieve multifunctionality in Bio-ABS offering improved strategies in performance over conventional buildings.

Keywords: biomimetics; building skin; multifunctionality; architectural design; building envelope; adaptability; design framework; building performance simulation

1. Introduction

Building design has advanced through technological developments, making the design and realization of innovative structures possible using new materials and construction techniques [1]. These breakthroughs in architecture helped building parts to host diverse functions, such as external walls shifting from being mostly load-bearing structures to having alternative roles [2]. The building skin now primarily acts as a thermal, acoustic, and visual barrier between the exterior and interior [3]. This change has seen the building skin heavily influence the building performance [4]. In response, new technologies including biomimetic adaptive building skins (Bio-ABS) are emerging that can adapt to changing environmental conditions, foster increased comfort, and reduce operational carbon emissions [5,6].

Using biological functions to design technical systems is called biomimetics [7,8]. Biomimetics is used in developing Bio-ABS, due to the opportunities in transferring biological adaptations into the design of climate-adaptable buildings [9–11]. There is extensive research in transferring biological adaptations into Bio-ABS with multiple frameworks and case studies [12–18]. Some of these frameworks are supported with resources such as databases or linguistic approaches [16,18–23]. Despite such developments, there is still a need for defining strategies to translate multifunctional strategies from nature into Bio-ABS, because environmental regulation in buildings often requires having multiple functions [5,24,25]. Most published works explore developing biomimetic strategies focusing on environmental adaptability, but are limited in addressing multifunctionality [10,26].

Multifunctional mechanisms in nature is achieved through the "natural design principles" including heterogeneity and hierarchy [27–29]. Organisms have developed multi-layered structures (hierarchy) with different morphologies of basic components (heterogeneity) to achieve multifunctionality [29–32]. This paper aims at structuring and mapping the existing knowledge in biological adaptations from the perspective of multifunctionality and developing a design framework to achieve multifunctionality in Bio-ABS. To do so, we provide (1) an understanding of how to achieve multifunctionality in building skin design through hierarchy and heterogeneity, (2) a systematic classification of multifunctional properties in biological systems, (3) a framework for developing multifunctional Bio-ABS, and (4) a case study of *Echinocactus grusonii* demonstrating the framework.

2. Existing Biomimetic Design Frameworks to Develop Bio-ABS

Several biomimetic design frameworks to develop Bio-ABS have been developed supported with case studies [16,18–23]. Most of these frameworks have focused on biological adaptations, being the main purpose of Bio-ABS [9,12,13,33]. However, it is important to fulfill multiple functional requirements in buildings for improved performance. There is limited exploration of this in the literature. However, nature offers ways to achieve multifunctionality, named the "natural design principles", yet to be significantly investigated in biomimetic design [30]. In this section, we review and discuss the existing biomimetic design frameworks to develop Bio-ABS regarding to what extent multifunctionality is addressed.

Most biomimetic design frameworks are applicable for use in architecture and some specialized to design Bio-ABS. All frameworks comprise of phases either starting with a technical problem followed by a biological solution (TD) or starting with a biological solution followed by a technical problem (BU), and all ending with developing a biomimetic strategy (Table 1). Some include databases to search for biological adaptations. The ones with no databases provide other means of investigation including linguistic approaches to define commonalities between nature and design, systematic analogical translation to find biological systems [18,22,23,34–37].

Biomimetic principles for the development of adaptive architectural envelopes, is based on understanding plants adaptations at macro- and micro-scales to be abstracted into adaptive architectural envelopes [13]. The framework focuses on dynamic mechanisms responding to external stimuli through movement and static strategies with surface properties. A case study developed using the framework is an adaptive architectural envelope inspired by plants genus *Mesembryanthemum* where seeds swell with rain [9]. The design is triggered by rainwater as an opening system [13]. The opportunity in this framework is addressing diversity in biological scales relating to hierarchy. This framework only considers plant adaptations; therefore, it is exclusive of other living systems in different kingdoms. Moreover, the mechanisms investigated are only triggered by external environmental influences limiting the diversity in biological domain. Most importantly, this framework has not specified how to combine multiple functions, either a combination of static and dynamic or otherwise.

Table 1. List of existing biomimetic design frameworks (TD: Top-down, BU: Bottom-up).

Reference	Biomimetic Design Framework	TD	BU	Database
[20]	BioTRIZ	✓		✓
[36]	Design Spiral-Helix Model	✓		
[37]	Computational Architectural Design Based on Biological Principles	✓		
[38]	Bioinspired Environmental Architectural Design	✓		✓
[39]	BioGEN	✓		✓
[34]	BioMAPS	✓		
[40]	Ecomimetic	✓		
[41]	New Product Development Process	✓		
[42]	The Law of System Completeness	✓		✓
[43]	Idea-Inspire	✓		✓
[44]	Design by Analogy to Nature Engine (DANE)	✓		✓
[45]	Biomimicry Theoretical Model		✓	
[13]	Biomimetic Principles for the Development of Adaptive Architectural Envelopes		✓	
[46]	Transferring Plant Movements to Elastic Systems in Architecture		✓	
[17]	Biomimicry 3.8	✓	✓	✓
[35]	Typological Analysis	✓	✓	
[22]	Analogical Translation	✓	✓	
[23]	Nature Studies Analysis	✓	✓	
[47]	Ecosystem for Biomimetic Design	✓	✓	
[18]	Engineering-to-Biology Thesaurus	✓	✓	

Transferring plant movements to elastic systems is used to develop kinetic elastic architectural systems inspired by plant movements [46]. A case study developed using this framework is Flectofin™, a shading prototype inspired by the petal movements of the bird of paradise flower [48]. The device is a hingeless louver system unfolding through a force. The success of this framework is investigating and transferring dynamic adaptations in nature into architecture. This framework only focused on plant movements and mechanical efficiency, it is limited in terms of multifunctionality. It could be, if it investigates environmental triggers informing movements in nature, to develop Bio-ABS. However, currently, the biological domain and biomimetic output can be triggered by the same stimulus, but they may not host the same function. For example, petals of the bird of paradise flower move for pollination, but Flectofin™ moves to provide shading. Moreover, this framework does not address how to integrate multiple functions together. For example, would it be possible to combine two types of plant(s) movements hosting different purposes and how?

"BioTRIZ" is a problem-solving method undertaken via a database of biological data [20,21]. The database's purpose is to make biological information available in a language specific to engineering. However, the process is out of biological context, meaning the strategies do not exist in nature, but as theoretical ideas. For example, the surface segmentation trend is observed in nature as naturally transparent ridged surfaces refracting light and becoming colored due to nanostructures. Such as the Menelaus butterfly (*Morpho menelaus*) has transparent wings which look blue with sunlight. This property could be transferred into functional surface structures. The success of this framework is introducing heterogeneity through generalizing concepts. The limitation is the absence of the biological solution and multifunctionality is not addressed. It could be, if this framework integrates how to achieve multifunctionality in addition to identifying heterogeneity in nature.

"Biomimicry 3.8" offers a framework accompanied with an online database of biological strategies, called AskNature [17]. AskNature contains over 1600 biological strategies and provides a taxonomy to organize biological adaptations [19]. The steps applied in the process are called the"Design Spiral". The success of this framework is the database accessible to everyone, as a key aspect in investigating biological strategies. The difficulty in adopting this framework is the lack of clarification in some stages; abstract, emulate, and evaluate. The abstraction stage is described as brainstorming ideas from nature;

emulation as seeking expert advice and evaluation as integrating life's principles. These measures are often complex and may be challenging for every design. The limitation of this framework is the lack of integrating multifunctionality. It could be, if the AskNature database provides multifunctional properties of organisms be used to design multifunctional strategies. This may require adding a step in the 'Design Spiral' on achieving multifunctionality.

'BioGen' is another biomimetic design framework [10,41]. It creates an exploration model mapping functional aspects, relevant processes and influencing factors. The biological entities called pinnacles are presented as examples for a specific function or process. An example of a case study developed by using BioGen is the Shading/Energy Generating Skin, inspired by plants tracking sun radiation, designed to tilt according to the angle of the sun [49]. While the creator of BioGen has proposed a discussion on multi-regulation of the four environmental factors (heat, light, water, and air) this is a theoretical approach and lacks further development. Multi-regulation is addressed by mapping biological functions of the same living system regulating multiple environmental factors dependently, i.e., ventilation is associated with heat regulation. This presents a success in addressing multifunctionality, but its applicability is limited and must be fulfilled with examples. It may also benefit addressing multifunctionality through combining functions of different living systems (inter)dependently.

DANE is another framework exploring analogical reasoning in biomimetics [44]. DANE follows the steps of defining the problem, searching for the biological solution, and applying the principle [50]. An example to this approach is the Shark Attack Project, designed to protect swimmers from sharks. It is inspired by pistol shrimp's snapping mechanism that creates a sound. The design was a shark repellant emitting sounds. The success in this framework is the identification of biological functions that gives clues on the heterogeneous properties of organisms. The limitation is addressing multifunctionality, either analogically or as design outcomes.

3. Achieving Multifunctionality through Natural Design Principles

During the course of evolution, living systems adapted their character to meet changing environmental conditions by multifunctional strategies [30]. The result is a compromise satisfying partially conflicting requirements; known as trade-offs [51,52]. A trade-off is having multiple functions through diminishing one property in return for gains in others [53]. The functional requirements of conventional buildings are complex and often contradictory and during the life cycle of a building, they must be adapted to changes [54,55]. Some examples of these functional requirements include decreasing solar gains, while providing views and improving illuminance levels [26,56,57]. These strategies are still mostly handled as isolated components that are monofunctional. One way to address this challenge is using biomimetics, as in nature, multifunctional trade-off strategies have been developed through evolution [58]. Multifunctionality is well understood in biology; however, its translation into architecture remains limited [24]. In this section, we present and discuss opportunities to achieve multifunctionality in biomimetics through the "natural design principles" (Figure 1). We include hierarchy (multi-scale) and heterogeneity (multidimension) and adaptability in this study. Beyond this, there remain opportunities to explore multifunctionality through other processes including anisotropy, redundancy, and more [59].

3.1. Adaptability in Nature and Bio-ABS

Adaptability is the ability of an organism to respond to changing internal or external conditions [60]. Most organisms have evolved various morphological, physiological, and behavioral adaptations to survive in their habitats. For example, plants exchange water vapor to decrease their internal temperatures during hot periods through their stomata. This is a means of diffusing useful gases for photosynthesis and to discharge excess materials to dissipate heat [61]. Depending on the natural habitat, stomata show various adaptations [62] (Figure 2). In some plants such as cacti and succulents, stomata are positioned on the epidermal (outmost) layer of the stem as they have minimal leaves to maintain a low surface to volume ratio. While in most plants, stomata are located on the leaves to ease the process of photosynthesis. The number of stomatal openings varies in different species according to the climatic influences as stoma loses water. For instance, desert plants lack in stomata while tropical plants have higher numbers of stomata [62–64].

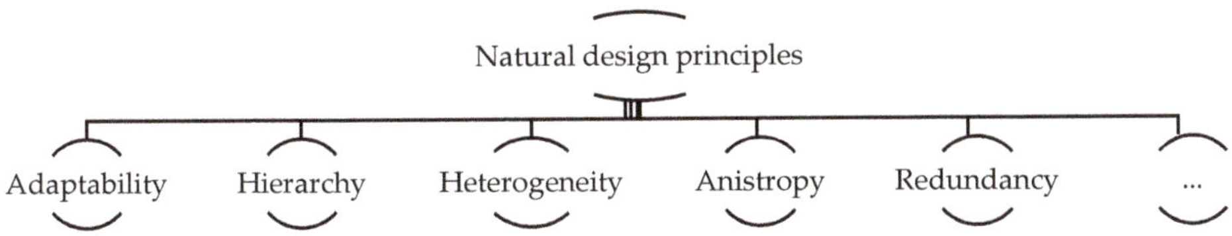

Figure 1. Some of the natural design principles as an ever-growing classification, where the last item in the list presents continuity.

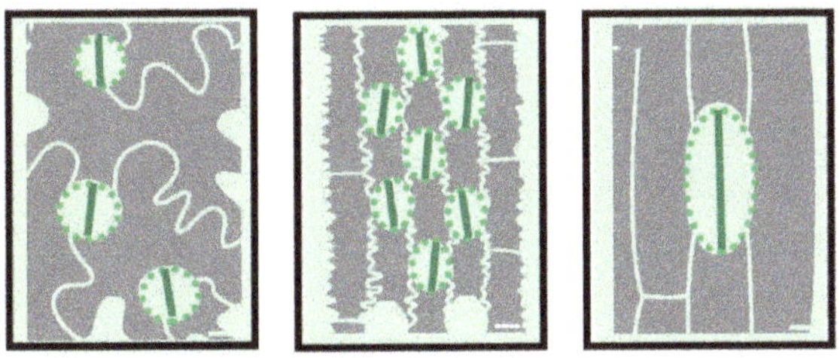

Figure 2. Stomatal adaptations in plants by changing size and density.

In engineering and design this is different, where a system is often designed to achieve one function and it does not reconfigure itself. Therefore, the strategy is static, where a design is made and must satisfy all needs during the lifetime [13]. However, adaptability in building skins learnt from nature is an emerging topic with many examples. Some examples of Bio-ABS include the Homeostatic Façade System that is an adaptive shading component, changing its shape according to solar radiation levels [65]. Also, the Ocean Pavilion, a full-scale application of the shading device Flectofin™, regulates solar irradiance levels by an automated hingeless mechanism [66] (Figure 3). Other examples, the HygroScope and HygroSkin, can change their morphologies through the changes in humidity by hygroscopic material properties of wood [67,68].

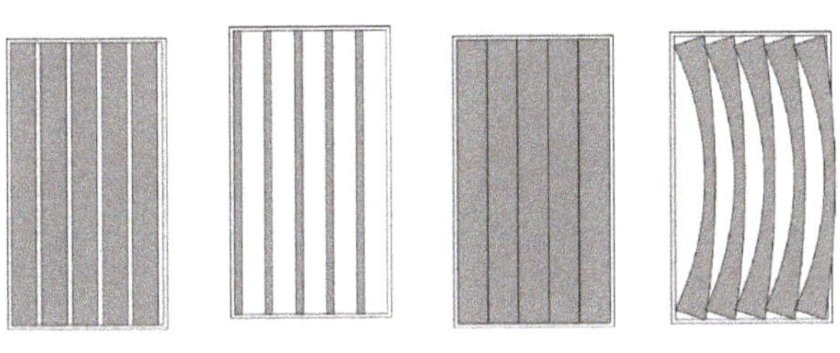

Figure 3. Schematic illustrations of some Bio-ABS examples, from left to right; the Ocean Pavilion in closed and open configurations, and Flectofin™ in closed and open configurations.

3.2. Achieving Multifunctionality through a Multi-Scaled Hierarchy

Hierarchy is having functional features in a multi-level structure from nano-to-macro scales. Having differentiated scales allows organisms to develop multiple functional adaptations at various levels. For example, adapting the form of an organ such as a tree branch is one level of adaptation, and a second is the microstructures of the tree bark's material [69] (Figure 4). This dual optimization is well known as a strategy for solving engineering problems. This approach can be combined with adaptability and achieve multifunctionality in Bio-ABS. An example to hierarchy in nature is seen in trees. The shape of a branch is created by the assembly of molecules to cells and of cells to wood with a specific shape. Therefore, at every scale, the branch is both form and material, where the materials compose the form; therefore, the biological structure becomes hierarchical [70]. It is important to note that tree bark and tannins is categorized under hierarchy due to being a material at a small scale covering the larger scaled tree bark. The adaptations of tannins are dependent on its scale, not its form.

Figure 4. A tree trunk and tannins. Photo by Madeleine Maguire on Unsplash.

Hierarchical biological structures can be adopted to achieve multifunctionality in architecture through multi-scaled structures with multiple uses. For example, building systems at a larger hierarchical level can host one function and its material at a smaller scale can host another. This can be explained through the hierarchical components of a conventional façade (Figure 5). The external walls are at the large hierarchical scale as the enclosure of a building. The windows on the external walls are smaller and provide external views. The window openings are even smaller and provide ventilation. The glazing material of the windows is at the smallest scale, limiting solar gains or light. The combination of these components at difference scales creates a façade with multiple functions. Therefore, through using hierarchical scales in conventional systems with adaptability, it is possible to achieve multifunctionality in developing Bio-ABS.

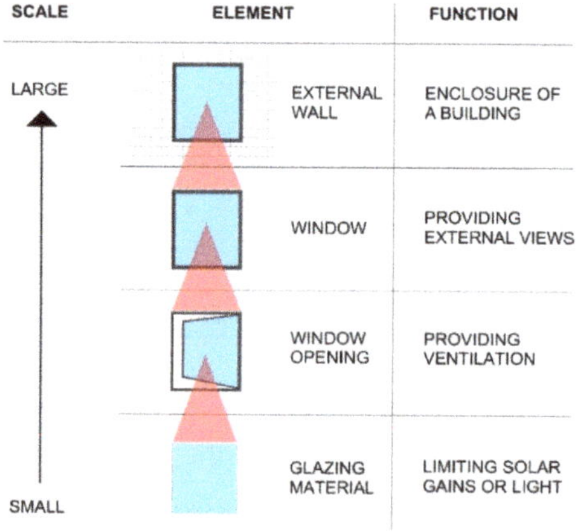

Figure 5. Hierarchical components on a conventional façade system at diverse scales including external walls, windows, window openings, and glazing material.

3.3. Achieving Multifunctionality through a Heterogeneous Geometric Differentiation

Heterogeneity is characterized by a geometric differentiation of elements in multi-dimensional structures hosting different functions with various forms. [30]. Form and function are interrelated in nature. As such, it is almost impossible to separate one from another. Nature uses shape or form, rather than added material and energy, to meet functional requirements. This allows the organism to accomplish its needs using a minimum of resources [7]. In addition, chemical and structural heterogeneities play an important role in allowing local adjustments to be integrated.

There is often a functional reason behind a form in nature. For example, the carnivorous plant Venus flytrap (*Dionaea muscipula*) has a cup-shaped leaf that catches insects with a mechanism trapping prey in (Figure 6). There are small hairs or cilia that are sensory organelles on the leaves. A slight disturbance of the cilia by an insect triggers the collapse of the leaf. This mechanism, called snap instability, is achieved by its shape, and its kinetics has been studied by researchers and transferred into shading devices [30,71,72]. This shape of leaves helps the plant save energy when moving. It is important to note that as the function of the cilia is achieved by their form rather than their scale, this adaptation of the Venus flytrap is categorized under heterogeneity.

Figure 6. Venus flytrap (*Dionaea muscipula*) with its specialized cilia. Photo by Gabriel on Unsplash.

Heterogeneity can be adopted into architecture to achieve multifunctionality through having various morphologies. For example, different geometrical forms with diverse functions can be integrated in a system. This can be explained through the heterogeneous components of a conventional roof (Figure 7). The roof itself functions as the top covering of a building and the eaves are the edges of the roof which project beyond the side of a building. The eaves function as shadings and through their geometric differentiation from the roof, the system becomes multifunctional. Therefore, through combining the use of heterogeneous scales in conventional systems with adaptability, it is possible to achieve multifunctionality in developing Bio-ABS.

3.4. Integrating Hierarchy and Heterogeneity Together

Hierarchy through scales and heterogeneity through form in nature is developed in various complexities much further than the human eye can detect. In many cases, color and light transmittance or refraction is achieved by nano-scaled ridges, combining the principles of hierarchy and heterogeneity together. For instance, *Morpho menelaus* has transparent wings that are covered with micro scaled structures forming a textured pattern that diffuse light to achieve its color (Figure 8) [73]. This is a more efficient solution than having pigments, which in time can lose their properties or require energy to maintain. The wings function as means of flight bodies. The micro-scaled ridges add the wings an additional function, by having hierarchical elements. The wings being the larger-scale flight function and the ridges being the micro-scaled colorists therefore becoming heterogeneous [74,75].

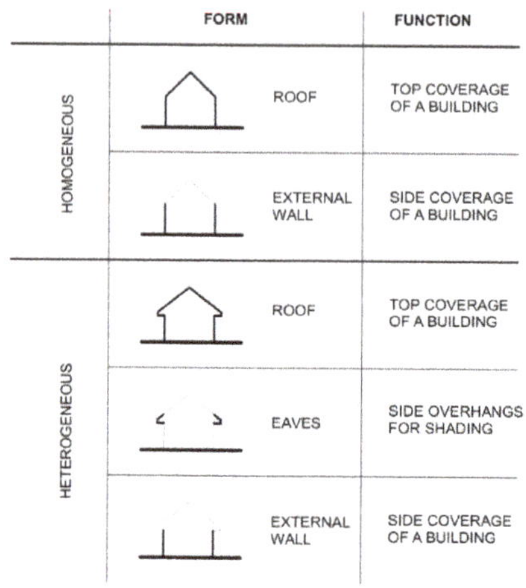

Figure 7. Heterogeneity in a conventional roof structure showing the heterogeneous forms of the roof and eaves.

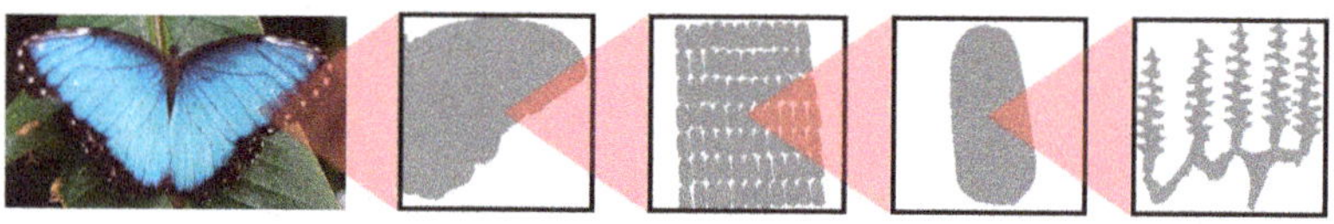

Figure 8. From left to right: *Morpho Menelaus*, its wings, micro-scaled structures forming a pattern over the wings, a single scale of *Morpho Menelaus* and a scale showing a few ridges giving color. Photo by Damon On Road on Unsplash.

Hierarchy and heterogeneity can be integrated together in achieving multifunctionality into architecture through differentiating functions at diverse scales and geometries. Similar to *Morpho menelaus*, a building can host a function on its different elements situated at diverse scales and on its functional surfaces differentiated by various geometries. For example, in conventional structures, this is achieved through having the hierarchy of system, component, material at diverse scales added together with surface structures as differentiated geometries. In this way, the traditional hierarchical parts of a building are formed. Then, the material properties of the specialized surfaces, i.e., nanostructured surfaces, are used at as small scale of the hierarchy. The material property adds heterogeneous forms in the hierarchy and therefore the whole system achieves multifunctionality through hierarchy and heterogeneity (Figure 9).

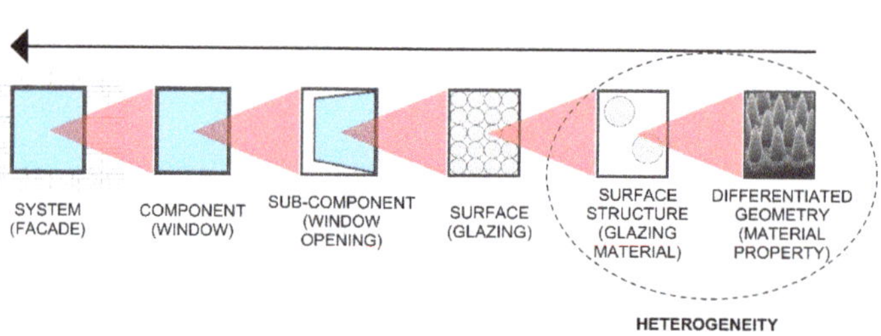

Figure 9. How buildings can achieve multifunctionality through integrating hierarchical building components with heterogeneous forms of material properties.

4. Methodology

Several research methods are used in this study to achieve multifunctionality in Bio-ABS and to measure the benefit of implementing multifunctional Bio-ABS in buildings. These methods include the following.

(1) Literature review, synthesis, and comparative analysis of the existing biomimetic design frameworks.
(2) The case study of *Echinocactus grusonii* demonstrating a new biomimetic design framework to achieve multifunctionality in Bio-ABS.
(3) Building performance simulation of a digital base-case building model and the case study of a multifunctional Bio-ABS.
(4) Comparative analysis of the simulation results showing the improvement in thermal comfort multifunctional Bio-ABS offers.

The literature review, synthesis, and comparative analysis of the existing design frameworks aim at identifying the research gaps in developing Bio-ABS. In doing so, they draw attention to the limitations and successes of existing frameworks to develop Bio-ABS and outline whether multifunctionality is addressed. However, it is found that achieving multifunctionality with existing frameworks remains a gap, and to address this, a new framework is proposed. Therefore, as a method, framework is used in order to address the challenges faced in developing multifunctional Bio-ABS. Frameworks, in general, provide a holistic approach to address specific problems or challenges and they present the opportunity to be further improved as validated methodologies. To aid the framework developed, a preliminary database is created by mapping multifunctional mechanisms of organisms through a systematic classification as a growing source of multifunctional biological mechanisms. The new framework presents a process to develop multifunctional Bio-ABS using the database. Furthermore, a case study of *Echinocactus grusonii* (golden barrel cactus) implemented on a naturally ventilated educational building modeled digitally is developed through the demonstration of the new framework. This case study provides an example to use the database and framework to develop multifunctional Bio-ABS by translating the natural design principles of hierarchy and heterogeneity.

The digital base-case building is taken from the repository of U.S. Department of Energy that provides models for various building types. The building type selected is educational to investigate problems associated with comfort in schools, as most existing studies focus on office and commercial buildings [76–82]. Building performance simulation (BPS) of the digital base-case building targeting thermal comfort through the Adaptive Comfort Model applicable to naturally ventilated buildings is performed. The multifunctional Bio-ABS case study is implemented on the base-case building, replacing its windows. The software used to model and simulate the base-case building and case study is EnergyPlus. This was chosen as it provides a flexible input-output interface to model multifunctional facades using the built-in tool Energy Management System (EMS). EMS has customizable input objects that is suitable to model adaptive facades [79,83,84]. A comparative analysis for the BPS results of the base-case building and Bio-ABS case study is made to draw out the improvements in comfort multifunctional Bio-ABS offer over conventional buildings.

5. Classification of Biological Mechanisms for Multifunctionality Bio-ABS

This section presents a database mapping multifunctional properties of organisms. To do so, we classified biological mechanisms in terms of multifunctionality. The classification is comprised of three layers in Table 2:

(1) Named species presenting the biological system,
(2) adaptability presenting the type of biological adaptations, and
(3) multifunctionality presenting biological functional strategies divided into hierarchy and heterogeneity sub-categories.

Table 2. The classification parameters of biological mechanisms for achieving multifunctionality in developing Bio-ABS.

Layer	Scope	Parameters	
1. Species	Biological system	-	
	Scientific name	-	
2. Adaptability	Stimulus	Physical	
		Chemical	
	Process	Dynamic	
		Static	
3. Multifunctionality	Environmental factor	Heat	
		Light	
		Air	
		Water	
		Energy	
	Function	Maintain	
		Exchange	
		Gain	
		Lose	
	Mechanism	Biological mechanism	
		Functional strategy	
		Performance measure	
3.a. Hierarchy	Biological organization scale	Pre-cellular	Atom, molecule
		Sub-cellular	Organelle
		Cellular	Cell
		Multicellular	Tissue, organ, organism
		Ecological	Ecosystem, biome
3.b. Heterogeneity	Adaptation type and level	Morphological	Form Structure Texture
		Physiological	Chemical response Trait
		Behavioral	Kinetic response Tropism Nastic movement

5.1. *Species and Adaptability Layers*

The species layer comprises the name of the biological system and its scientific name. The stimulus presents the set of triggers biological mechanisms respond to, defined as physical and chemical. Physical stimuli refer to the internal and external environmental triggers including the changes in environmental factors. These range anywhere from heat, light, to water content. Chemical stimuli refer to physiological triggers detected by a receptor. Chemical stimuli promote internal responses including changes in pH. Adaptive features of living systems occur as either dynamic or static processes, demonstrating the kinds of changes. A dynamic process relates to motion such as behavioral adaptations. In static processes, no change is present. Examples include nanostructured textural features of living systems such as waxy surface structures of water lily (Nymphaeaceae) and lotus flower (*Nelumbo nucifera*) keeping them dry off the water [85,86].

5.2. *Multifunctionality Layer*

5.2.1. Environmental Factor, Function, and Mechanism

The environmental factor refers to climatic variables that biological mechanisms control, categorized as heat, light, air, water, and energy. For example, heat includes surface or body temperature (referred as thermoregulation in living systems), absorption, and dissipation of heat. Light is related to the impact and management of solar irradiance and radiation including light intensity, reflection, refraction, and absorption. Air is related to air and gas management including oxygen intake, air flow and gas exchange. Water is related to water content in the body or in the air or the surrounding environment of a living system including absorbing the moisture from the air, diffusion of moisture, and waxy surfaces to hold moisture over the surface. Energy is related to the generation and conservation of energy.

The function describes the control that biological mechanisms host over the environmental factors. These include gain, lose, maintain, and exchange. Gain refers to absorbing, warming up, and enhancing; describing an increase. Lose is the opposite of gain, referring to reflecting, refracting, cooling down, and evaporation; describing a decrease. Maintain refers to managing, intensifying, and thermal regulation; describing a certain variable is remained as a constant. Last, exchange refers to absorbing, taking, and filtering; describing the simultaneous emergence of gain and lose.

The biological mechanism is the driving force of the classification that presents the functional characteristics of living systems. Examples of biological mechanisms include self-shading areoles and spines of cacti, and shrinking and swelling stem structures of succulents [87]. A functional strategy refers to the performative role played by an organism's adaptations. Examples include maximizing the amount of light reflected, passive ventilation through altering air pressure and thermoregulation through retaining absorbed heat [88].

5.2.2. Hierarchy

Biological organization is the hierarchical order of biological systems, extending from atoms to biospheres Each level in the hierarchy represents an increase in organizational complexity, with each 'object' being composed of the previous level's basic unit. The basic principle behind the organization is the concept of emergence: the properties found at a hierarchical level are not present and irrelevant at the lower levels. In most simple terms, the biological organization level relates to the level at which the biological mechanism is part of a living system (Figure 10). Organisms achieve multifunctionality through having multiple functions located at their hierarchical multi-level biological structures. To promote biomimetic strategies with multiple functions, the biological levels of those biological mechanisms are identified. This categorization aims at presenting how diverse functions are combined in living systems and gives clues to creating corresponding technical systems. For instance, if a biological mechanism is situated at a cell level, it can be combined with another mechanisms that is located at a different level such as organ or organism [34].

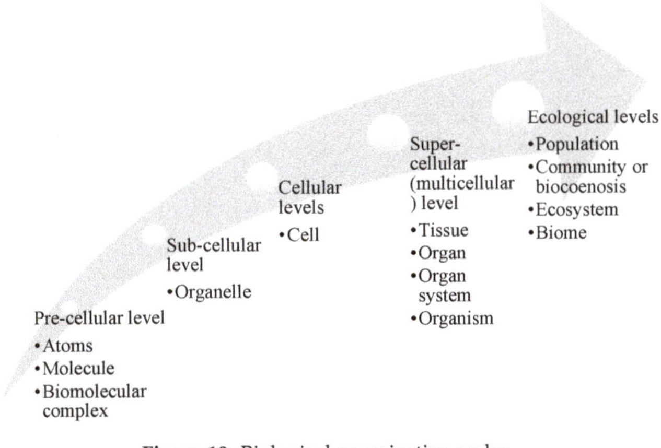

Figure 10. Biological organization scales.

5.2.3. Heterogeneity

Morphological adaptions occur through the changes in the morphology of an organism. Examples include the movement of wooden-like scales of the big pine cone (*Pinus coultieri*) when the water content in the air fluctuates to manage moisture. In morphological adaptations, form is related to the size or shape of an organism. Structure relates to a structure of an organism, as biological spatial and structural material-systems. Examples include shell structures such as the sea urchin (*Echnoidea*) and sand dollar (*Clypeasteroida*). Texture results from segmentation trends of biological surfaces of an organism. Examples include thorny fruit trees such as durian, to provide self-shading with scale structures [89].

Physiological adaptations occur through internal changes of an organism. Examples include Crassulacean Acid Metabolism in cacti, a carbon fixation pathway where the stomata remain shut during the day to reduce evapotranspiration, while open at night to collect carbon dioxide to diffuse into cells. In physiological adaptations, trait relates to phenotypic characteristics of an organism as results of the evolutionary processes. Examples include the eye color as a character of humans, while blue, brown, and green colors are traits. Chemical processes result from internal chemical processes of an organism. Examples include the physiology of stomatal openings facilitating gas exchange through a pair of specialized guard cells [61].

Behavioral adaptations are changes in the behavior of an organism, mostly resulting in movement. Kinetic response is the movement in living systems other than plants. Examples include kangaroos (Macropodidae) licking their paws for evaporation and crouching into smaller position for decreased heat gain. Nastic movement is plants' response to an external stimulus (i.e., temperature, light, and chemicals) independent from its direction. Examples include thermonasty of tulips (Tulipa), closing and openings according to the changes in temperature and thigmonasty of wood sorrels (*Oxalidaceae*) as a response to vibration. Tropism is plants' response to an external stimulus (i.e., oxygen, sun, and humidity) depending on its direction, as opposed to nastic movement. Examples include hydrotropism such as tomato roots, the tendency to grow towards higher moisture content (Figure 11).

Figure 11. Examples of adaptations or characteristics from left to right; in morphology—succulent's form, sea urchin's structure, durian's thorns; in physiology—eye color trait, photosynthesis of stoma; in behavior—licking paws for evapotranspiration, thermonasty, and phototropism.

5.3. Examples of Classified Biological Systems

In an attempt to demonstrate this systematic classification, we mapped biological systems in a 'preliminary database' for achieving multifunctionality in biomimetic designs (Table 3). The biological systems are organized to present their multifunctional properties, giving insight to their hierarchical or heterogeneous structures. The classification and the database are to be used as part of the framework proposed in this paper further. The table presented below provides the information on an extract (sixteen entries) from the database. The database in its current format consists of 43 entries and it is yet to be complete as a growing source of multifunctional mechanisms of biological systems.

Table 3. A preliminary mapping of several biological systems through the systematic classification. (in stimulus P: physical, C: chemical; in process D: dynamic, S: static; in biological organization scales O: organ, T: tissue, C: cell; in adaptation types M: morphological, P: physiological, B: behavioral; in adaptation levels F: form, S: structure, TEX: texture, CR: chemical response, KR: kinetic response, NM: nastic movement, TRO: tropism; N/A: Not available). Information is gathered from various sources [6,8,12,13,50,89–110].

Species	Adaptability		Multifunctionality				
Biological System	Stimulus	Process	Function	Mechanism	Biological Organization Scale (Hierarchy)	Adaptation Level (Heterogeneity)	Adaptation Type (Heterogeneity)
Australian Banksia Seeds (*Banksia attenuata*)	P	D	Maintain/Exchange light and heat	Managing UV radiation, high ambient summer temperatures by crack openings	T	P	Trait
Stone Plant (*Lithops*)	N/A	S	Gain/Maintain/Lose light	Translucent and colored patterned epidermal windows on the leaves	T	P	Trait
	P	D	Gain/Maintain waterLoose/Maintain heat	Shrinking and swelling leaves	O	M	F
Saharan Silver Ant (*Cataglyphis bombycina*)	P	D	Maintain/Lose light and heat	Triangular reflective hair and grooves reducing heat absorption, reflection, and refraction	T	M	TEX
Desert Snail (*Sphincterochila boisseri*)	N/A	S	Maintain/Lose light and heat	Highly reflective shell surface allowing conduction	O	P	Trait
	P	D	Maintain heat	Layer of insulating air cushion	O	B	KR
Glass Snail (*Oxychilus draparnaudi*)	N/A	S	Maintain/Lose light and heat	Glossy translucent shell called glass house allowing reflectance	O	P	Trait
Butterfly—Menelaus Blue Butterfly (*Morpho menelaus*)	P	D	Maintain light and heat	Wings scale structures allowing structural coloration	O	M	TEX
	N/A	S	Gain energy	Microscopically thin layers of film (chitin) on wings absorbing energy/infrared light	O	P	Trait
Bark of Trees	N/A	S	Maintain light and heat	Tannins on the bark surface managing optical properties through nanostructures	T	P	Trait
	N/A	S	Maintain air, light, and heat Lose light and heat	Rough bark surface producing shadowed areas amongst the illuminated ones, stimulating convection of air	T	M	TEX
Cactus-Barrel cactus (*Echinocatus grusonii*)	P	D	Maintain water Lose light and heat	Swelling and shrinking cortex achieving high surface to volume ratio through the ribs structured stem	O	M	F
	N/A	S	Lose light and heat	Self-shading areoles and spines over the cortex epidermal layer	T	M	TEX
Stoma	C	D	Exchange heat, air, and light	Microscopic and permeable stomatal openings	C	P	CR
Camel (Camelus)	P	D	Maintain/Exchange light and water	Open/closed configurations of the shell	T	B	KR
	P	D	Maintain/Lose heat Exchange water	Hygroscopic nasal passages cooling exhaled air during night and extracting water vapor from air	T	P	Trait
Spurge (Euphorbias)	N/A	S	Maintain heat and water Lose heat Gain water	Managing water content and heat through waxy surface covering the stem	T	P	Trait
Rhododendron Leaves	P	D	Maintain/Gain heat Maintain water Lose light and heat	Curling movement of the leaves triggered by heat allowing the reduction of the total quantity of light absorbed by the leaf	O	B	TRO
Fern Leaves	N/A	S	Maintain/Lose light Gain energy	Special structures absorbing solar radiation, and managing light by reflection and refraction	C	P	CR
	C	D	Exchange air and water	Permeability of outer leaf surface allowing diffusion	C	P	CR

Table 3. *Cont.*

Species	Adaptability		Multifunctionality				
Biological System	Stimulus	Process	Function	Mechanism	Biological Organization Scale (Hierarchy)	Adaptation Level (Heterogeneity)	Adaptation Type (Heterogeneity)
Polar Bear (*Ursus maritimus*)	N/A	S	Gain heat light Maintain/Lose light	Long, transparent, hollow guard hairs scattering and reflecting sunlight	T	M	TEX
	N/A	S	Maintain light and heat	Dense underfur, darkly pigmented skin, and blubbering	T	P	Trait
Big pinecone (*Pinus coulteri*)	P	D	Exchange water, air, and light	Movement through the material capacity of wooden scales	T	M	TEX
Succulents	P	D	Gain/Maintain water Lose light and heat	Swelling and shrinking cortex achieving high surface to volume ratio	O	M	F

6. Developing a Framework to Achieve Multifunctionality in Bio-ABS

To achieve multifunctionality in Bio-ABS, a framework called the 'Multi-Biomechanism Approach' is proposed. This is a top-down approach focusing on technical problems to be solved through biological inspiration. It is comprised of four stages (Figure 12):

(1) Identifying a technical problem.
(2) Selecting a biological solution.
(3) Achieving multifunctionality.
(4) Developing a biomimetic strategy.

Each stage is comprised of sub-stages, which are facilitated by the classification of multifunctional biological mechanisms as outlined in Table 3. Stage one is comprised of the two sub-stages of selecting a base-case scenario that includes a location, the climate, and the performance analysis of that base-case scenario and identifying functional requirements to improve the performance. Stage two is comprised of matching functional requirements of the base-case scenario with a corresponding biological system found in the database. A suitable biological system with multifunctional properties is selected to serve as a case study. Stage three is comprised of outlining the properties of the chosen biological system as hierarchical and heterogeneous structures to achieve multifunctionality. This includes the identification of biological organization scales, adaptation levels and types of the chosen biological mechanisms. Stage four is comprised of designing a Bio-ABS with functions at diverse scales and with different geometries. In doing so, several configurations of the Bio-ABS are produced, and actuation mechanisms are presented that deliver climate-adaptability. Further details on how to perform the stages of the framework are described thoroughly in the following sections of this paper with a case study demonstrating its use.

6.1. *Stage 1: Identifying Technical Problems*

The identification of technical problems involves identifying optical, acoustic, and energetic controls over a base-case scenario. This is achieved through selecting location, climate, and a base-case building to identify functional requirements. A performance analysis of the base-case is proposed to define the functions required, such as through building performance simulation. The functions defined are suggested to use a simplified language as maintain, exchange, lose, and gain; of the environmental factors as heat, light, air, water, and energy. For instance, results of a performance analysis may suggest that cooling energy loads are relatively high. This indicates heat regulation through the function lose can be investigated as thermoregulation in organisms.

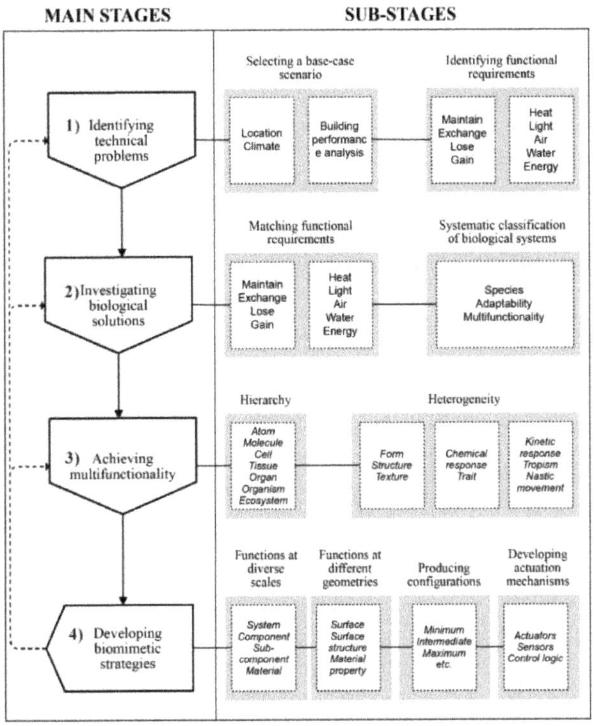

Figure 12. The multi-biomechanism approach and its stages. (1) Identifying technical problems, (2) investigating biological solutions, (3) achieving multifunctionality, and (4) developing biomimetic strategies.

6.2. Stage 2: Investigating Biological Solutions

The second stage is to find solutions in nature that respond to similar problems as identified in Stage 1. This stage is comprised of matching functional requirements, investigating corresponding biological mechanisms, and selecting biological models. To match functional requirements between the base-case building and nature, the same terms for functions (maintain, exchange, lose, and gain) and environmental factors (heat, light, air, water, and energy) are suggested to be used.

6.3. Stage 3: Achieving Multifunctionality

The third stage is to achieve multifunctionality through selecting multiple biological mechanisms, situating mechanisms at diverse scales, and developing actuation mechanisms. Multiple biological mechanisms whether they belong to the same biological model or not, should be selected to develop a multifunctional system. The significance in this stage is to employ hierarchy or heterogeneity as drivers. This means either situating the selected mechanisms at diverse scales, or selecting diverse types of morphological, physiological, or behavioral adaptations in a heterogeneous structure.

6.4. Stage 4: Developing Biomimetic Strategies

The final stage is developing biomimetic strategies, comprised of developing a façade design, selecting smart materials as actuators, and producing configurations. For example, the Stone Plant *(Lithops)* maintains light levels through its translucent and colored structures. This mechanism can be transferred as a texture changing its light transmittance. Moreover, the opening movement of stomata is a dynamic process that relates to a motion happening at a cellular scale. These mechanisms can be combined together and translated into a potential Bio-ABS design.

7. The Case Study of *Echinocactus grusonii*

This section describes the case study of a multifunctional Bio-ABS following the four stages of the Multi-Biomechanism Approach. The case study of *Echinocactus grusonii* is implemented on a digital reference building through translating its multifunctional properties using the concepts of hierarchy and heterogeneity. Further on, building performance simulation of the base-case building before and after implementing the multifunctional Bio-ABS case study is conducted. A comparative analysis of the simulation results is presented showing the performance improvements.

7.1. Stage 1: Identifying Technical Problems

To determine the technical problems in a base-case scenario, a climatic context with a location and reference building must be selected. A base-case scenario is selected in Sydney, Australia, with humid warm temperate climate characterized by warm summers and cool winters [111]. A digital reference educational building from the United States Department of Energy repository was selected to serve as a base-case model [112]. The building type selected is educational to investigate problems associated with comfort in schools, as most existing studies focus on offices and commercial buildings [78–84]. The reference building was located in Atlanta, USA, as Atlanta shows climatic similarity to Sydney. It is anticipated that the geometry of the reference building is suitable for a similar climate.

The reference building is simulated using the software EnergyPlus, in which the building was already modeled. The simulation results are presented further in this section in '7.5. Comparative analysis of environmental performance evaluation'. As results of the simulation, technical problems in the building are identified. The technical problems are defined as excessive heat, need for cooling, and high solar gains. The problems are revised using the simplified language specified in the framework. The translation of excessive heat is described as to lose and maintain heat, the need for cooling as to gain and exchange air, and high solar gains as to lose and maintain light (Table 4).

Table 4. Defining technical problems as functional requirements.

Technical Problem	Functional Requirement
Excessive heat	Lose/maintain heat
Need for cooling	Gain/exchange air
High solar gains	Lose/maintain light

7.2. Stage 2: Investigating Biological Solutions

A search for the functional requirements resulted with several biological systems including *Echinocactus grusonii*, *Pinus coultieri*, and succulents. The search for three different functions in the database is performed to investigate various biological mechanisms. The first function (air regulation) results with 11 entries including *Echinocactus grusonii*, big pine cone (*Pinus coulteri*), stomata, succulents, and barnacles (*Chthamalus stellatus*). The second function (light regulation) results with 33 entries including *Echinocactus grusonii*, *Pinus coulteri*, succulents, *Mimosa pudica*, *Lithops*, and Saharan silver ant (*Cataglyphis bombycina*). The third function (heat regulation) results with 43 entries including *Echinocactus grusonii*, *Pinus coulteri*, stomata, succulents, *Mimosa pudica*, *Lithops*, and *Cataglyphis bombycina*. The functions are as listed below.

(1) Function IN ('Exchange', 'Gain') AND 'Environmental Factor' = 'Air',
(2) Function IN ('Lose', 'Maintain') AND 'Environmental Factor' = 'Light'
(3) Function IN ('Lose', 'Maintain') AND 'Environmental Factor' = 'Heat'

As results of the search, three biological systems are found in the intersection with corresponding strategies. These are *Echinocactus grusonii*, *Pinus coulteri*, and succulents. In addition, cacti and succulents have their stomatal openings situated on the external layer of their epidermis, which forms a hierarchical adaptation at a smaller scale of the stem. In *Pinus coulter*, stomata are situated on the leaves that are part of the tree instead. *Pinus coulteri* exchanges water, air, and light by a response to moisture. It presents a movement through the movement by the wooden scales, which results in one type of morphological adaptation as a texture. Succulents gain and maintain water; lose light and heat through swelling and shrinking cortex, similar to cacti. However, this results in one type of morphological adaptation as a form of an organ. However, *Echinocactus grusonii* not only achieves what succulents do through a similar adaptation with a different form, and have their stomata on their stem different to pine cone, but it also loses light and heat through its self-shading areoles and spines. This additional morphological adaptation is a texture. Therefore, *Echinocactus grusonii* is chosen among the three biological systems as it provides two morphological adaptations at diverse hierarchical scales (organ and tissue) and has different textural heterogeneous structures (areoles and spines).

Some of the adaptations of *Echinocactus grusonii* involve the swelling and shrinking movement and high surface-to-volume ratio of the stem, self-shading areoles, and spines over the cortex as a morphology; and the microscopic stomatal openings as physiological mechanisms (Table 5). The cactus stem swells, shrinks, and maintains a high surface-to-volume ratio, through the unique ribs structure. Studies show that *Echinocactus grusonii* can expand up to 54% of its initial surface area [105,109]. At the shrunk state, self-shaded areas in between the ribs help cool the surface temperatures down. The surface-to-volume ratio increases as the cactus gets larger, losing heat and light. There are areoles on the cortex out of which grow spines, self-shading and creating cooler microclimate. Studies show that *Echinocactus grusonii* can achieve up to a difference of 17 °C in winter and 25 °C in summer between the surface and air temperatures through these morphological adaptations [64]. Stomata are microscopic pores on leaves to transpire water and exchange air and heat, but in cacti they are placed directly on the stem. *Echinocactus grusonii* has 15 to 70 stomata per square millimeter [103].

Table 5. *Echinocactus grusonii* and its properties. (in stimulus P: physical, C: chemical; in process D: dynamic, S: static; in biological organization scales O: organ, T: tissue, C: cell; in adaptation types M: morphological, P: physiological; in adaptation levels F: form, TEX: texture, CR: chemical response; N/A: Not available).

Species	Adaptability		Multifunctionality				
Biological System	Stimulus	Process	Function	Mechanism	Biological Organization Scale (Hierarchy)	Adaptation Type (Heterogeneity)	Adaptation Level (Heterogeneity)
Cactus-Barrel cactus (*Echinocatus grusonii*)	P	D	Maintain water Lose light and heat	Swelling and shrinking cortex achieving high surface to volume ratio through the ribs structured stem	O	M	F
	N/A	S	Lose light and heat	Self-shading areoles and spines over the cortex epidermal layer	T	M	TEX
Stoma	C	D	Exchange heat, air, and light	Microscopic and permeable stomatal openings	C	P	CR

7.3. Stage 3: Achieving Multifunctionality

The mechanisms of *Echinocactus grusonii* are situated at diverse scales of biological organization and differentiated morphologies. For example, the swelling and shrinking cortex is a morphological adaptation that hosts the function of losing heat by the differentiated form of the ribs' structure covering the cortex over the spherical stem. This is an example of heterogeneity in nature. Moreover, the self-shading areoles and spines are morphological adaptations as a differentiated form of texture over the ribs presenting heterogeneity. On the other hand, the microscopic stomatal openings operate as physiological adaptation. All three adaptations are situated at different scales of biological organization: stomatal openings at the cellular level and areoles and spines at the tissue and ribs structured cortex at the organ levels. This shows an example for hierarchy in nature. Therefore, the translation of the cortex can be activated by heat and light. Being at the largest level among the mechanisms, it may be transferred into a larger spatial scale. The translation of the stomatal openings may work in conjunction with the ribbed stem, regulating heat and light. This mechanism can be transferred as openings at a medium-level spatial scale for air intake, activated by temperature. The areoles and spines regulate light through creating a texture over the cortex, activated by light. This mechanism can be translated at a smaller scale such as sub-component (Figure 13).

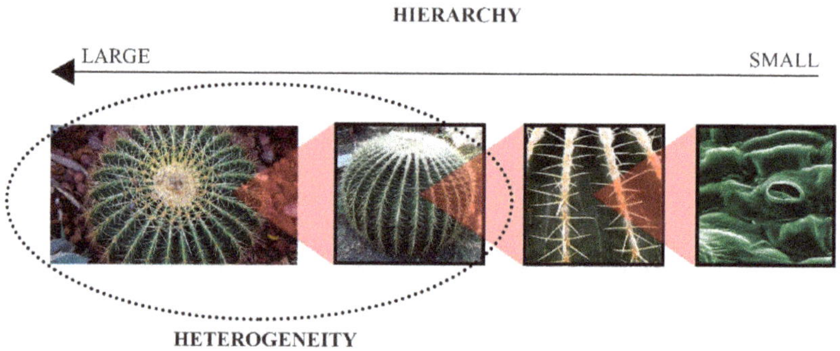

Figure 13. Multifunctionality through hierarchy and heterogeneity in *Echinocactus grusonii.*

7.4. Stage 4: Developing Biomimetic Strategies

The selected mechanisms of *Echinocactus grusonii* are translated into a façade design to improve the performance of the base-case scenario. The ribbed stem is translated into a morphology regulating heat by expanding and contracting triggered by temperature difference. As the form of a stem is a sphere but through giving it dimension by the ribs, it becomes geometrically heterogeneous and hosts another function of shading. This achieves cooling between the ribbed surfaces instead of reaching extreme temperatures otherwise without ribs. The self-shading areoles and spines texture is translated into an opacity-changing glazing (photochromic) regulating light and solar gains triggered by solar radiation. The photochromic glazing chemically changes its properties various solar heat gain coefficient and visible light transmittance values to different levels of solar irradiance [113]. This is similar to having an additional function through heterogeneous surface properties. The heterogeneous morphology and the hierarchical shading material together achieve multifunctionality. The stomata are translated into openings regulating heat and air triggered by heat for ventilation. This presents hosting another function by hierarchy and combine with the other two mechanisms interdependently, as it is linked to the ribbed morphology. Therefore, the design hosts two functions at its hierarchical and heterogeneous structure (Figure 14).

The Bio-ABS design is a folding module with an expanding and contracting mechanism activated by temperature difference through the use of thermally restrictive smart material of shape memory alloys (SMAs). SMAs change their length when exposed to differences in solar radiation levels [114,115]. Through this morphology, the design forms a ribbed structure while creating openings for ventilation. The opening's size is controlled by the actuator's displacement, as a percentage of contraction in length. The component is formed by isosceles triangle shaped creases connected by mountain folds and divided into two identical creases by valley folds. The design is a symmetric double-line vertex of degree 6-case rigid origami with a hexagonal base, as hexagon offers improved mechanical properties. The double-line technique allows the creases to have a gap for material thickness. Folds allow the component to change its shape while keeping the triangular creases rigid. A selected material placed in the central point can trigger the system with by a pull and push force into a pattern similar to the biomechanics of the rib structure of *Echinocactus grusonii*. The second function is achieved by color changing smart materials called chromogenics that present an example for functional surface material properties. A study on visualizing the façade on the base-case reference building through replacing its windows with 188 modules is presented in Figure 15.

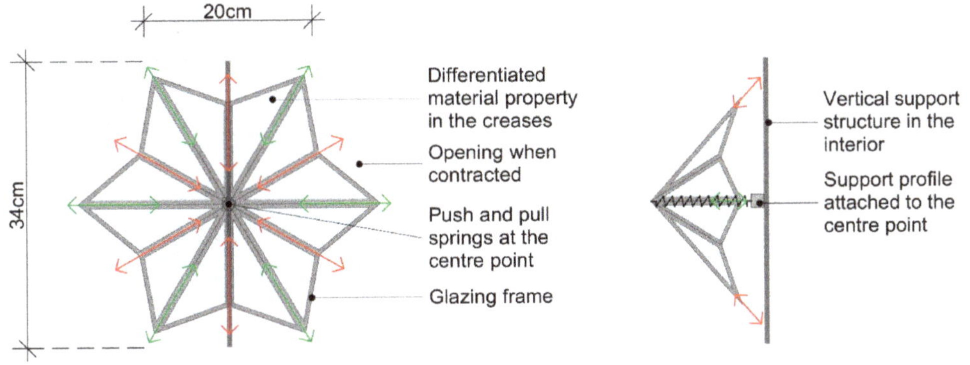

Figure 14. Movement and dimensions of the façade module in elevation and section, where red arrows represent the displacement of valley folds and green arrows represent the displacement of mountain folds.

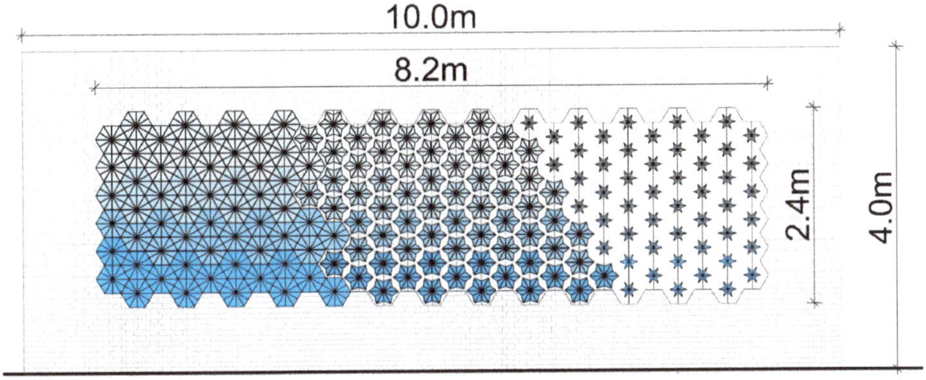

Figure 15. The pattern the Bio-ABS module, with various changing configurations, when implemented on the reference building's windows consisting of 188 modules; from left to right opening ratio increases and from bottom to top glazing opacity decreases.

The aesthetics of this system exist in the built and designed realm, for instance, the dynamic shading façade on the Al Bahar Tower in Abu Dhabi [116]. However, the operation and performance of this design differ significantly, in that they manage multiple parameters (ventilation and shade). As another difference, this design integrates the smart materials of shape memory alloys and photochromic cells within its mechanisms and therefore having a passive operation of the system. As such, Al Bahar Towers are automated through building management system (BMS) that uses electricity [116]. Therefore, most designs in the area with similar constructions are active, meaning they are operated by electricity [116,117]. Those designs are mostly programmed to operate at certain situations and therefore they are not considered as fully climate adaptable. The morphological movement of the origami folds is known, while the parameters that control the movement are unique, including the thermally activated pull and push through the central point. Another difference is the integration of hierarchical scales and heterogeneous material properties into this design. Most existing similar designs perform a single function and do not consider the integration of multi-scale and multi-dimension transferred from the natural design principles of hierarchy and heterogeneity, respectively.

7.5. Comparative Analysis of Environmental Performance Evaluation

To quantify the performance improvement Bio-ABS offer, a comparative analysis with the base-case building is done through building performance simulation. Only one thermal zone, a classroom is modeled to simplify the process (Figure 16). The ventilation type is switched to natural ventilation from mechanical ventilation to determine thermal comfort through the Adaptive Model. Glazing type and its thermal properties, aperture ratio, and ventilation rate are compliant with National Construction Code (NCC) of Australian Building Codes Board (Table 6) [118,119].

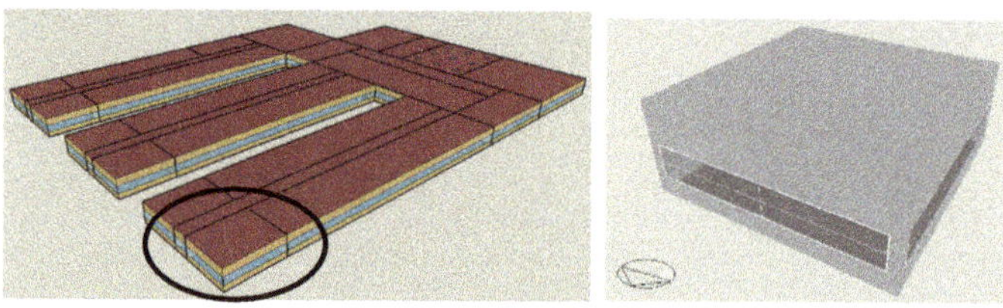

Figure 16. Base-case building.

Table 6. Performance descriptors of the base-case building.

Performance Descriptor	Value	Reference
Lighting load	8 W/m	
Equipment load	5 W/m	
Occupants density	0.4 people/m	
Ventilation operation schedule	Temperature	[118,119]
Window-to-wall ratio	40%	
Glazing opening ratio	25%	
Glazing thermal transmittance	1.786 W/m^{2}K	
Glazing solar heat gain coefficient	0.39	
Floor area	97 m	
Zone volume	388 m	[112]
Floor-to-ceiling height	4 m	
External walls thermal resistance	1.469 W/mK	
Air change rate	7.5 ac/h	[120]

Building performance simulation is performed to determine thermal comfort analysis of the base-case building using EnergyPlus. The climate file used is available at EnergyPlus Weather for Sydney IWEC (International Weather for Energy Calculations) station number 947,670. Thermal comfort is calculated through the adaptive model and found that according to 90% Acceptability Limits (A.L.), 74.03% of the time occupied does not fall in the comfort zone with 1588 h of discomfort and according to 80% A.L., it is 38.14% with 818.25 h of discomfort (The calculation follows the presence of occupants on a daily basis in weekdays from 08:00 am to 16:00 pm by a fraction of 0.75 and from 16:00 pm to 21:00 pm by a fraction of 0.15) (Table 7).

Table 7. Comfort analysis of the base-case building.

Analysis Type	Acceptability Limits	Discomfort Hours	Discomfort Ratio
Adaptive thermal comfort	90%	1588 h	74.03%
	80%	818.25 h	38.14%

The simulation results are analyzed to understand the causes for increased discomfort ratio. The results show that maximum values for indoor temperatures are calculated as 37 °C for mean radiant and 38 °C for operative and air temperatures. This suggests that the maximum temperatures are above the limits for temperature, that is, 33–33.5 °C. Moreover, the windows and infiltration are identified as the causes for heat loss with values of 18 kWh/m^2 and 41 kWh/m^2, respectively. The evaluation shows that the base-case building suffers high temperatures and excessive heat gains. This suggests that solar gains negatively impact comfort in the building and the regulation of heat and light must be addressed to improve the performance (Figure 17).

The windows of the base-case building are replaced with the Bio-ABS design and the performance is analyzed through simulation. The multifunctional Bio-ABS is modeled in EnergyPlus using its feature EMS, that integrates customizable input–output objects and allows using if statements. The photochromic (PC) glazing is modeled with its four states changing the solar heat gain coefficient (SHGC = 0.508, 0.396, 0.325, 0.238) and visible light transmittance (VLT = 0.595, 0.446, 0.341, 0.238) with a fixed thermal transmittance (U-value = 1.786 W/m^{2}K). The U-value of the base-case building and the case study are the same. The SHGC and VLT of the base-case building are calculated as the average values of the four states of the case study. The properties of the PC glazing are taken from a previous study outlining the performance improving PC glazing systems for the chosen climatic context [82]. The properties of the morphology triggering Shape Memory Alloys (SMAs) are set to demonstrate a comparable case against the base-case, which provides the same window opening ratio (25%) with a corresponding SMA displacement ratio. Other properties of the SMA including the actuation (18 °C) and de-actuation temperatures (60 °C) are determined to provide an adaptable system operated by changes in temperature. The simulation results after replacing the base-case building's windows with the Bio-ABS show a decrease in discomfort hours by 23.18% for 90% A.L. and 5.09% for 80% A.L. (Figure 18).

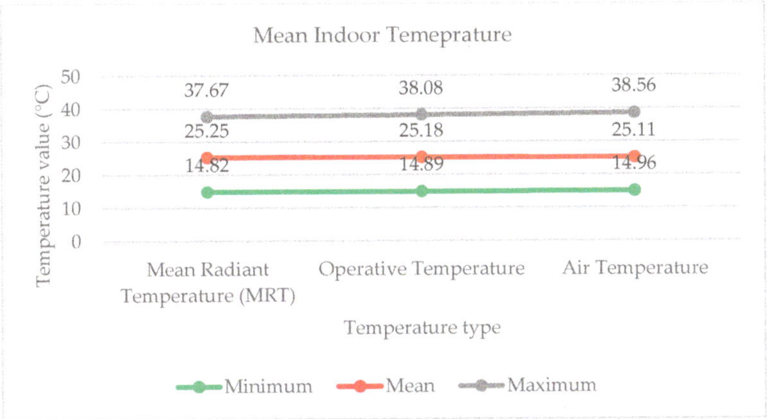

Figure 17. Indoor temperature analysis of the base-case building, where the gray line presents the maximum, red line presents the mean and the green line presents the minimum values.

The mean values for the indoor temperatures of mean radiant, mean operative, and mean air temperatures are decreased (Figure 19). The mean value for MRT is decreased by 2.23 °C, the mean value for operative temperature is decreased by 2.76 °C, and the mean value for air temperature is decreased by 3.28 °C. Overall, the implementation of this multifunctional Bio-ABS improves the thermal comfort in an educational building in Sydney. This study has focused on the integration of a PC glazing and SMA activated ventilation with set values for its performance descriptors (i.e., actuation temperature, SHGC, VLT). Further work could investigate different living systems, their functional transfer into multifunctional engineered designs, consider different performance descriptors of Bio-ABS, and simulate their environmental performance.

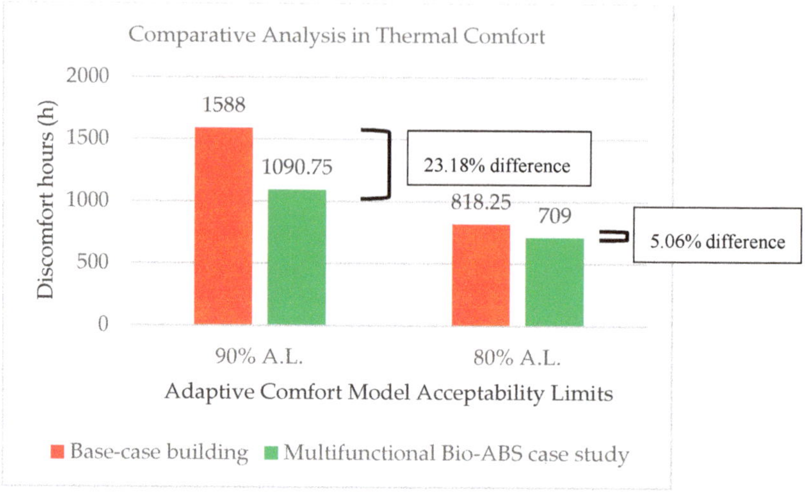

Figure 18. Comparative analysis in thermal comfort between the base-case building and the multifunctional Bio-ABS case study.

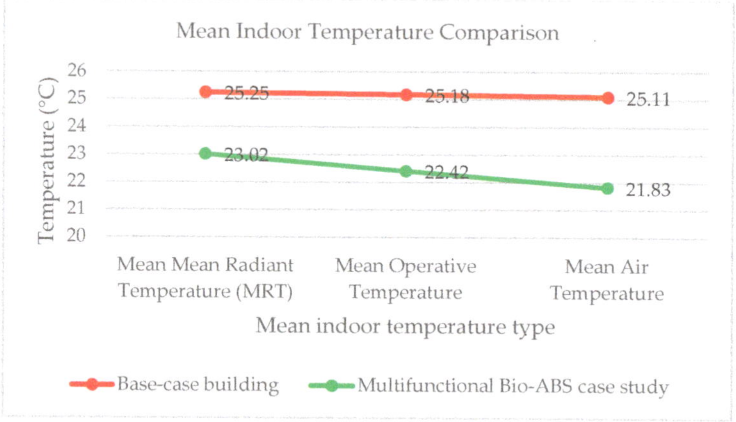

Figure 19. Mean indoor temperature comparison between the base-case building, and the case study.

8. Conclusions

This paper presents a framework for achieving multifunctionality in Bio-ABS. It does so by translating hierarchy and heterogeneity from nature into architecture. A systematic classification to map biological systems from the perspective of how they host multiple functions in their heterogeneous multi-level structures is presented. Several biological systems are mapped using the classification to define a "preliminary database" to categorize biological data. This database could be expanded over time to create a systemic collection of biological information as a resource for biomimetic design.

Hierarchy and heterogeneity in nature are described and their transfer into designs can achieve multifunctionality. Hierarchy is described as having multiple scales and heterogeneity as multiple geometric differentiations. Their transfer into Bio-ABS is proposed through a framework named the "Multi-Biomechanism Approach", that uses the systematic classification. The framework is demonstrated through the case study of translating *Echinocactus grusosnii* and three of its many biological adaptations: rib structured cortex, self-shading areoles and spines, and microscopic stomatal openings. A Bio-ABS design is presented showing how the hierarchical features of the areoles, spines, and stomatal openings, as well as the heterogeneous form of the ribs structure can be translated into a biomimetic strategy integrated in a building skin.

The base-case building and the case study of the multifunctional Bio-ABS replacing the windows of the base-case building are simulated to calculate their performance. The comparative analysis of the results show that Bio-ABS offer improved comfort both for 80% and 90% Acceptability Limits according to Adaptive Comfort Model. The mean value for MRT is decreased by 2.23 °C, the mean value for operative temperature is decreased by 2.76 °C, and the mean value for air temperature is decreased by 3.28 °C. Overall, the results after replacing the base-case building's windows with the Bio-ABS show a decrease in discomfort hours by 23.18% for 90% Acceptability Limits and 5.09% for 80% A.L. for adaptive thermal comfort.

As results of the work carried out and the findings, this research draws attention to multifunctionality in nature and in engineered designs, particularly of Bio-ABS, and promotes biomimetic design as a promising approach to be taken to develop environmentally sustainable building systems. This study attempts to point out the significance of the "natural design principles" and their limited application in architecture. However, it is limited to the translation of hierarchy and heterogeneity, excluding others that may help achieving multifunctionality, which further work can focus.

Author Contributions: A.K.: Conceptualization, methodology, software, formal analysis, investigation, resources, data curation, writing—original draft, writing—review & editing, visualization, funding acquisition. P.O.: Conceptualization, methodology, formal analysis, investigation, resources, data curation, writing—original draft, writing—review & editing, visualization, funding acquisition, supervision. S.B.: Conceptualization, methodology, formal analysis, investigation, resources, data curation, writing—review & editing, funding acquisition, supervision. F.F.: Conceptualization, methodology, software, formal analysis, investigation, resources, data curation, writing—review & editing, funding acquisition, supervision. All authors have read and agreed to the published version of the manuscript.

Funding: This work was supported by the Faculty of Built Environment at UNSW Sydney under Grant Wightman PG School Architecture (PGA1005).

Acknowledgments: The authors acknowledge the support provided by UNSW Sydney.

Conflicts of Interest: The authors declare no conflict of interest.

References

1. Hasselaar, B.L. Climate adaptive skins: Towards the new energy-efficient façade. In Proceedings of the 1st International Conference on the Management of Natural Resources, Sustainable Development and Ecological Hazards; 2006; pp. 351–360. Available online: https://www.scribd.com/document/458674117/WIT-transactions-on-ecology-and-the-environment-volume-99-C-A-Brebbia-Management-of-Natural-Resources-Sustainable-Development-And-Ecological-Ha (accessed on 1 June 2020).
2. Del Grosso, A.E.; Basso, P. Adaptive building skin structures. *Smart Mater. Struct.* **2010**, *19*, 124011. [CrossRef]
3. Loonen, R.C.; Trcka, M.; Costola, D.; Hensen, J.L. Climate adaptive building shells: State-of-the-art and future challenges. *Renew. Sustain. Energy Rev.* **2013**, *25*, 483–493. [CrossRef]
4. Aelenei, D. Adaptive Façade: Concept, Applications, Research Questions. *Energy Procedia* **2016**, *91*, 269–275. [CrossRef]
5. Kuru, A.; Oldfield, P.; Bonser, S.; Fiorito, F. Biomimetic adaptive building skins: Energy and environmental regulation in buildings. *Energy Build.* **2019**, *205*, 109544. [CrossRef]
6. Gruber, P. *Biomimetics in Architecture*; Springer Verlag Wien: Wien, Austria, 2011.
7. Vincent, J.F.; Bogatyreva, O.A.; Bogatyrev, N.R.; Bowyer, A.; Pahl, A.K. Biomimetics: Its practice and theory. *J. R. Soc. Interface* **2006**, *3*, 471–482. [CrossRef] [PubMed]
8. Bar-Cohen, Y. (Ed.) *Biomimetics: Nature-Based Innovation*; CRC Press, Taylor and Francis Group Publishing: Boca Raton, FL, USA, 2012.
9. López, M.; Rubio, R.; Martín, S.; Croxford, B.; Jackson, R. Active materials for adaptive architectural envelopes based on plant adaptation principles. *J. Facade Des. Eng.* **2015**, *3*, 27–38. [CrossRef]

10. Badarnah Kadri, L. Towards the LIVING Envelope: Biomimetics for Building Envelope Adaptation. Ph.D. Thesis, Technical University Delft, Delft, The Netherlands, 2015.
11. Kuru, A.; Fiorito, F.; Oldfield, P.; Bonser, S.P. Multi-functional biomimetic adaptive façades: A case study. In Proceedings of the FACADE 2018 Final Conference of COST TU1403 Adaptive Facades Network, Lucerne, Switzerland, 26–27 November 2018; pp. 241–250.
12. Badarnah, L. Form follows environment: Biomimetic approaches to building envelope design. *Buildings* **2017**, *7*, 40. [CrossRef]
13. López, M.; Rubio, R.; Martín, S.; Croxford, B. How plants inspire façades. From plants to architecture: Biomimetic principles for the development of adaptive architectural envelopes. *Renew. Sustain. Energy Rev.* **2017**, *67*, 692–703. [CrossRef]
14. Vincent, J.F.V.; Mann, D.L. Systematic technology transfer from biology to engineering. *Philos. Trans. Math. Phys. Eng. Sci.* **2002**, *15*, 159–173. [CrossRef] [PubMed]
15. Shu, L.H. A natural-language approach to biomimetic design. *Artif. Intell. Eng. Des. Anal. Manuf.* **2010**, *24*, 507–519. [CrossRef]
16. Helms, M.; Vattam, S.S.; Goel, A.K. Biologically inspired design: Process and products. *Des. Stud.* **2009**, *30*, 606–622. [CrossRef]
17. Biomimicry Institute. Biomimicry.net: Biomimicry 3.8. Available online: http://biomimicry.net/ (accessed on 25 June 2020).
18. Hirtz, J.; Stone, R.B.; McAdams, D.A.; Szykman, S.; Wood, K.L. A functional basis for engineering design: Reconciling and evolving previous efforts. *Res. Eng. Des. Theory Appl. Concurr. Eng.* **2002**, *13*, 65–82. [CrossRef]
19. Biomimicry Institute. AskNature: A Project of Biomimicry 3.8. Available online: www.asknature.org (accessed on 24 June 2020).
20. Vincent, J.F.V. *Research and Practice on the Theory of Inventive Problem Solving (TRIZ)*; Springer: Berlin, Germany, 2016.
21. Bogatyrev, N.; Bogatyreva, O. TRIZ evolutionary trends in biology and technology: Two opposites. In Proceedings of the CIRP Design Conference, Cranfield, UK, 30–31 March 2009; p. 293.
22. Wilson, J.O. A Systematic Appraoch to Bio-Inspired Conceptual Design. Ph.D. Thesis, Georgia Institute of Technology, Atlanta, GA, USA, 2008.
23. Gamage, R.S.D.; Wickramanayake, A.U. Parallels between nature and design teaching through nature studies. *Built Environ. Sri Lanka* **2005**, *5*, 1–12.
24. Svendsen, N.; Torben, L.A. How does biologically inspired design cope with multi-functionality? In Proceedings of the International Conference on Engineering Design ICED19, Delft, The Netherlands, 5–8 August 2019; pp. 349–358.
25. Cruz, E.; Raskin, K.; Aujard, F. Biological strategies for adpative building envelopes. In Proceedings of the FACADE 2018 Final Conference of COST TU1403 Adaptive Facades Network, Lucerne, Switzerland, 26–27 November 2018; pp. 223–229.
26. Kuru, A.; Fiorito, F.; Oldfield, P.; Bonser, S.P. Multi-functional biomimetic adaptive façades: Developing a framework. In Proceedings of the FACADE 2018 Final Conference of COST TU1403 Adaptive Facades Network, Lucerne, Switzerland, 26–27 November 2018; pp. 231–240.
27. Fratzl, P.; Dunlop, J.; Weinkamer, R. *Materials Design Inspired by Nature: Function through Inner Architecture*; RSC Publishing—Royal Society of Chemisstry: Cambridge, UK, 2013.
28. Fratzl, P.; Weinkamer, R. Nature's hierarchical materials. *Prog. Mater. Sci.* **2007**, *52*, 1263–1334. [CrossRef]
29. Dunlop, J.W.C.; Fratzl, P. Biological composites. *Ann. Rev. Mater. Res.* **2010**, *40*, 1–24. [CrossRef]
30. Knippers, J.; Speck, T. Design and construction principles in nature and architecture. *Bioinspir. Biomim.* **2012**, *7*, 15002. [CrossRef] [PubMed]
31. Jeronimidis, G.; Atkins, A.G. Mechanics of biological materials and structures: Nature's lessons for the engineer. *Proc. Inst. Mech. Eng.* **1995**, *209*, 221–235. [CrossRef]
32. Burgert, I.; Fratzl, P. Actuation systems in plants as prototypes for bioinspired devices. *Philos. Trans. R. Soc. A Math. Phys. Eng. Sci.* **2009**, *367*, 1541–1557. [CrossRef]
33. Pedersen Zari, M. Ecosystem processes for biomimetic architectural and urban design. *Archit. Sci. Rev.* **2014**, *58*, 106–119. [CrossRef]

34. John, G.; Clements-Croome, D.; Jeronimidis, G. Sustainable building solutions: A review of lessons from the natural world. *Build. Environ.* **2005**, *40*, 319–328. [CrossRef]
35. Garcia-Holguera, M.; Clark, O.G.; Sprecher, A.; Gaskin, S. Ecosystem biomimetics for resource use optimization in buildings. *Build. Res. Inf.* **2016**, *44*, 263–278. [CrossRef]
36. Appio, F.P.; Achiche, S.; Martini, A. On designers' use of biomimicry tools during the new product development process: An empirical investigation. *Technol. Anal. Strateg. Manag.* **2017**, *29*, 775–789. [CrossRef]
37. Pedersen Zari, M. Biomimetic Approaches To Architectural Design for Increased Sustainability. In Proceedings of the SB07 NZ Sustainable Building Conference, Auckland, New Zealand, 14–16 November 2017.
38. Hastrich, C. The biomimicry design spiral. *Biomimicry Newsl.* **2006**, *4*, 5–6.
39. Ahmar, S.E.; Fioravanti, A.; Hanafi, M. A Methodology for Computational Architectural Design Based on Biological Principles. In Proceedings of the 31st eCAADe Conference, Delft, The Netherlands, 18–20 September 2013.
40. Sheta, A.A.A.A. Biomimicry in Environmental Architecture: Exploring the Concept and Methods of the Bio-Inspired Environmental Architecrural Design. 2010. Available online: http://www.cpas-egypt.com/pdf/Ayat_El-Jawhary/MS.c/MS.c.pdf (accessed on 25 June 2020).
41. Badarnah, L.; Kadri, U. A methodology for the generation of biomimetic design concepts. *Archit. Sci. Rev.* **2014**, *8628*, 1–14. [CrossRef]
42. Cohen, Y.H.; Reich, Y.; Greenberg, S. What can we learn from biological systems when applying the law of system completeness? *Procedia Eng.* **2015**, *131*, 104–114. [CrossRef]
43. Chakrabarti, A.; Siddharth, L.; Dinakar, M. Idea Inspire 3.0—A tool for analogical design. In *Research into Design for Communities*; Smart Innovation, Systems and Technologies; 2017; Volume 2, pp. 475–485. Available online: https://link.springer.com/chapter/10.1007%2F978-981-10-3521-0_41 (accessed on 25 June 2020).
44. Goel, A.; Rugaber, S.; Vattam, S. Structure, Behavior and Function of complex systems: The SBF Modeling Language. *Artif. Intell. Eng. Des. Anal. Manuf.* **2009**, *23*, 23–25. [CrossRef]
45. Gamage, A.; Hyde, R. A model based on biomimicry to enhance ecologically sustainable design. *Archit. Sci. Rev.* **2012**, *55*, 224–235. [CrossRef]
46. Schleicher, S.; Lienhard, J.; Poppinga, S.; Speck, T.; Knippers, J. A methodology for transferring principles of plant movements to elastic systems in architecture. *Comput. Des.* **2015**, *60*, 105–117. [CrossRef]
47. Pedersen Zari, M. Ecosystem services analysis for the design of regenerative built environments. *Build. Res. Inf.* **2012**, *40*, 54–64. [CrossRef]
48. Lienhard, J.; Schleicher, S.; Poppinga, S.; Masselter, T.; Milwich, M.; Speck, T.; Knippers, J. Flectofin: A hingeless flapping mechanism inspired by nature. *Bioinspir. Biomim.* **2011**, *6*, 045001. [CrossRef]
49. Badarnah, L.; Knaack, U. Shading/Energy generating skin inspired from natural systems. In Proceedings of the 2008 World Sustainable Building Conference: SB08, Melbourne, Australia, 21–15 September 2008; pp. 305–312.
50. Goel, A.G.; Mcadams, D.A.; Stone, R.B. (Eds.) *Biologically Inspired Design: Computational Methods and Tools*; Springer: London, UK, 2014.
51. Martone, P.T.; Boller, M.; Burgert, I.; Dumais, J.; Edwards, J.; Mach, K.; Rowe, N.; Rueggeberg, M.; Seidel, R.; Speck, T. Mechanics without muscle: Biomechanical inspiration from the plant Wworld. *Integr. Comp. Biol.* **2010**, *50*, 888–907. [CrossRef] [PubMed]
52. Rowe, N.P.; Speck, T. Hydraulics and mechanics of plants: Novelty, innovation and evolution. In *The Evolution of Plant Physiology*; Academic, A.R., Poole, I., Eds.; Elsevier: London, UK, 2004; pp. 301–329.
53. Garland, T. Quick guide: Tradeoffs. *Curr. Biol.* **2014**, *24*, 60–61. [CrossRef] [PubMed]
54. Knaack, U.; Klein, T.; Bilow, M.; Auer, T. *Facades: Principles of Construction*, 2nd ed.; BIRKHAUSER: Berlin, Germany, 2014.
55. Prieto, A.; Knaack, U.; Auer, T.; Klein, T. Passive cooling & climate responsive façade design: Exploring the limits of passive cooling strategies to improve the performance of commercial buildings in warm climates Energy & Buildings Passive cooling & climate responsive façade design exploring the. *Energy Build.* **2018**, *175*, 30–47.
56. Gruber, P. Has biomimetics arrived in architecture? *Bioinspir. Biomim.* **2012**, *7*, 010201. [CrossRef]
57. Fiorito, F.; Sauchelli, M.; Arroyo, D.; Pesenti, M.; Imperadori, M.; Masera, G.; Ranzi, G. Shape morphing solar shadings: A review. *Renew. Sustain. Energy Rev.* **2016**, *55*, 863–884. [CrossRef]

58. Palomo, I.; Dujardin, Y.; Midler, E.; Robin, M.; Sanz, M.; Pascual, U. Modeling trade-offs across carbon sequestration, biodiversity conservation, and equity in the distribution of global REDD+ funds. In Proceedings of the National Academy of Sciences, 21 October 2019; Volume 116, pp. 22645–22650. [CrossRef]
59. Speck, T.; Rowe, N. How to become a successful climber–mechanical, anatomical, ultra-structrural and biochemical variations during ontogeny in plants with different climbing strategies. In Proceedings of the 5th International PLant Biomechanics Conference, Lausanne, Switzerland, 21–23 May 2007; Volume 1, pp. 103–108.
60. Al-Obaidi, K.M.; Ismail, M.A.; Hussein, H.; Abdul, A.M. Biomimetic building skins: An adaptive approach. *Renew. Sustain. Energy Rev.* **2017**, *79*, 1472–1491. [CrossRef]
61. Levitt, J. Physiological basis of stomatal response. In *Water and Plant Life*; Springer: Berlin/Heidelberg, Germany, 1976; Volume 19.
62. Bertolino, L.T.; Caine, R.S.; Gray, J.E. Impact of stomatal density and morphology on Water-Use efficiency in a changing world. *Front. Plant Sci.* **2019**, *10*, 1–11. [CrossRef]
63. Ye, H.; Yuan, Z.; Zhang, S. The heat and mass transfer analysis of a leaf. *J. Bionic Eng.* **2013**, *10*, 170–176. [CrossRef]
64. Nobel, P.S. Water relations and photosynthesis of a barrel cactus, Ferocactus acanthodes, in the Colorado desert. *Oecologia* **1977**, *27*, 117–133. Available online: https://link.springer.com/article/10.1007/BF00345817 (accessed on 25 June 2020). [CrossRef]
65. Decker, M. Emergent Futures: Nanotechnology and emergenct materials in architecture. In Proceedings of the BTES Conference 2013—Tectonics of Teaching Roger Williams University, New Jersey Institite of Technology, Bristol, RI, USA, 11–13 July 2013.
66. Schinegger, K.; Rutzinger, S.; Oberascher, M.; Weber, G. One Ocean: Theme Pavilion EXPO 2012 Yeosu, Residenz Verlag. Available online: http://www.soma-architecture.com/index.php?page=theme_pavilion&parent=2 (accessed on 25 June 2020).
67. Reichert, S.; Menges, A.; Correa, D. Meteorosensitive architecture: Biomimetic building skins based on materially embedded and hygroscopically enabled responsiveness. *Comput. Des.* **2015**, *60*, 50–69. [CrossRef]
68. Menges, A. HygroScope: Meteorosensitive morphology. In Proceedings of the Project Catalogue of the 32nd Annual Conference of the Association for Computer-Aided Design in Architecture ACADIA, San Francisco, CA, USA, 18–21 October 2012.
69. Henrion, W.; Tributsch, H. Optical solar energy adaptations and radiative temperature control of green leaves and tree barks. *Sol. Energy Mater. Sol. Cells* **2009**, *93*, 98–107. [CrossRef]
70. Jeronimidis, G. Design and function of structural biological materials. *Pergamon Mater. Ser.* **2000**, *4*, 19–29.
71. Schleicher, S. *Bio-Inspired Compliant Mechanisms for Architectural Design: Transferring Bending and Folding Principles of Plant Leaves to Flexible Kinetic Structures*; University of Stuttgart: Stuttgart, Germany, 2016.
72. Speck, T.; Knippers, J.; Speck, O. Self-X materials and structures in nature and technology: Bio-inspiration as a driving force for technical innovation. *Archit. Des.* **2015**, *85*, 34–39. [CrossRef]
73. Vukusic, P.; Sambles, J.R.; Lawrence, C.R.; Wootton, R.J. Quantified interference and diffraction in single Morpho butterfly scales. *Proc. R. Soc. B Biol. Sci.* **1999**, *266*, 1403–1411. [CrossRef]
74. Yoshioka, S.; Kinoshita, S. Wavelength–selective and anisotropic light–diffusing scale on the wing of the Morpho butterfly. *Proc. R. Soc. Publ. Biol. Sci.* **2004**, *271*, 581–587. [CrossRef] [PubMed]
75. Prum, R.O. Anatomically diverse butterfly scales all produce structural colours by coherent scattering. *J. Exp. Biol.* **2006**, *209*, 748–765. [CrossRef] [PubMed]
76. Bui, D.K.; Nguyen, T.N.; Ghazlan, A.; Ngo, N.T. Enhancing building energy efficiency by adaptive façade: A computational optimization approach. *Appl. Energy* **2020**, *265*, 114797. [CrossRef]
77. Sheikh, W.T.; Asghar, Q. Adaptive biomimetic facades: Enhancing energy efficiency of highly glazed buildings. *Front. Archit. Res.* **2019**, *8*, 319–331. [CrossRef]
78. Park, B.R.; Hong, J.; Choi, E.J.; Choi, Y.J.; Lee, C. Improvement in energy performance of building envelope incorporating electrochromic windows (ECWs). *Energies* **2019**, *12*, 1181. [CrossRef]
79. Giovannini, L.; Favoino, F.; Pellegrino, A.; Lo Verso, V.R.M.; Serra, V.; Zinzi, M. Thermochromic glazing performance: From component experimental characterisation to whole building performance evaluation. *Appl. Energy* **2019**, *251*, 113335. [CrossRef]

80. Svetozarevic, B.; Begle, M.; Jayathissa, P.; Caranovic, S.; Shephard, R.F.; Nagy, Z.; Hischier, I.; Hofer, J.; Schluter, A. Dynamic photovoltaic building envelopes for adaptive energy and comfort management. *Nat. Energy* **2019**, *4*, 671–682. [CrossRef]
81. Gao, Y.; Dong, J.; Isabella, O.; Santbergen, R.; Tan, H.; Zeman, M.; Zhang, G. A photovoltaic window with sun-tracking shading elements towards maximum power generation and non-glare daylighting. *Appl. Energy* **2018**, *228*, 1454–1472. [CrossRef]
82. Fiorito, F.; Cannavale, A.; Santamouris, M. Development, testing and evaluation savings potentials of photovoltachromic windows in office buildings. A perspective study for Australian climates. *Sol. Energy* **2020**, *205*, 358–371. [CrossRef]
83. Loonen, R.C.G.M.; Favoino, F.; Hensen, J.L.M.; Overend, M. Review of current status, requirements and opportunities for building performance simulation of adaptive facades. *J. Build. Perform. Simul.* **2016**, *1493*, 1–19. [CrossRef]
84. Loonen, R. *Approaches for Computational Performance Optimization of Innovative Adaptive Façade Concepts*; Issue 153 Bouwstenen series of the Department of the Built Environment; Eindhoven University of Technology: Eindhoven, The Netherlands, 2018.
85. Bhushan, B.; Jung, Y.C. Natural and biomimetic artificial surfaces for superhydrophobicity, self-cleaning, low adhesion, and drag reduction. *Prog. Mater. Sci.* **2011**, *56*, 1–108. [CrossRef]
86. Gorb, S.; Speck, T. Biological and biomimetic materials and surfaces. *Beilstein J. Nanotechnol.* **2017**, *8*, 403–407. [CrossRef]
87. Garrett, T.Y.; Huynh, C.; North, G.B. Root contraction helps protect the "Living rock" cactus Ariocarpus fissuratus from lethal high temperatures when growing in rocky soil temperatures when growing in rocky soil 1. *Am. J. Bot.* **2010**, *97*, 1951–1960. [CrossRef]
88. Worall, M. Intelligent thermoregulation and homeostasis: Lessons from nature. In Proceedings of the CIB World Congress, Salford, UK, 11–13 May 2010.
89. Badarnah, L. Light Management Lessons from Nature for Building Applications. In Proceedings of the International Conference on Sustainable Design, Engineering and Construction, Tempe, AZ, USA, 18–20 May 2016; Volume 145, pp. 595–602.
90. Helfman, Y.C.; Reich, Y. *Biomimetic Design Method for Innovation and Sustainability*; Springer: Berlin, Germany, 2016.
91. Gruber, P. Transfer of nature to architecture—Analysis of case studies. In Proceedings of the Biological Approaches for Engineering Conference, University of Southampton, Southampton, UK, 17–19 March 2008.
92. Mazzoleni, I. *Architecture Follows Nature*; Taylor & Francis: Oxfordshire, UK, 2013.
93. Pawlyn, M. *Biomimicry in Architecture*; RIBA Publishing: London, UK, 2011.
94. Harman, J. *The Shark's Paintbrush: Biomimicry and How Nature Is Inspiring Innovation*; Knopf Doubleday Publishing Group: New York, NY, USA, 2013.
95. Kapsali, V. *Biomimetics for Designers*; Thames and Hudson: London, UK, 2016.
96. Knippers, J.; Nickel, K.G. (Eds.) *Biomimetic Research for Architecture and Building Construction: Biological Design and Integrative Structures*; Springer International Publishing: Cham, Switzerland, 2016.
97. Pacheco, F.-T.; Labrincha, J.A.; Diamanti, M.V.; Yu, C.P.; Lee, H.K. (Eds.) *Biotechnologies and Biomimetics for Civil Engineering*; Springer International Publishing: Cham, Switzerland, 2015.
98. Pohl, G.; Nachtigall, W. *Biomimetics for Architecture and Design: Nature—Analogies—Technology*; Springer: Cham, Switzerland, 2015. [CrossRef]
99. Shi, N.N.; Tsai, C.C.; Camino, F.; Bernard, G.D.; Yu, N.; Wehner, R. Keeping cool: Enhanced optical reflection and radiative heat dissipation in Saharan silver ants. *Science* **2015**, *349*, 298–301. [CrossRef]
100. Bhasin, D.; McAdams, D. The characterization of biological organization, abstraction, and novelty in biomimetic design. *Designs* **2018**, *2*, 54. [CrossRef]
101. Field, K.J.; George, R.; Fearn, B.; Quick, W.P.; Davey, M.P. Best of both worlds: Simultaneous high-light and shade-tolerance adaptations within individual leaves of the living stone lithops aucampiae. *PLoS ONE* **2013**, *8*, e75671. [CrossRef]
102. Badarnah, L. Environmental adaptation of buildings through morphological differentiation. In Proceedings of the 15th Conference on Advanced Building Skins, Bern, Switzerland, 1–2 October 2018.
103. Gibson, A.; Nobel, P. *The Cactus Primer*; Harvard University Press: Cambridge, MA, USA, 1986.

104. Bar-Cohen, Y. Biomimetics—Using nature to inspire human innovation. *Bioinspir. Biomim.* **2006**, *1*, 1–12. [CrossRef]
105. Bhushan, B. Biomimetics: Lessons from nature—An overview. *Philos. Trans. R. Soc. A Math. Phys. Eng. Sci.* **2009**, *367*, 1445–1486. [CrossRef]
106. Bhushan, B. *Biomimetics: Bioinspired Hierarchical-Structured Surfaces for Green Science and Technology*; Springer: Berlin/Heidelberg, Germany, 2012.
107. Brownell, B.; Swackhamer, M. *Hyper-Natural: Architecture's New Relationship with Nature*; Princeton Architectural Press: New York, NY, USA, 2015.
108. Laver, J.; Clifford, D.; Vollen, J. High performance masonry wall systems: Principles derived from natural analogues. *WIT Trans. Ecol. Environ.* **2008**, *114*, 243–252.
109. Dawson, C.; Vincent, J.F.V.; Rocca, A.M. How pine cones open. *Nature* **1997**, *39*, 668. [CrossRef]
110. Mott, K.A.; Gibson, A.C.; Leary, J.W.O. The adaptive significance of amphistomatic leaves. *Plant Cell Environ.* **1982**, *5*, 455–460. [CrossRef]
111. Bureau of Meteorology. Climate Statistics for Australian Locations. 2016. Available online: www.bom.gov.au (accessed on 24 June 2020).
112. U.S.D. of Energy. US DoE Documentation. Available online: https://www.environment.gov.au/energy (accessed on 27 June 2020).
113. Casini, M. Active dynamic windows for buildings: A review. *Renew. Energy* **2018**, *119*, 923–934. [CrossRef]
114. Formentini, M.; Lenci, S. An innovative building envelope (kinetic façade) with Shape Memory Alloys used as actuators and sensors. *Autom. Constr.* **2018**, *85*, 220–231. [CrossRef]
115. Pesenti, M.; Masera, G.; Fiorito, F. Exploration of Adaptive Origami Shading Concepts through Integrated Dynamic Simulations. *J. Archit. Eng.* **2018**, *24*, 4018022. [CrossRef]
116. Attia, S. Evaluation of adaptive facades: The case study of Al Bahr Towers in the UAE. *QSci. Connect* **2017**, *6*. [CrossRef]
117. Suralkar, R. Solar responsive kinetic facade shading systems inspired by plant mvoements in nature. In Proceedings of the People and Buildings, the offices of Arup UK, Netowkr for Comfort and Energy Use in Buildings, London, UK, 23 September 2011.
118. ABCB. NCC 2019 BCA Volume 1. 2019. Available online: https://ncc.abcb.gov.au/ncc-online/NCC/2019/NCC-2019-Volume-One (accessed on 24 June 2020).
119. ABCB. NCC 2019 BCA Volume 2. 2019. Available online: https://ncc.abcb.gov.au/ncc-online/NCC/2019/NCC-2019-Volume-Two (accessed on 24 June 2020).
120. Australian Standards. AS 1668 2 Supplement 1—2002 the Use of Ventilation and Airconditioning in Buildings—Ventilation Design for Indoor Air Contaminant Control. 2016. Available online: https://www.saiglobal.com/PDFTemp/Previews/OSH/as/as1000/1600/N16682S1.pdf (accessed on 24 June 2020).

© 2020 by the authors. Licensee MDPI, Basel, Switzerland. This article is an open access article distributed under the terms and conditions of the Creative Commons Attribution (CC BY) license (http://creativecommons.org/licenses/by/4.0/).

buildings

MDPI

Article

Non-Commercial Air Purifier—The Effectiveness and Safety

Anna Mainka [1,*], Walter Mucha [1], Józef S. Pastuszka [1], Ewa Brągoszewska [2] and Agnieszka Janoszek [1]

1 Faculty of Energy and Environmental Protection, Department of Air Protection, Silesian University of Technology, 44-100 Gliwice, Poland; Walter.Mucha@polsl.pl (W.M.); Jozef.Pastuszka@polsl.pl (J.S.P.); ajanoszek@vp.pl (A.J.)

2 Faculty of Energy and Environmental Protection, Department of Technologies and Installations for Waste Management, Silesian University of Technology, 44-100 Gliwice, Poland; Ewa.Bragoszewska@polsl.pl

* Correspondence: Anna.Mainka@polsl.pl; Tel.: +48-32-237-1060

Received: 30 April 2020; Accepted: 27 May 2020; Published: 31 May 2020

Abstract: (1) Background: On the Internet, we can find the guidelines for homemade air purifiers. One of the solutions includes the use of a low-cost ozone generator to decrease the level of odors and biological contaminants. However, the authors do not notify about hazardous effects of ozone generation on human health; (2) Methods: We elaborated our test results on the bacterial and fungal aerosol reduction by the use of two technical solutions of homemade air purifiers. First, including a mesh filter and ozone generator, second including an ozone generator, mesh filter, and carbon filter. (3) Conclusions: After 20 min of ozone generation, the concentration of bacteria decreased by 78% and 48% without and with a carbon filter, while fungi concentration was reduced in the lower range 63% and 40%, respectively. Based on our test results, we proposed a precise periodical operation of homemade air purifier to maintain the permissible level of ozone for the occupants.

Keywords: IEQ; bioaerosols; airborne bacteria; airborne fungi; ozone; portable air purifier; ozone generation

1. Introduction

As we spend about 90% of our lifetime inside, indoor air pollution appears as one of the most harmful threats these days. One of the crucial indicators of indoor air pollution is bioaerosols. These biological particles are one of the risk factors bringing adverse health effects [1,2]. Biological aerosols are a broad category of airborne particles, comprising all particles having a biological source. One group includes bacteria, fungi, viruses and pollen suspended in the air, while the second includes biomolecules (toxins, debris from membranes such as lipids and proteins) [3]. This kind of aerosol has been linked to various health effects from allergic, through infections, to toxic reactions [4–8].

In order to improve air quality by reducing bioaerosols in the environment, air purifiers are used. There are many air purifiers on the market. In particular, mobile devices have the advantage of being independent of integrated installations and can also be used in buildings where no air conditioning has been installed. Furthermore, they are considered to be an alternative way of treating hazardous and/or odorous pollutants in buildings that cannot be fully remediated at reasonable costs [9]. The typical method of air cleaning is filtering. The most common material is high-efficiency particle-arresting (HEPA) filters or carbon, often used in activated modifications [9]. Generally, air purifiers recommended by the Association of Home Appliance Manufacturers (AHAM) meet the 80% effectiveness criterion for small particles (e.g., environmental tobacco smoke—ETS). Shaughnessy et al. [10] underlined that meeting this criterion for all particles is difficult because specific air cleaner effectiveness depends on three key elements—room size, clean air delivery rate (CADR), and particle-size category. The authors pointed to three particle types crucial for effectiveness evaluations: environmental tobacco smoke (ETS) (particle size range, 0.09–1.0 μm diameter); dust (particle size range, 0.5–3.0 μm diameter); and paper mulberry pollen (particle size range, 5–11 μm diameter). The particle removal effectiveness depending on these three parameters can vary from 19% to 99%. Portable air purifiers can run on various technologies, including mechanical filters, electrostatic precipitators, ion generators, hybrid filters, gas phase filters, and ozone generators [11]. Many devices and appliances dedicated to indoor use release ozone either intentionally or unintentionally [12]. Siegel [13] pointed out that a device intended as an air cleaner that intentionally emits any compound into indoor air should not be considered a true air cleaner because the contamination can outweigh any air cleaning benefits. However, very often, commercial ozone generators are marketed as air purifiers [14]. Among various mechanisms, the ozone generators aim to improve the microbiological quality of air, reducing the levels of fungi, bacteria, and viruses present in the environment by over 80% [15]. Some have claimed that ozone can oxidize airborne gases, and even particulates, to simple carbon dioxide and water vapor [16].

The literature reports several tests on air purifiers reducing levels of bioaerosols in indoor environments (Table 1). In these studies, the effectiveness of air purifiers is defined as the percentage of reduction in pollutant concentration in a room of interest [17,18]. Some studies have examined the efficacies of air purifiers in indoor environments with a single dominating emission source, such as cat/dog allergens [19–21]. Others investigated the effectiveness of air filtration in lowering concentrations of air pollution in educational [22,23] and residential buildings [24,25].

Air purifiers generally cost about USD 200 to 500, but can go over USD 1000. Replacement filters and filter sets typically cost from USD 30 to 100, depending on the brand and model of air purifier [26]. Currently, technological progress and the availability of various technical solutions allow one to build a homemade air purifier and purchase for USD 25–50. On the internet (YouTube), movies on homemade air cleaners can be found—some recommend an ozone generator as an effective way to remedy the problem of odors and bacteria. However, we should put forward the question regarding the safety of such a solution.

As opposed to publications complimenting the ozone-generating devices to improve indoor air quality (IAQ), some studies cited US EPA the Consumers Union statement that "Air cleaners that generate ozone intentionally should not be used indoors" and focus on potentially deleterious consequences of overexposure to ozone as a public health concern [16,27]. For example, Britigan et al. [27] examined thirteen air purifiers and pointed out that ozone generators can produce O_3 levels above public health standards. Except for measurements indoors, in cars or airliner cabins [27,28], the tests with personal air purifiers (PAP) were done [27,29]. All skeptical publications underline that O_3 emission rates can maintain levels over public health standards, particularly in urban areas where ozone levels are already elevated as a result of outdoor emission. They also underline the susceptibility of the elderly, children, and persons with chronic diseases to ozone emissions.

Among the documented health effects of exposure, there is an increased risk of deaths and illnesses due to respiratory diseases (including asthma and chronic obstructive pulmonary disease—COPD) and cardiovascular diseases. A separate issue is occupational exposure to ozone, which arises, among others, for example, during working with office devices, such as photocopiers, printers, or projectors [12,30].

Due to the rapidly changing conditions, ozone concentrations are usually provided as 1 h, 8 h, 1 month, and annual averages [12]. In order to protect human health, the World Health Organization (WHO) has provided a guideline value of 100 μg/m^3 as the maximum 8 h mean ozone concentration [31]. However, governmental organizations have issued various recommendations or standards for ozone. According to current Polish regulations of the Ministry of Family, Work, and Social Policy defining the highest permissible concentrations and intensities of agents harmful to health in the work environment [32], the highest permissible concentration of O_3 by 8-h exposure is 150 μg/m^3 (0.06 ppm = 60 ppb). The U.S. Occupational Safety and Health Administration (OSHA) established a permissible exposure limit (PEL) of 100 ppb (241 μg/m^3) for an 8-h exposure and short-term exposure limit (STEL) of 300 ppb (723 μg/m^3) for a 15-min exposure [27]. The most restricted values are in

Canada. The Health Canada recommends a residential maximum exposure limit of 40 μg/m^3 (20 ppb) ozone, based on an averaging time of 8-h [33]. The WHO working group puts O_3 in the second group of potential interest as a pollutant, possibly carcinogenic to humans [34].

Table 1. Selected air cleaning devices.

Device	Mechanism	Reduction	References
Ozone generator	With prefilter, primary glass media filter, a 0.45 kg granular activated carbon filter, ion generating pulsating section, and ozone generating section	Dust 29% ± 1% Total spores 42% ± 14% Viable spores 38% ± 29%	[35]
AS150G(Wein Products, Inc. (Los Angeles, CA, USA)), AS150GX(Wein Products, Inc. (Los Angeles, CA, USA)), AS250B(Wein Products, Inc. (Los Angeles, CA, USA)), AS1250((Wein Products, Inc. (Los Angeles, CA, USA), VI-2500 (Wein Products, Inc. (Los Angeles, CA, USA)	Ionic air purifiers	Respirable bacteria 50% after 15 min almost 100% after 1.5 h	[11]
Not commercial device	Negative ion emitter and HVAC filter	Bacteria 75–85% (20% without ion emitter) Fungi 80–90% (40–60% without ion emitter)	[36]
Air purifier (model AX9000, Samsung, Suwon, Korea)	HEPA filters for air filtration and is reported by the manufacturer to have a Clean Air Delivery Rate (CADR) up to 1000 m^3/h.	Endotoxin 29%–37%	[17]
Not commercial device	HEPA filter device and UV filter	Bioaerosols 79%–82%	[37]
Aerobiotix Illuvia 500uv system (Aerobiotix, West Carrollton, OH, USA)	Combining HEPA filtration, zirconium-based Photochemical oxidation, and germicidal UV irradiation	Bioaerosols 41%	[38]
Philips AP (AC3256)	Equipped with a HEPA and an activated carbon filter	Cat and dog allergen 60%	[21]
Pure Air XL® PU6020F0 model Rowenta	HEPA filters	Cat allergen 35%	[20]

On the other hand, some authors considered the ozone levels from 0.08 (190 μg/m^3) to 3 ppm (7230 μg/m^3) in the environment as toxic and it has been associated with various adverse health effects, including the formation of reactive oxygen species (ROS), reduction in lung function induced airway inflammation through the infiltration of neutrophils and macrophages in both healthy individuals, and those who already have some kind of respiratory disease, exacerbation of respiratory illnesses, and increased rates of hospital admissions [15,39].

The review of available literature data confirms well-known knowledge that the emission of ozone is a very effective sterilization process, killing viable bacteria and fungi suspended in indoor air. Unfortunately, ozone is a toxic gas. Therefore, the exposure to ozone must be strongly limited. It should be underlined that the sterilization of indoor air using ozone, especially when people are present in the indoor environment, can be performed only using a device continuously sampling air and sterilizing it inside this instrument. It is easy to understand that although the idea of such a device is quite simple, it is difficult to construct such an instrument which should meet different expectations.

The main objective of our study was to verify whether low-cost ozone generator equipped with a mesh and carbon filter will meet the following expectations: (1) high sterilization efficiency, (2) low emission of ozone, guaranteeing the concentration of ozone in the indoor environment meeting health standards and, (3) relatively low cost of the designed air purifier. In this paper, we report our work in this field. The following sections describe the experimental procedure, key results, and implications of these measurements.

2. Materials and Methods

2.1. Research Room and Experimental Conditions

The challenges associated with the non-uniform distribution of bioaerosols point to the experimental measurements being an appropriate method to evaluate the efficiency of portable air purifiers. For comparison of bioaerosol reduction with the use of a mobile air cleaning device, multiple experiments need to be conducted for the same environmental boundary conditions. For a residence in a house, they vary significantly due to the stochastic nature of external weather conditions, but in the laboratory environment, these parameters can be controlled [40]. Generally, airtight test chambers have mostly been used for the ranking of the individual air purifiers. Unfortunately, the mixtures of typical air pollutants introduced in the chamber at selected concentrations have higher concentrations than those of indoor environments [41]. In Polish apartments, an average bedroom, primarily a child bedroom, is characterized by a surface area of approx. 9 m^2, so in our study, we used a laboratory room with a similar area equipped with a fume hood, both characterized in Table 2. During the test, environmental parameters, including temperature, relative humidity, PM_{10}, and CO_2 concentrations, were reported. An automatic portable monitor (model 77535, Az Instruments International Ltd., Hong Kong, China) connected to a PC with RS232 software installed was used to monitor the temperature, relative humidity, and CO_2 concentrations. For PM_{10} monitoring, the TSI SidePak AM510 Personal Aerosol Monitor (TSI, Inc., Shoreview, MN, USA) was used. The SidePak is a portable, battery-operated device using a built-in sampling pump that continuously measures PM_{10}. It uses the method of laser scattering and has proven useful in measuring exposure particles within the sampling area [42]. According to Mahyuddin et al. [43], one sensor in a room with a <100 m^2 floor area has significant p-value relationships. Both monitors displayed and recorded in real-time the measurements of environmental parameters in the laboratory room (Table 2), which allows recorded data to be downloaded for analysis. The sampling interval was 1 min.

Table 2. Characteristics of laboratory room, fume hood, and environmental conditions.

Parameter	Fume Hood	Laboratory Room
Length, m	1.4	3.8
Height, m	1.3	3.2
Width, m	0.7	2.5
Volume, m^3	1.3	30.4
Environmental conditions	**Average**	**SD**
Temperature, °C	18.6	0.5
Relative humidity, %	39.1	3.9
PM_{10}, $\mu g/m^3$	20.0	5.0
CO_2, ppm	363.6	22.5

2.2. Air Purifier

The air purifier used in the study (Figure 1) includes VENTS, TT 150 duct fan (1), mesh filter (2), ozone generator (3), carbon filter (4). For the measurements of ozone concentrations, two air sampling pumps model 224-PCMTX8–SKC (SKC Inc., UK) (5), and set of scrubbers for collecting air samples for ozone determination according to [44] (6) were used. The fan (1) is used to induce constant flow through the air purifier. The mesh filter (2) is used to decrease the concentration of particulate matter (PM), it provides good filtration efficiency down to sizes of 2 to 10 μm [45].

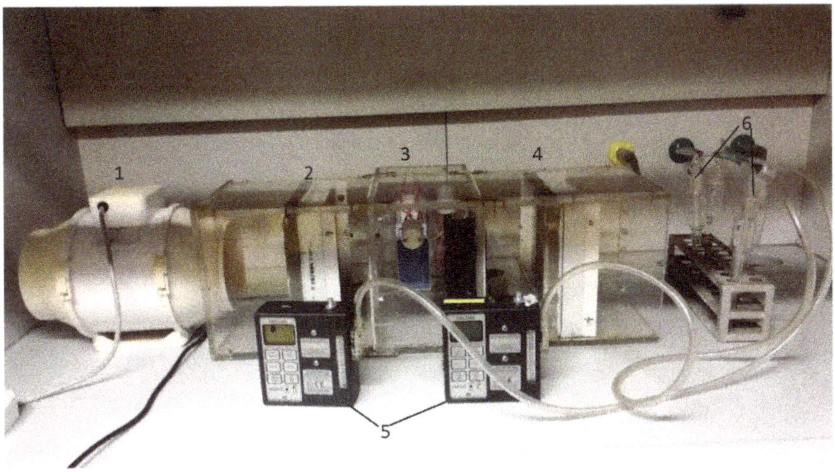

Figure 1. Construction of air purifier used in the study.

Next is the ozone generator to sterilize the air from bacteria and fungi. Ozone was emitted by generator model OG-CP-7G with yield 7000 mg/h. It has to be underlined that commercial air purifiers are characterized by ozone emission rates from 16 µg/h to 220 mg/h [27,46]. Our low-cost solution is characterized by a significantly higher emission rate of ozone, so undoubtedly will be characterized by a high reduction in bioaerosols, but its high ozone emission can have a negative impact on the health of occupants.

The air purifier was operated in two modes. The first mode included the airflow through the fan (1), mesh filter (2), and ozone generator (3), while the second mode included the assembling of carbon filter (4) after ozone generator (3). The carbon filter was used to reduce the ozone emission into the laboratory room. The standard parameters of the carbon filter included: the mass of the activated carbon filter (90 g), the granulation of the activated carbon filter (0.5–1.0 mm), the specific surface of the activated carbon filter (900 m^2/g), and the cross-sectional area of the activated carbon filter (0.042 m^2). The carbon filter was added to reduce the O_3 emission into the laboratory room, but the use of a carbon filter substantially decreases the air velocity (Table 3).

Table 3. Characteristics of air purifiers with and without carbon filter.

Parameter	Air Purifier	
	without Carbon Filter	with Carbon Filter
Average air velocity through the air purifier, m/s	1.43	1.18
Average airflow through the air purifier, m^3/h	218	180
The number of the air exchange rate in the room, h^{-1}	7	6

2.3. Bioaerosol Determination

The bacterial and fungal aerosol concentrations were measured by using an Air Ideal (AI) one-stage impactor (Figure 2). The air flow throughout AI was 100 dm^3/minute, and the sampling time was 3 min. Every measurement was conducted before the air purifier was active—to determine the background, and also after 5 and 20 min of air purifier work.

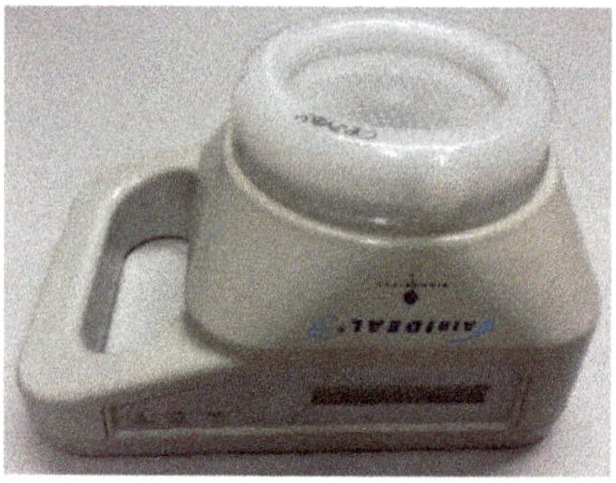

Figure 2. Air Ideal (AI) one-stage impactor used during the study.

Between each sampling, AI was disinfected. The disinfection was performed by the use of cotton balls immersed in 70% ethanol. Tryptic soy agar (TSA) with cycloheximide added to inhibit fungal growth was used for bacteria growth. At the same time, malt extract agar (MEA 2%) with chloramphenicol added to inhibit bacterial growth was used for fungal bioaerosol growth. Results were calculated as total colony counts which were corrected for multiple impactions by the positive hole method and expressed as colony-forming units per cubic meter of air (CFU/m^3). As presented in our previous studies [47,48], to control the quality of the bioaerosols sampling and determination the authors followed PN-EN 12322 and ISO 11133 standards. The Shapiro–Wilk test checked the normality of distribution of the bacterial and fungal concentrations in construction without or with a carbon filter. The distributions were lognormal, so comparison at 0, 5, and 20 min without and with carbon filter were calculated using Student's t-test. The Statistica Software v. 13 was used to perform all statistical analyses. All data are reported as means ± standard deviation (SD), and p-values less than 0.05 are accepted as statistically significant.

2.4. Ozone Determination

In the study, two methods of ozone determination were used. During measurements of bioaerosol reduction, the method was based on the ozone absorption in the potassium iodide solution and a colorimetric determination (SHIMADZU UV/VIS 2101PC) of iodine-based on the violet color of the reaction product with dimethyl-p-phenylenediamine. In order to collect the sample, 10 dm^3 of the air is passed through two scrubbers containing 10 cm^3 of a 1% KI solution at a rate of 1 dm^3/min. To determine the ozone concentration in the samples the operator should follow the following instruction [49]:

- take 5 cm^3 of the absorbing solution from the scrubber and transfer it to a colorimetric tube,
- add 0.5 cm^3 of 0.02% dimethyl-p-phenylenediamine hydrochloride and mix it.
- after 15 min, test the color of the solution in visible light (λ = 550 nm) in compare to earlier prepared scale of standards.

However, during ozone emission rate testing, the portable indoor air quality monitor Aeroqual series 500 (Aeroqual Limited Auckland, New Zealand) was used. It enables accurate real-time measurement of ozone in the air. Two sensor heads (OZL and OZH) were used. OZL (gas-sensitive semiconductor—GSS method) sensor measuring head enables to monitor ozone levels in the range 0–0.5 ppm, while the OZH sensor (gas-sensitive electrochemical—GSE method) determines the ozone level from 0.5 to 20 ppm. The resolution is 0.001 and 0.01 ppm, respectively [50]. The procedure of ozone decay rate measurement included the following steps:

- in both cases (without or with carbon filter) the laboratory room was not ventilated,
- ozone generation was turned on for 20 min (116 mg/min × 20 min), which corresponds to 2320 mg of generated O_3,
- after 20 min, the ozone generator was turned off,

- in the case of an activated carbon filter, the flow remained constant of 180 m^3/h,
- time in which the ozone concentration dropped below the lowest limit value of 40 $\mu g/m^3$ (Health Canada) was measured.

The measurement of mechanical ventilation use was also performed.

3. Results

3.1. Bacterial And Fungal Aerosol Reduction with the Ozone Generator

Figure 3 shows changes in the number of CFU of bacteria and fungi per cubic meter of air at the beginning of measurements as well as after 5 and 20 min of work of the ozone generator. In the first 5 min, the concentration of bacterial aerosol decreased from 500 to 180 CFU/m^3, which corresponds to the reduction speed of approx. 64 CFU/min. Between the 5th minute and 20th minute of ozone generation, the concentration of bacterial aerosol decreased significantly ($p = 0.01$) from 180 to 110 CFU/m^3; the reduction speed was approx. 5 CFU/min.

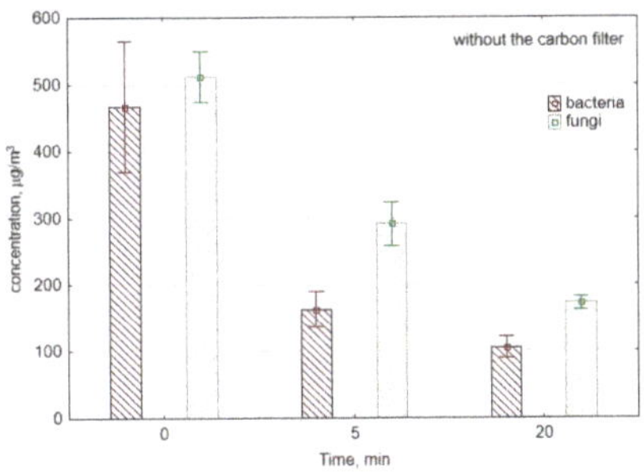

Figure 3. The concentration of bacterial and fungal aerosols during the work of air purifier consisting of mesh filter and ozone generator. The data represent the average values and the standard deviations of six measurements.

In the case of culturable fungal spores, the concentration decreased significantly from 477 to 253 CFU/m^3 ($p = 0.01$) during the first 5 min of ozone generation. So, during the first period of measurements, the reduction speed was approx. 45 CFU/min. Meanwhile, during the second period of measurements between the 5th minute and 20th minute, the reduction speed was significantly lower than approx. 5 CFU/min. This corresponds to the decrease from 253 to 179 CFU/m^3 ($p = 0.04$), with and without a carbon filter.

Usually, for both bacteria and fungi, the reduction speed is higher at the beginning of ozone generation. In the beginning, there is substantial damage and deformity of the surface structure of ozone-treated microorganisms. The progressive degradation involves the changes in membrane permeability and cell integrity. During the next stage, there is the lysis reaction, which corresponds to the biological effect on cell viability [51].

3.2. Ozone Concentrations

As can be seen in Figure 4, after 5 min of ozone generation, the concentration of O_3 increased to the hazardous level of 420.7 $\mu g/m^3$. After 20 min of ozone generation, the concentration of ozone decreases significantly ($p = 0.02$) compared to the concentration level consistent with 5 min of work. This decrease is due to the decomposition time of ozone, which is approx. 20 min. Britigan et al. [27] pointed that O_3 lifetime is moderately dependent on temperature variation and highly dependent on the presence of many reactive surfaces; for example, inside the car, the lifetime was only 2 min. In our study, after 20 min of air purification, the concentration of ozone remained at the level exceeding the acceptable O_3 concentration (150 μg/m^3) more than twice. Therefore, we installed a carbon filter after the ozone generator to decrease the emission of ozone into the room. Figure 4 presents the reduction in ozone concentration with the use of a carbon filter. The installation of carbon filer significantly ($p < 0.01$) decreased the level of O_3 in the air from 420.7 to 222.6 μg/m^3 and from 382.2 to 239.8 μg/m^3 after 5 min and 20 min of the ozone generator work without and with a carbon filter, respectively. The concentration of ozone after 5 and 20 min of the ozone generator maintenance with carbon filter is not significantly different ($p = 0.13$), because the installation of carbon filter stabilizes the concentration of ozone.

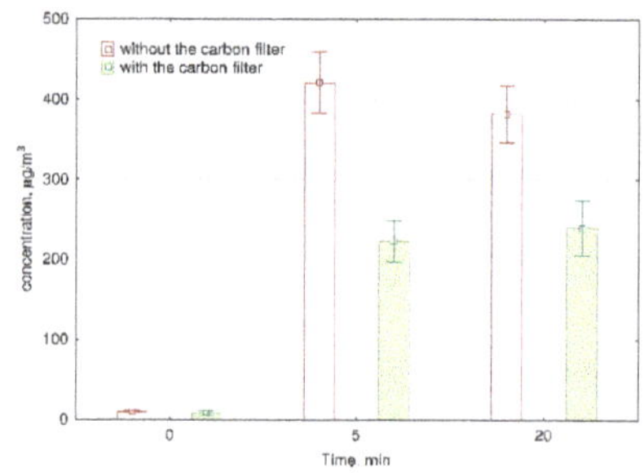

Figure 4. The concentration of ozone during the work of air purifiers with and without carbon filter. The data represent the average values and the standard deviations of six measurements.

3.3. Bacterial and Fungal Aerosol Reduction with Ozone Generator and Carbon Filter

Although the ozone levels between the 5th minute and 20th minute decrease without carbon filter or remain on the constant level with a carbon filter, respectively, the concentrations of bacterial and fungal aerosols significantly decrease.

Figure 5 presents changes in concentrations of fungi and bacteria in the laboratory room if the ozone generator was equipped with a carbon filter to reduce ozone emissions into the room. The use of carbon filter decreased the viable bacterial cells concentration with a similar reduction speed like without carbon filter approx. 6 CFU/min, the decrease was from 330 to 245 CFU/m^3 ($p = 0.001$). The difference in the reduction speed was crucial in the first 5 min of ozone generation. It was 64 and 28 CFU/min without and with a carbon filter, respectively, which corresponds to the decrease from 500 to 180 CFU/m^3 ($p = 0.007$) and 470 to 330 CFU/m^3 ($p = 0.001$), respectively.

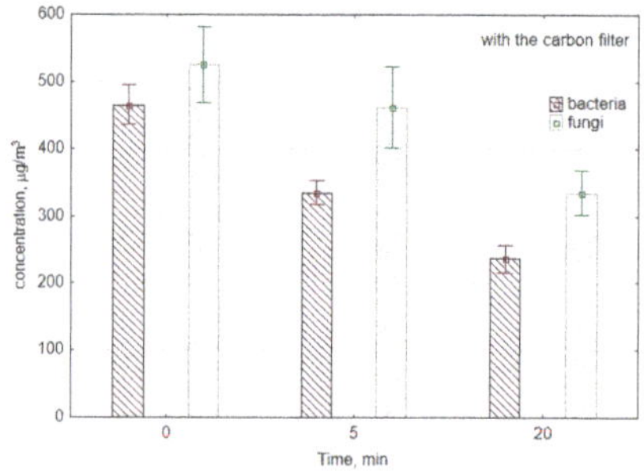

Figure 5. The average concentration of bacterial and fungal aerosols during the work of air purifier consisting of mesh filter, ozone generator, and carbon filter. The data represent the average values and the standard deviations of six measurements.

In the case of culturable fungal spores, the concentration decreased from 530 to 500 CFU/m^3, so it was not as significant a difference ($p = 0.51$) as without a carbon filter. During the first period of measurements, the reduction speed was higher without carbon filter approx. 45 and 6 CFU/min with a carbon filter. Nevertheless, during the second period of measurements between the 5th minute and 20th minute, the reduction speed was higher with the carbon filter 12 CFU/min and approx. 5 CFU/min without a carbon filter. It corresponds to the decrease from 500 to 320 CFU/m^3 (p = 0.03) and 253 to 179 CFU/m^3 (p = 0.04), with and without a carbon filter, respectively.

3.4. The Efficiency of Bacterial and Fungal Aerosol Removal

During the use of ozone generator both without and with a carbon filter, the number of air changes in the room is not significantly different 7 and 6 h^{-1}, respectively. However, the efficiency of reduction is significantly different. After just 5 min of work without a carbon filter, we observed (Table 4) a 64% reduction in bacterial particles and a 47% reduction in fungal spores. After 20 min of O_3 generation, the concentrations of bacterial and fungal aerosols decreased by 78% and 63%, respectively. With a carbon filter, the reduction in bacterial and fungal aerosols is significantly lower, for bacteria 30% and 50% as well as 6% and 40% for fungi, after 5 and 20 min of ozone generation, respectively. However, in both cases, such a reduction level is sufficient for the users. Faster reduction in bacteria levels compared to fungi levels was observed because the fungal aerosol is generally more resistant to environmental and mechanical stress than bacterial aerosol [52–54].

Table 4. The average efficiency of air purifier with and without carbon filter (n = 24).

Parameter	Air Purifier			
	without Carbon Filter		with Carbon Filter	
Time of work, min	5	20	5	20
Reduction in bacterial aerosol, %	64.0 ± 8.9	78.0 ± 6.0	29.8 ± 7.0	47.9 ± 9.9
Reduction in fungal aerosol, %	47.1 ± 20.7	62.6 ± 7.5	5.7 ± 20.3	39.6 ± 22.1

3.5. Ozone Decay Rate

Since the use of carbon filter did not allow to reduce O_3 concentration to the acceptable level below 150 µg/m^3, we have decided to perform the measurements of the ozone decay rate. They were performed in two series at the same laboratory room (approx. 30 m^3). The first series included ozone concentration measurements with ozone generator without a carbon filter; in the second series, the carbon filter was used. In the first case, the OZH sensor head with detection limit 0.01 ppm was used, while the second series was performed with the use of the OZL head with a detection limit of 0.001 ppm.

For both measurement sessions, the background concentration level was first monitored. Next, the air purifier was switched on, and the ozone generation started. The rapid increase in ozone concentration was observed. After 20 min, the homemade purifier was switched off, and the ozone concentration decay was further monitored for about 1.5 h. The highest concentrations of ozone have exceeded 3500 and 500 µg/m^3, without and with a carbon filter, respectively (Figure 6). The decrease in ozone concentration provides the natural decay of ozone in the environmental test conditions. As can be seen from Figure 6, the ozone concentration in the case of a system without a carbon filter meets the acceptable Polish level 33 min after turning off the device, and, in the case of a system with a carbon filter, after 8 min.

The ozone decomposition rate was determined from O_3 concentration changes in the reaction time. The reaction orders and rate constants were determined by the standard integral technique (1).

$$r = -\frac{d[O_3]}{dt} = k_1[O_3]^n \quad (1)$$

We also checked the decrease in ozone concentration with mechanical ventilation use. At the lowest available airflow of 820 m^3/h and an air exchange rate of 27 h^{-1}, the ozone concentration drops below the acceptable level after 2 min.

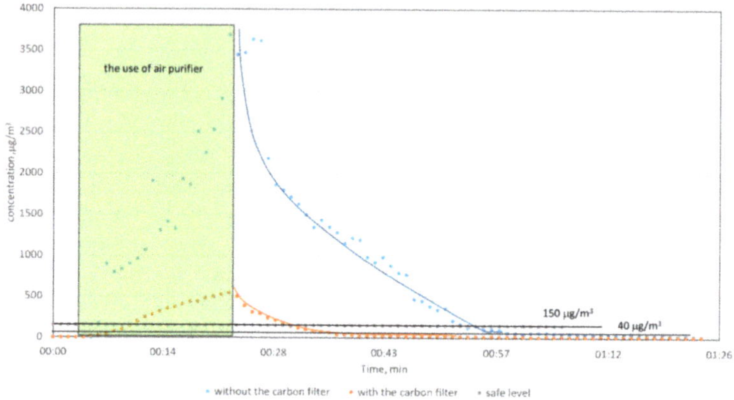

Figure 6. The rate of ozone reduction during the work of air purifier with and without a carbon filter.

To interpret the monitored data of a decay mechanism of ozone during two separate experiments performed without and with a carbon filter, we calculated the empirical reaction order. The highest correlation coefficients point to first–order of the reaction (R^2 = 0.95 and 0.99) with experimental decay constants 5.45 and 10.56 h^{-1} without and with a carbon filter, respectively. The first-order decay model was first used in the 1970s to interpret outdoor-originated ozone decay in the indoor environment [55–57].

4. Discussion

4.1. Effectiveness of Bioaerosol Removal by a Low-Cost Air Purifier

This study focuses on the efficiency of a homemade portable bioaerosol purifier consisting of an ozone generator preceded by mesh filter in the basic construction and in the second construction additionally equipped with a carbon filter to decrease the concentration of generated ozone. Both constructions proved effective in the removal of bacterial aerosol. However, the reduction in fungal aerosol levels with a carbon filter was effective after more extended use (20 min).

Commercial devices allow for reducing bioaerosol concentration with the effectiveness of 30% to 60% (Table 1). Our construction without carbon filter provides >60% of bacterial aerosol and 47.1% ± 20.7% to 62.6% ± 7.5% of fungal aerosol reduction. The achieved levels of reduction are similar to other, non-commercial devices. Huang et al. [36] reached 75–85% reduction in bacterial aerosol and 80–90% reduction in fungi, while Lee et al. [37] reported 79–82% reduction in bioaerosols. The reported levels are higher than in our more effective construction without a carbon filter; however, both reported devices used HEPA filters, which increase the cost of such an air purifier. Our construction is a low-cost device, the total value of which is comparable to the price of a single HEPA filter.

The construction with carbon filter allowed us to reduce bacterial and fungal aerosols to the level of 47.9% ± 9.9% and 39.6% ± 22.1%, respectively. Our levels of reduction are similar to the results of Shaughnessy et al. (42% ± 14%) [35], who used analogous construction elements: ozone generator, prefilter, and activated carbon. On the other hand, the bioaerosol reduction by commercial air purifiers allows for achieving 50% of bioaerosol reduction with a much higher cost of the device [11,17,20,21]. However, higher efficiency and lower costs of our construction require particular attention because of the accompanying emission of ozone into the room.

4.2. Ozone Threat

It is well known that ozone due to its oxidative power supports the biological decontamination of environments. However, the careless usage of any ozone generator available on the market can be hazardous to the consumers. Our study assessed the effectiveness of the ozonolysis air purification system in reducing the overall bacterial and fungal burden load throughout the laboratory room with an ozone generator and in the next step with the use of carbon filter. Both technical solutions ensure a significant reduction in bioaerosols. However, without a carbon filter, the concentration of O_3 after 20 min of generation reached 160 ppb. Even though it does not exceed OSHA STEL (15-min exposure) limit 300 ppb, it is above OSHA PEL (8-h) exposure limit (100 ppb), and Polish PEL limit (60 ppb).

Our research included determination of O_3 concentration decay due to decomposition; we based this on the ozone generation calculation method [58], based upon a generation-decay model:

$$V\frac{dC}{dt} = E - kVC \tag{2}$$

where V is the chamber volume (m^3), C the ozone concentration (mg/m^3), t the time (s), E the ozone generation rate by the unit tested (mg/s), k the decay constant due to various decomposition mechanisms (s^{-1}). When no ozone emission source is present, i.e., when an air cleaner is off, we have $E = 0$, and then Equation (1) becomes:

$$\frac{dC}{C} = -kdt \tag{3}$$

where k represents the natural decay constant in the chamber when an air cleaner is not operating. We have obtained two ozone decay constants $k_{without\ carbon\ filter}$ and $k_{with\ carbon\ filter}$, which are 1.5×10^{-3} s^{-1} and 2.9×10^{-3} s^{-1}. The solution is:

$$\ln\frac{C}{C_0} = -k(t - t_0) \tag{4}$$

$$t - t_0 = -\frac{\ln\frac{C}{C_0}}{k} \tag{5}$$

where C_0 (mg/m^3) is the initial ozone concentration in the laboratory room, C (mg/m^3) is the maximum ozone concentration that can be in the laboratory room, (e.g., C_{OSHA}= 0.241 mg/m^3, C_{Poland} = 0.150 mg/m^3, and C_{Canada} = 0.04 mg/m^3), t (s) is time, and $t_0 = 0$, k (s^{-1}) represents the natural decay constant in the laboratory room when the air cleaner is not operating; we used separately $k_{without\ carbon\ filter}$ and $k_{with\ carbon\ filter}$.

The question is how long after turning-off a homemade air purifier can the occupants enter the room without a health threat. If we consider three permissible exposure limits for 8-h exposure (C_{OSHA}= 241 $\mu g/m^3$, C_{Poland} = 150 $\mu g/m^3$, and C_{Canada} = 40 $\mu g/m^3$), after 20 min of ozone generation the air will meet the OSHA standard in 29 and 4 min without and with a carbon filter, respectively. The Polish standard will be preserved after 35 and 7 min, while the most restrictive Canadian standard will be preserved after 74 and 15 min without and with a carbon filter, respectively. As it can be seen, the installation of carbon filer decreases the O_3 level below 100 ppb (241 $\mu g/m^3$) in less than 5 min, but the decrease in O_3 concentration to the safe level of 40 $\mu g/m^3$ without using a carbon filter requires turning-off the ozone generator approx. 1.5 h before occupancy. Following Britigan et al. [27], it should be underlined that ozone decrease depends on the surface area of the room—in this case, the loss is dominated by heterogeneous removal on surfaces as well as on the total volume of the room, when the O_3 decrease is dominated by air exchange.

4.3. Implications for Health

To effectively control indoor air pollution, the flux of generated O_3 should exceed permissible exposure limits, but to improve IAQ, health benefits are superior. As written in the introduction, high ozone levels (0.08–3 ppm) in indoor spaces are considered toxic and are associated with various health effects. Thiele et al. [59] reported that following a single 2 h exposure to 10 ppm of O_3 and a 30 min exposure to air, the outermost skin layers of mice were significantly depleted of vitamins C and E. Furthermore, there was a dramatic increase in lipid peroxidation, as evidenced by increased concentrations of epidermal malondialdehyde (MDA). The research on pulmonary toxicity on the lungs of rats pointed out that rats exposed to 0.06 ppm of O_3 for 5 days presented increased alveolar epithelium volume and cellular injury [15]. Beyond detrimental health effects of breathing O_3-polluted air, ozone can react with indoor surfaces, such as carpets, linoleum, clothing, and furniture, releasing volatile oxidation products that may have adverse health effects. One of the examples is the reaction of O_3 with volatile organic compounds (VOCs) found in many detergents, which produces aldehydes [27]; others include reactions with nitric oxide and some free radicals. There is evidence that the by-products may often be more irritating than the original reactants. Clausen et al. [60] exposed mice to a high concentration of ozone (3 ppm) and limonene (48 ppm) for 2 h. The mice experienced a 33% reduction in respiratory rate, while the individual compounds that were measured in the chamber could not explain this reduction. Although ozone generated by air purifiers in concentration about 0.05 ppm is defined as safe, either in the use of 3 h/day as well as 24 h/day [15], the health impacts of lower O_3 levels are possible, particularly in more susceptible individuals such as children and persons with asthma or other respiratory diseases or allergies, and depend on the duration of exposure and breathing rate. The epidemiological studies document a dose—response relationship between the long-term exposure of O_3 concentrations below 0.05 ppm, particularly in ambient air and human health, leading to a decrease in pulmonary function and broncho-provocation [31].

As our device is based on O_3 emission, below please find a scheme (Figure 7) and instruction with the steps required for safe use. The authors recommend the use of the device with a carbon filter. The crucial feature is that the operation of the device in occupied places is forbidden. The device is designed to be used periodically, and the occupants can enter the room 1.5 h after turning it off. Intended or unintended entry to the room during the operation of the device is highly dangerous, due to O_3 exposure.

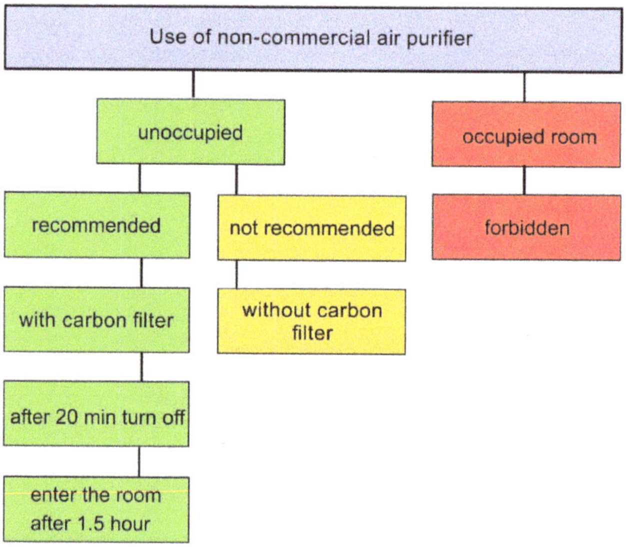

Figure 7. The scheme with maintenance instruction of non-commercial air purifier.

Not following the instructions mentioned above can lead to pulmonary system effects, including inflammation, reflexes, and reduction in pulmonary defenses. Ozone affects pulmonary defenses by several mechanisms: the impairment of mucociliary clearance, decreased macrophage activity, and effects on circulating lymphocytes. These effects of short time exposure on immunity are acute, and recovery might take more than three days. Longer exposures can cause significant structural alterations to the lungs, characterized by diffuse mucus hyperplasia, bronchiolar narrowing and alveolar fibrosis. These alterations revert partially after finishing the exposure. Although uptake of ozone is almost exclusively by inhalation, there is also a possibility that some effects can be detected in the tear duct epithelial cells [31]. If the user must enter the room during air ozonolysis, breathing through the nose is recommended, because ozone removal in the upper respiratory tract is lower for oral than for nasal inhalation.

5. Conclusions

The following conclusions regarding the use of low-cost portable air cleaner for removing bacterial and fungal aerosols can be drawn based on laboratory tests reported here.

To summarize our results, we demonstrated the purification of air by a low-cost ozone generator. However, it is highly effective but requires an ozone reduction solution, e.g., carbon filter, as well as periodical operation to ensure a safe environment for the users.

Generally, there is a belief among users that if something is easy to buy, it is checked and safe for the user. Unfortunately, many easily accessible technologies can be harmful to users. Hence, further studies are necessary to compare different low-cost air filtration and purification products to develop meaningful outcomes regarding their impact on human health.

Following Hashimoto and Kawakami [24], we have to underline the limitations of our study. The experiments were performed in a laboratory room to minimize fluctuations in the number of airborne microbes due to external factors. Houses are less airtight and contain many obstacles; however, the amount of in-house microbe generation (CFU/h) and ventilation (m^3/h), which affect the microbial concentration, differ among houses.

Although our ozone generator is dedicated to killing bacteria and fungi indoors, a new global epidemic of coronavirus requires consideration of whether this device can also be used to destroy coronavirus. The new coronavirus is considered to be an "enveloped virus". Enveloped viruses are usually more sensitive to physicochemical challenges than naked viruses (without an envelope). In addition, ozone has been shown to kill the SARS coronavirus. Since the structure of the new 2019-nCoV coronavirus is almost identical, we believe that it will also work on the new coronavirus. It is also known that ozone destroys this type of virus, breaking through the outer shell to the core, causing damage to the viral RNA. Ozone can also damage the outer layer of the virus in a process called oxidation. However, this problem requires further investigation.

Author Contributions: Conceptualization A.M., W.M. and J.S.P.; methodology A.M. and W.M.; formal analysis E.B.; investigation A.J.; data curation W.M.; writing—original draft preparation A.M.; writing—review and editing A.M. and J.S.P.; visualization A.M. and W.M.; supervision, A.M. and W.M. All authors have read and agreed to the published version of the manuscript.

Funding: This research was funded by the Faculty of Energy and Environmental Engineering, Silesian University of Technology, Poland (statutory research).

Acknowledgments: The authors would like to thank Jan Kaczmarczyk and Monika Blaszczok for technical support (sharing the portable indoor air quality monitor Aeroqual—series 500 for O_3 decay rate measurements).

Conflicts of Interest: The authors declare no conflict of interest.

References

1. Jiayu, C.; Qiaoqiao, R.; Feilong, C.; Chen, L.; Jiguo, W.; Zhendong, W.; Lingyun, C.; Liu, R.; Guoxia, Z. Microbiology Community Structure in Bioaerosols and the Respiratory Diseases. *J. Environ. Sci. Public Health* **2019**, *3*, 347–357. [CrossRef]

2. Brągoszewska, E.; Biedroń, I.; Hryb, W. Microbiological air quality and drug resistance in airborne bacteria isolated from a waste sorting plant located in poland—A case study. *Microorganisms* **2020**, *8*, 202. [CrossRef] [PubMed]
3. Wéry, N. Bioaerosols from composting facilities-a review. *Front. Cell. Infect. Microbiol.* **2014**, *4*, 42. [PubMed]
4. Pearson, C.; Littlewood, E.; Douglas, P.; Robertson, S.; Gant, T.W.; Hansell, A.L. Exposures and health outcomes in relation to bioaerosol emissions from composting facilities: A systematic review of occupational and community studies. *J. Toxicol. Environ. Health B Crit. Rev.* **2015**, *18*, 43–69. [CrossRef] [PubMed]
5. Fung, F.; Hughson, W.G. Health Effects of Indoor Fungal Bioaerosol Exposure. *Appl. Occup. Environ. Hyg.* **2003**, *18*, 535–544. [CrossRef]
6. Douwes, J.; Thorne, P.; Pearce, N.; Heederik, D. Bioaerosol health effects and exposure assessment: Progress and prospects. *Ann. Occup. Hyg.* **2003**, *47*, 187–200. [PubMed]
7. Kim, K.H.; Kabir, E.; Jahan, S.A. Airborne bioaerosols and their impact on human health. *J. Environ. Sci. China* **2018**, *67*, 23–35. [CrossRef]
8. Górny, R.L. Microbial Aerosols: Sources, Properties, Health Effects, Exposure Assessment—A Review. *KONA Powder Part. J.* **2020**, *37*, 264–285. [CrossRef]
9. Gunschera, J.; Markewitz, D.; Bansen, B.; Salthammer, T.; Ding, H. Portable photocatalytic air cleaners: Efficiencies and by-product generation. *Environ. Sci. Pollut. Res.* **2016**, *23*, 7482–7493. [CrossRef]
10. Shaughnessy, R.J.; Sextro, R.G. What is an effective portable air cleaning device? A review. *J. Occup. Environ. Hyg.* **2006**, *3*, 169–181. [CrossRef]
11. Grinshpun, S.A.; Mainelis, G.; Trunov, M.; Adhikari, A.; Reponen, T.; Willeke, K. Evaluation of ionic air purifiers for reducing aerosol exposure in confined indoor spaces. *Indoor Air* **2005**, *15*, 235–245. [CrossRef] [PubMed]
12. Salonen, H.; Salthammer, T.; Morawska, L. Human exposure to ozone in school and office indoor environments. *Environ. Int.* **2018**, *119*, 503–514. [CrossRef] [PubMed]
13. Siegel, J.A. Primary and secondary consequences of indoor air cleaners. *Indoor Air* **2016**, *26*, 88–96. [CrossRef] [PubMed]
14. Hubbard, H.F.; Coleman, B.K.; Sarwar, G.; Corsi, R.L. Effects of an ozone-generating air purifier on indoor secondary particles in three residential dwellings. *Indoor Air* **2005**, *15*, 432–444. [CrossRef] [PubMed]
15. Cestonaro, L.V.; Marcolan, A.M.; Rossato-Grando, L.G.; Anzolin, A.P.; Goethel, G.; Vilani, A.; Garcia, S.C.; Bertol, C.D. Ozone generated by air purifier in low concentrations: Friend or foe? *Environ. Sci. Pollut. Res.* **2017**, *24*, 22673–22678. [CrossRef] [PubMed]
16. Boeniger, M.F. Use of ozone generating devices to improve indoor air quality. *Am. Ind. Hyg. Assoc. J.* **1995**, *56*, 590–598. [CrossRef]
17. Cheng, K.C.; Park, H.K.; Tetteh, A.O.; Zheng, D.; Ouellette, N.T.; Nadeau, K.C.; Hildemann, L.M. Mixing and sink effects of air purifiers on indoor PM2.5 concentrations: A pilot study of eight residential homes in Fresno, California. *Aerosol Sci. Technol.* **2016**, *50*, 835–845. [CrossRef]
18. Zhang, Y.; Mo, J.; Li, Y.; Sundell, J.; Wargocki, P.; Zhang, J.; Little, J.C.; Corsi, R.; Deng, Q.; Leung, M.H.K.; et al. Can commonly-used fan-driven air cleaning technologies improve indoor air quality? A literature review. *Atmos. Environ.* **2011**, *45*, 4329–4343. [CrossRef]
19. Custovic, A.; Smith, A.; Green, R.; Woodcock, A. Distribution and aerodynamic characteristics of major cat allergen Fel d 1. *Thorax* **1996**, *51*, 33–38.
20. Gherasim, A.; Jacob, A.; Schoettel, F.; Domis, N.; de Blay, F. Efficacy of air cleaners in asthmatics allergic to cat in ALYATEC® environmental exposure chamber. *Clin. Exp. Allergy* **2020**, *50*, 160–169. [CrossRef]
21. Punsmann, S.; Van Der Graaf, T.; Zahradnik, E.; Czibor, C.; Raulf, M.; Sander, I.; Punsmann, S.; Zahradnik, E.; Czibor, C.; Raulf, M.; et al. Effectiveness of a portable air filtration device in reducing allergen exposure during household chores. *Allergo J. Int.* **2019**, *28*, 299–307. [CrossRef]
22. Polidori, A.; Fine, P.M.; White, V.; Kwon, P.S. Pilot study of high-performance air filtration for classroom applications. *Indoor Air* **2013**, *23*, 185–195. [CrossRef] [PubMed]
23. Scheepers, P.T.J.; De Hartog, J.J.; Reijnaerts, J.; Beckmann, G.; Anzion, R.; Poels, K.; Godderis, L. Influence of combined dust reducing carpet and compact air filtration unit on the indoor air quality of a classroom. *Environ. Sci. Process. Impacts* **2015**, *17*, 316–325. [CrossRef]
24. Hashimoto, K.; Kawakami, Y. Effectiveness of Airborne Fungi Removal by using a HEPA Air Purifier Fan in Houses. *Biocontrol Sci.* **2018**, *23*, 215–221. [CrossRef] [PubMed]

25. Vyas, S.; Srivastav, N.; Spears, D. An experiment with air purifiers in delhi during winter 2015–2016. *PLoS ONE* **2016**, *11*, 1–20. [CrossRef] [PubMed]
26. How to Buy the Best Air Purifier. CHOICE. Available online: https://www.choice.com.au/home-and-living/cooling/air-purifiers/buying-guides/air-purifiers (accessed on 5 September 2019).
27. Britigan, N.; Alshawa, A.; Nizkorodov, S.A. Quantification of ozone levels in indoor environments generated by ionization and ozonolysis air purifiers. *J. Air Waste Manag. Assoc.* **2006**, *56*, 601–610. [CrossRef] [PubMed]
28. Nagda, N.L. *Air Quality and Comfort in Airliner Cabins*; ASTM: Chelsea, MI, USA, 2000; ISBN 0803128665. Available online: https://books.google.pl/books?id=41mdymdg0cwC&pg=PA208&lpg=PA208&dq=28.+Nagda,+N.L.+Air+Quality+and+Comfort+in+Airliner+Cabins;+ASTM&source=bl&ots=YubnoN5U7k&sig=ACfU3U0VONopy27atgomJMihPXBdJXu_Cg&hl=pl&sa=X&ved=2ahUKEwiWq7mOjtnpAhUrBhAIHd1WCZMQ6AEwAHoECAcQAQ#v=onepage&q=28.%20Nagda%2C%20N.L.%20Air%20Quality%20and%20Comfort%20in%20Airliner%20Cabins%3B%20ASTM&f=false (accessed on 30 August 2019).
29. Phillips, T.J.; Bloudoff, D.P.; Jenkins, P.L.; Stroud, K.R. Ozone emissions from a "personal air purifier". *J. Expo. Anal. Environ. Epidemiol.* **1999**, *9*, 594–601. [CrossRef]
30. Kowalska, M.; Zajusz-Zubek, E. Occupational exposure to ozone in workers using photocopiers and printers. *Medycyna Pracy* **2010**, *61*, 549–551. (In Polish)
31. WHO. *Air Quality Guidelines for Particulate Matter, Ozone, Nitrogen Dioxide and Sulfur Dioxide*; WHO: Geneva, Switzerland, 2006.
32. Dz.U.1286 Regulation of the Minister of Family, Labor and Social Policy regarding the highest permissible concentrations and intensities of harmful factors in the work environment. *Pol. J. Laws* **2018**. Available online: http://www.ilo.org/dyn/natlex/natlex4.detail?p_lang=en&p_isn=99664 (accessed on 26 August 2019). (In Polish).
33. Health Canada Residential Indoor Air Quality Guideline: Ozone. Available online: www.canada.ca/content/dam/canada/health-canada/migration/healthy-canadians/publications/healthy-living-vie-saine/ozone/alt/ozone-eng.pdf (accessed on 31 October 2019).
34. Penney, D.; Benignus, V.; Kephalopoulos, S.; Kotzias, D.; Kleinman, M. Agnes Verrier Guidelines for indoor air quality. *WHO Guidel.* **2010**, *9*, 454.
35. Shaughnessy, R.J.; Levetin, E.; Blocker, J.; Sublette, K.L. Effectiveness of Portable Indoor Air Cleaners: Sensory Testing Results. *Indoor Air* **1994**, *4*, 179–188. [CrossRef]
36. Huang, R.; Agranovski, I.; Pyankov, O.; Grinshpun, S. Removal of viable bioaerosol particles with a low-efficiency HVAC filter enhanced by continuous emission of unipolar air ions. *Indoor Air* **2008**, *18*, 106–112. [CrossRef] [PubMed]
37. Lee, J.H.; Kim, J.Y.; Cho, B.B.; Anusha, J.R.; Sim, J.Y.; Raj, C.J.; Yu, K.H. Assessment of air purifier on efficient removal of airborne bacteria, Staphylococcus epidermidis, using single-chamber method. *Environ. Monit. Assess.* **2019**, *191*, 1–10. [CrossRef] [PubMed]
38. Bischoff, W.; Russell, G.; Willard, E.; Stehle, J. Impact of a novel mobile high-efficiency particulate air-ultraviolet air recirculation system on the bacterial air burden during routine care. *Am. J. Infect. Control* **2019**, *47*, 1025–1027. [CrossRef] [PubMed]
39. Alexis, N.E.; Becker, S.; Bromberg, P.A.; Devlin, R.; Peden, D.B. Circulating CD11b expression correlates with the neutrophil response and airway mCD14 expression is enhanced following ozone exposure in humans. *Clin. Immunol.* **2004**, *111*, 126–131. [CrossRef] [PubMed]
40. Novoselac, A.; Siegel, J.A. Impact of placement of portable air cleaning devices in multizone residential environments. *Build. Environ.* **2009**, *44*, 2348–2356. [CrossRef]
41. Pichat, P. A brief survey of the practicality of using photocatalysis to purify the ambient air (indoors or outdoors) or air effluents. *Appl. Catal. B Environ.* **2019**, *245*, 770–776. [CrossRef]
42. Travers, M.J.; Kulak, J.A.; Vogl, L. Waterpipe cafés are hazardous to your health: Determination of a waterpipe specific calibration factor. *Int. J. Hyg. Environ. Health* **2018**, *221*, 48–53. [CrossRef]
43. Mahyuddin, N.; Awbi, H.B. A Review of CO_2 measurement procedures in ventilation research. *Int. J. Vent.* **2012**, *10*, 353–370. [CrossRef]
44. PN-Z-04007-2. *Protection of Air Purity—Ozone Content Tests—Determination of Ozone in the Presence of Nitrogen Oxides at Workplaces Using Spectrophotometric Method in Visible Light*; Polish Committee for Standarization: Warsaw, Poland, 1994. (In Polish)

45. Liu, G.; Xiao, M.; Zhang, X.; Gal, C.; Chen, X.; Liu, L.; Pan, S.; Wu, J.; Tang, L.; Clements-Croome, D. A review of air filtration technologies for sustainable and healthy building ventilation. *Sustain. Cities Soc.* **2017**, *32*, 375–396. [CrossRef]
46. Guo, C.; Gao, Z.; Shen, J. Emission rates of indoor ozone emission devices: A literature review. *Build. Environ.* **2019**, *158*, 302–318. [CrossRef]
47. Brągoszewska, E.; Mainka, A.; Pastuszka, J.S. Bacterial aerosols in an urban nursery school in Gliwice, Poland: A case study. *Aerobiologia (Bologna)* **2016**, *32*, 469–480.
48. Brągoszewska, E.; Mainka, A.; Pastuszka, J. Bacterial and Fungal Aerosols in Rural Nursery Schools in Southern Poland. *Atmosphere (Basel)* **2016**, *7*, 142. [CrossRef]
49. Pastuszka, J.S.; Mucha, W.; Wlazło, A.; Lis, D.; Marchwińska-Wyrwał, E.; Mainka, A. The study of the sterilization of the indoor air in hospital/clinic rooms by using the electron wind generator. *Int. J. Environ. Res. Public Health* **2019**, *16*, 4935. [CrossRef] [PubMed]
50. Aeroqual Limited, Aeroqual Aeroqual Series 200, 300 & 500 User Guide MRK-D-0022V3.1. Auckland New Zeland. Available online: https://www.aeroqual.com/wp-content/uploads/Series-200-300-500-Portable-Monitor-User-Guide-11-14.pdf (accessed on 5 September 2019).
51. Thanomsub, B.; Anupunpisit, V.; Chanphetch, S.; Watcharachaipong, T.; Poonkhum, R.; Srisukonth, C. Effects of ozone treatment on cell growth and ultrastructural changes in bacteria. *J. Gen. Appl. Microbiol.* **2002**, *48*, 193–199. [CrossRef] [PubMed]
52. Tang, J.W. The effect of environmental parameters on the survival of airborne infectious agents. *J. R. Soc. Interface* **2009**, *6*, S737–S746. [CrossRef] [PubMed]
53. Lin, W.H.; Li, C.S. Associations of fungal aerosols, air pollutants, and meteorological factors. *Aerosol Sci. Technol.* **2000**, *32*, 359–368. [CrossRef]
54. Moore, G.; Griffith, C.; Peters, A. Bactericidal properties of ozone and its potential application as a terminal disinfectant. *J. Food Prot.* **2000**, *63*, 1100–1106. [CrossRef]
55. Sabersky, R.H.; Sinema, D.A.; Shair, F.H. Concentrations, decay rates, and removal of ozone and their relation to establishing clean indoor air. *Environ. Sci. Technol.* **1973**, *7*, 347–353. [CrossRef]
56. Mueller, F.X.; Loeb, L.; Mapes, W.H. Decomposition rates of ozone in living areas. *Environ. Sci. Technol.* **1973**, *7*, 342–346. [CrossRef]
57. Shair, F.H.; Heitner, K.L. Theoretical model for relating indoor pollutant concentrations to those outside. *Environ. Sci. Technol.* **1974**, *8*, 444–451. [CrossRef]
58. Niu, J.; Thung, T.C.W.; Burnett, J. Ozone emission rate testing and ranking method using environmental chamber. *Atmos. Environ.* **2001**, *35*, 2143–2151. [CrossRef]
59. Thiele, J.E.N.S.J.T.; Raber, M.A.G.T.; Sang, K.E.T.; Ross, C.A.E.C.; Acker, L.E.P. In vivo exposure to ozone depletes vitamins c and e and induces lipid peroxidation in epidermal layers of murine skin. *Free Radic. Biol. Med.* **1997**, *23*, 385–391. [CrossRef]
60. Clausen, P.A.; Wilkins, C.K.; Wolkoff, P.; Damga, G. Chemical and biological evaluation of a reaction mixture of R-(+) -limonene/ozone Formation of strong airway irritants. *Environ. Int.* **2001**, *26*, 511–522. [CrossRef]

© 2020 by the authors. Licensee MDPI, Basel, Switzerland. This article is an open access article distributed under the terms and conditions of the Creative Commons Attribution (CC BY) license (http://creativecommons.org/licenses/by/4.0/).

buildings

MDPI

Article

Modeling of an Aerogel-Based "Thermal Break" for Super-Insulated Window Frames

Alessandro Cannavale [1,2,*], Francesco Martellotta [1], Umberto Berardi [3], Chiara Rubino [1], Stefania Liuzzi [1], Vincenzo De Carlo [4] and Ubaldo Ayr [1]

1 Department of Civil Engineering and Architecture (DICAR), Technical University of Bari, via Orabona 4, 70125 Bari, Italy; francesco.martellotta@poliba.it (F.M.); chiara.rubino@poliba.it (C.R.); stefania.liuzzi@poliba.it (S.L.); Ubaldo.Ayr@poliba.it (U.A.)
2 Istituto di Nanotecnologia, CNR Nanotec, Via Arnesano 16, 73100 Lecce, Italy
3 Department of Architectural Science, Ryerson University, 350 Victoria St., Toronto, ON M5B 2K3, Canada; uberardi@ryerson.ca
4 Dcs Group S.R.L., Via per Castellaneta Z. Ind. S. Basilio, 74017 Mottola (Ta), Italy; vincenzo.decarlo@decarlo.it
* Correspondence: alessandro.cannavale@poliba.it; Tel.: +39-080-5963718

Received: 7 February 2020; Accepted: 12 March 2020; Published: 18 March 2020

Abstract: Research activities in the field of innovative fixtures are continuously aiming at increasing their thermal and optical performances to offer optimal exploitation of daylight and solar gains, providing effective climate screen, according to increasing standards for indoor comfort and energy saving. Within this work, we designed an innovative aerogel-based "thermal break" for window frames, so as to consistently reduce the frame conductance. Then, we compared the performance of this new frame both with currently used and obsolete frames, present in most of the existing building stock. Energy savings for heating and cooling were assessed for different locations and confirmed the potential role played by super-insulating materials in fixtures for extremely rigid climates.

Keywords: window frame; granular aerogel; energy saving

---

1. Introduction

Awareness of the anthropogenic effects on the environment dramatically has increased in recent years, especially in terms of global warming and uncontrolled greenhouse gas production. For this reason, the control of energy consumption of buildings is a pivotal challenge. The latest European Directive 2018/844 reports that the building stock contributes to 36% of greenhouse gas emissions, considering that almost 50% of the Union's final energy consumption is used for heating and cooling, and 80% of this amount is employed in buildings [1]. The 2015 Paris Agreement on climate change, following the United Nations Conference on Climate Change (COP21), required the subscribing States to reduce carbon emissions in the building stock. To do this, in the EU and worldwide, the priority is to enhance energy efficiency by deploying low-cost renewable energies and innovative technologies, especially deriving from recent achievements in the field of nanomaterials research, with reference to building integration of novel technologies, spanning from chromogenics [2] to semi-transparent photovoltaics [3,4], super-insulating materials [5,6], and phase change materials [7,8]. One of the possible approaches, especially in countries with a more rigid climate, is to find new devices to increase the standards of insulation of buildings. This challenge can be taken up by some materials, defined as super-insulators [5], characterized by a relatively lower thermal conductivity compared with the typical insulating materials used in building envelopes. The main evaluation parameters in the choice of an innovative insulating material are undoubtedly the thermal conductivity, but also volume and cost [9–11]. For these reasons, interest in a nanostructured porous material called aerogel has been growing in recent years [12,13]. The name itself, aerogel, indicates a precise property of the material, namely, the substitution of the liquid component of the gel with a gas. The gel is generally made by means of low temperature sol–gel technique.

1.1. Aerogel for Super-Insulation

The synthesis of the aerogel involves the hydrolysis and condensation at room temperature of a solution, starting from an alkoxide precursor (tetraethyl-orthosilicate, tetramethyl orthosilicate, or poly-ethoxydisiloxane are the most used), in the presence of a convenient catalyst [5]. A continuous three-dimensional lattice forms inside the liquid. After the aging phase of the gel and the subsequent removal of water from the inside of the pores with solvent washes, the material is ready for the drying process. Although there are different ways to proceed with the drying of the silica gel, the most widespread is still the so-called supercritical drying method. The substitution gives rise to a structure made of silica gel containing pores with a diameter of variable size (between 5 and 70 nm) and a density between 70 and 150 kg/m^3 [14]. For this reason, aerogel can be considered an open-celled, mesoporous solid foam containing a network of nanostructures.

The resulting material is characterized by a thermal conductivity that can be very low (0.013 W/m·K) [15], in which the nanopores occupy more than 85% of the volume; the high porosity allows the obtaining of interesting properties from the optical, thermal, and physical point of view [16]. The very low thermal conductivity of the aerogel is due to the very small fraction of solid silica: the structure skeleton contains a large number of free ends that "complicate" the path of the thermal flow. The size of pores (~20 nm) hinders thermal conduction, which is inversely proportional to the diameter of the pores themselves. Moreover, the dimension of the pores prevents the Brownian motion of the air molecules and therefore prevents convective heat exchange [17]. As stated by Baetens [11], if one would be able to find a cheaper manufacturing process, aerogel might become a real alternative to existing building insulating materials.

1.2. Aerogel Products for Constructions

Currently, aerogel products are available for construction as monolithic panels with high transparency [18], small size granules [19,20], or thermal insulation blankets [21,22]. Each of these products can be employed to improve the thermal performance of mortars and concretes; plasters for opaque systems [16,23]; or translucent/transparent systems, such as insulating glazed units, suitably filled with monolithic or granular aerogels [10,24,25]. In terms of potential aerogel applications, windows represent the best candidates because they are a thermal "weak point" in the building envelope, due to their relatively higher global heat exchange coefficient compared to opaque surfaces.

1.3. Aerogel Insulation for Windows

Aerogel can be used in windows in various ways, as reported in the literature [26]. If aerogel were used to fill air gaps between glass panes, the consequent benefits would be represented by lower solar heat gain coefficient, as glazing containing aerogel is translucent and lower values of glazing thermal transmittance (U_g) would be obtained. Double- or triple-glazed units filled with aerogel—as granules or in a monolithic form—will output a diffuse light, which is sometimes desirable to achieve comfort in large commercial or office spaces, airports, museums, etc. However, it is in terms of reduction in energy consumption that aerogel may make a difference compared to conventional materials. It has been proven that granular aerogel may reduce the heat loss in office buildings [27] by 80% or cooling load in humid subtropical climates by 4% [28]. A study by Gao et al. also reported a yearly saving as high as 21% in a building's energy consumption [29]. Moretti and Buratti compared monolithic aerogel glass panes to conventional glazing systems, reporting a 55% reduction in heat losses, whereas granular aerogel windows showed a 25% reduction in heat losses. In the latter case, a 66% reduction in visible transmittance was observed [30,31]. A recent work investigated the effect of inserting an aerogel blanket in the air chambers of a PVC window frame, leading to a reduction in the frame U-value from

1.06 to 0.92 W/m^2·K [32]. In the present work, the energy benefits achievable by using granular aerogel inside the fixed and mobile frames of a high thermal performance window were studied. The window frame was made of wood and aluminium and an interposed aerogel thermal break was also used. A finite element method analysis of a window frame containing aerogel was carried out, obtaining a high performance level. This innovative window was compared with others, embodying different frames, as explained in the Methodology section, by means of simulations in test rooms within the Energyplus software platform [33]. Energy consumption for heating and cooling was tested within a typical office on all possible exposures, in two locations with very different climates: the Canadian city of Toronto and Bari, in Southern Italy.

2. Materials and Methods

2.1. Frame and Window Description

This innovative window (Figure 1), specifically designed to meet the highest thermal insulation standards, includes a three-pane glazing. The internal glazing embodied a low emittance coating (ε = 0.21) deposited on face 5, i.e., the outer face of the internal pane. The aerogel section (the red part in Figure 2) was conveniently surrounded by a thin acrylonitrile–butadienestyrene (ABS) skin (2 mm thick), and embodied granular aerogel within the frame section, thus behaving like a super-insulating "thermal break", between the external aluminium frame and the internal wooden frame. The same ABS skin is generally adopted in existing commercial frames using the same technology, to embody expanded polystyrene (Isover EPS 035) insulating material. Cabot Enova IC3100 granular aerogel was used, with granule size spanning between 2 and 40 µm, a thermal conductivity of 0.012 W/m·K at ambient temperature, and very low density, ranging from 120 to 150 kg/m^3 according to technical specifications of the material. The coupling between the aerogel section and the wooden frame was made by using a stripe of aerogel blanket, Pyrogel-XTE by Aspen, with comparable thermal conductivity (less than 0,02 W/m·K) suitably glued and mechanically fixed to the frame.

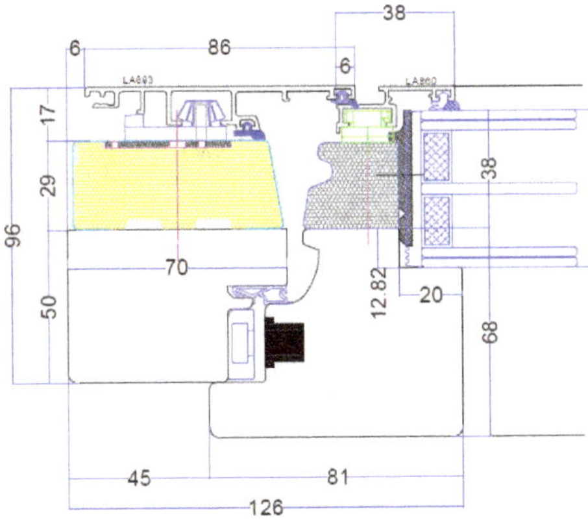

Figure 1. Cross section of the frame reporting dimensional data of the reference window frame, corresponding to a window designed in compliance with regulations. Above, the external aluminum section is clearly visible; in the intermediate part, the area devoted to thermal insulation (in yellow); below, the wooden profile, facing the indoor space.

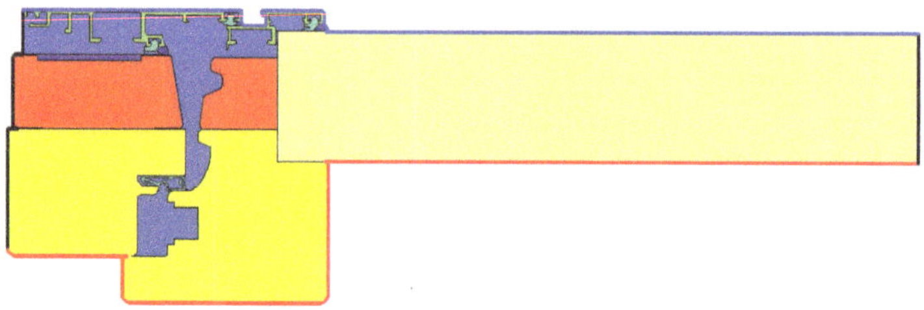

Figure 2. Cross section of the frame, as designed in the finite element method software Flixo, used for the simulation. In this case, each color represents a material, modeled with different thermal properties. The yellow section on the left represents soft wood; the light yellow area represents the replacement panel with given thermal resistance, replacing the triple glass in the numerical analysis; the aerogel is colored in red; and the aluminium is in light green. A different shade of green was used for EPDM and light blue for air within the frame.

2.2. Finite Element Method

The thermal heat loss due to the transmission through windows is strongly affected by the technical features of their transparent (glazing) and opaque (frames and spacers) components. Such figure of merit can be determined according to the standard ISO 10077-2:2017, "Thermal performance of windows, doors and shutters—Part 2: Numerical method of frames" [34]. To calculate the thermal transmittance for the two-dimensional window section, including the frame and the glazing, according to the standard, the thermal transmittance of the frame section (U_f) in W/m^2·K must be calculated according to the following formula, remembering that in the computational model, glazing must be replaced with a panel with given thermal insulating properties,

$$U_f = \frac{L_f^{2D} - U_p b_p}{b_f} \tag{1}$$

where U_p is the thermal transmittance of the replacement panel, in W/m^2·K, which is assumed to have a thermal conductivity of 0.035 W/m·K, according to EN ISO 10077-2; b_f is the width of the frame section in m; and b_p is the visible width of the panel, in m. L_f^{2D} is the thermal conductance of the section in W/m·K, calculated from the total heat flow rate per unit length through the section divided by the temperature difference between both adjacent environments. The value of U_f, according to current standards, was obtained by using the commercial software Flixo, allowing CAD-based input of constructions and to import DXF-files and materials databases, to make a detailed two-dimensional finite elements analysis of the frame, in steady-state conditions (Figure 2).

Afterwards, the global thermal exchange coefficient of the whole window (U_w) was calculated according to the equation reported in the standard ISO 10077-1:2017, "Thermal performance of windows, doors, and shutters—Calculation of thermal transmittance—Part 1: General" [34]:

$$U_w = \frac{A_g U_g + A_f U_f + \Psi_g l_g}{A_g + A_f} \tag{2}$$

where U_g is the thermal transmittance of the glass section (W/m^2·K); A_g and A_f are the surface areas of glazing and frame, respectively (m^2); Ψ_g is the linear thermal transmittance of the glazing joint (W/m·K); and l_g is its length (m), calculated according to ISO 10077-2.

To highlight the effect of the aerogel material, further FEM simulations were made to compare the thermal performance of the aerogel-based frame with those observed in a similar frame, embodying expanded polystyrene, instead of granular aerogel. Other materials used in this FEM analysis (Figure 2) were aluminium (λ = 160 W/m·K), ethylene propylene diene monomer (EPDM, with λ = 0.25 W/m·K), soft wood (λ = 0.12 W/m·K), Cabot 3120 granular aerogel (λ = 0.012 W/m·K), and expanded polystyrene (λ = 0.035 W/m·K).

The innovative window has been fabricated, and some images of the designed prototype are shown in Figure 3. Experimental campaigns will subsequently be carried out in order to assess the energy saving benefits obtainable with this type of window frame, in the context of a real building.

Figure 3. Aerogel-enhanced window (Aeroshield) embodying a triple-glazed unit, with external aluminium frame and internal wooden frame (**a**); Reference frame with expanded polystyrene insulation, in black (b); Aerogel-insulated frame with polymer skin (**c**).

2.3. *EnergyPlus Model*

To analyze the energy performance implications, the aerogel-based window was compared to different window configurations. The first one, identified as "Standard compliant scenario", was made of an aluminium frame with the so-called "thermal break", i.e., a continuous polymer barrier between internal and external parts of window frames, preventing conductive thermal energy loss and a triple-glazed unit with the same properties of the newly designed window, so as to assess performance benefits strictly deriving from the use of aerogel in frames. The second kind of window considered was made of an obsolete double-glazed unit and a simple aluminium frame, with no thermal break inside. The latter window was considered to envisage results in a "Refurbishment scenario", a useful comparison with typical window performance in the existing building stock.

A 3D geometrical model of the case study was first designed in SketchUp, using the OpenStudio plugin. Afterwards, it was imported in EnergyPlus v. 8.9, a free simulation tool by the U.S. Department of Energy's Building Technology Office, capable of performing dynamic simulations, providing detailed energy analysis.

An office test room, to be considered part of a multi-storey building, was consequently modeled. The floor surface was 20 m^2 and the internal height was 3.5 m. The gross surface of the glazed envelope was 7.5 m^2 and it was located on one of the vertical walls, which was assumed as the only thermally exchanging surface (made of wood cladding/thermal insulation/gypsum plaster with U-Factor = 0.35 W/m^2·K).

Glazing and frame properties for all the windows considered in this study were reported in Table 1. The window-to-wall ratio (WWR) was 60%, whereas the surface ratio between frame and glazed area of the window was 41%. Lateral walls (20 cm thick walls) and horizontal surfaces (floor and ceiling, 30 cm thick, U-Factor = 0.819 W/m^2·K) were shared with other offices, and thus considered adiabatic in their external surface. All possible exposures were considered for the windowed wall.

Table 1. Glazing, frame, and dividers properties. All the cavities between glass panes were considered filled with argon gas. Conductance was calculated excluding surface convection resistance.

Window Type	Glazing Structure	Glass U-Factor (W/m^2·K)	Frame and Divider Conductance (W/m^2·K)	Low-e Coating	Frame Materials
Refurbishment scenario	4/16/4	2.60	7.57	-	Aluminium without thermal break
Standard Compliant scenario	4/16/4/16/4	1.08	3.5	Face 5	Aluminium with thermal break

To better point out the role of the window (and its frame), the envelope properties were considered the same in Bari and Toronto, although their respective climates are significantly different. For "Refurbishment scenario" and "Standard compliant scenario", typical values of window frame conductance were considered, as specified in Table 1. A window with a triple glass pane and a low performing frame was not taken into account, in this work, for two main reasons: first of all, we aimed at excluding the effect of glazing when considering the comparison between the standard compliant window and the one equipped with the innovative aerogel. In fact, glazing might have affected simulations much more than frames, due to the larger surface area involved in heat transfer, compared to frames; second, a window with three panes would be technically incompatible with a thin frame for structural and dimensional reasons.

Two locations were taken into account: Bari (Southern Italy) and Toronto (Canada). Their climate specifications are provided in Table 2, showing the significant differences in terms of "extreme" winter conditions. Figure 4 reports yearly solar radiation available on a vertical plane, in Bari and Toronto. Graphs also show that total radiation is similar in the two locations, due to the similar latitude. This confirms that the radiative contribution does not represent an element of inhomogeneity for the analyses reported hereafter. Consequently, main differences in terms of heat transfer are due to frame conductance, rather than sun irradiance.

Table 2. Climate characteristics of the two locations considered in this work.

City	Latitude [°]	Koppen-Geiger Climate Class	Average Temperature [°C]	Winter Average Temperature [°C]	Winter Average Daily Minimum Temperature [°C]	Winter Average Daily Maximum Temperature [°C]
Toronto	43.70	Dfb	9.85	−3.00	−6.53	2.60
Bari	41.11	Csa	16.10	9.87	6.27	13.57

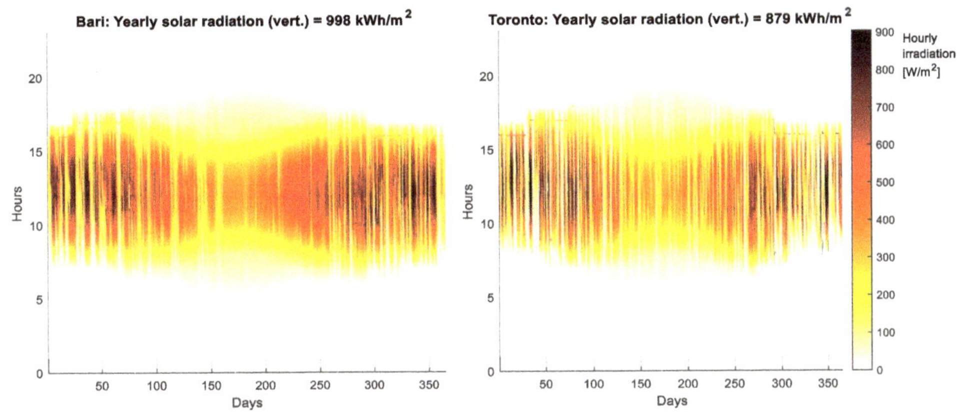

Figure 4. Solar radiation available on vertical planes, in Bari and Toronto.

The heating period was considered according to regulations in force. According to the Italian regulations, Bari falls in a climatic zone in which heating systems work from November 15th to March 31st, whereas in Toronto, heating works from October 1st to May 15th. The cooling period was considered to extend from July 1st to September 30rd both in Bari and Toronto, with average temperatures of 23.1 °C and 19.5 °C, respectively. In the proposed model, EnergyPlus provides heating and cooling energy required to meet the temperature at a set point of 20.5 °C in heating mode and 26 °C in cooling mode. To calculate energy use for heating and cooling with a simplified method an "IdealLoadAirSystem" was used, which gives the thermal energy strictly necessary to achieve the given set-point temperature. Zone ventilation was assumed to be 0.0025 m^3/s, according to ASHRAE 62.1/2013, Table 6.2.2.1. Infiltrations were considered 0.0003 $m^3/s{\cdot}m^2$ of exterior surface area, and an "always on" schedule was applied. Internal gains due to equipment were taken into account to the extent of 5 W/m^2, whereas artificial lighting was considered as 10.66 W/m^2.

3. Results

3.1. FEM Model

Numerical analysis, by means of FEM modeling, was made to investigate the thermal figures of merit of a frame containing granular aerogel, acting as a "thermal break" within an aluminium/wood frame. To obtain intelligible information, in this case, we compared a highly performing frame, containing expanded polystyrene insulator with the new aerogel-insulated window. In this way, ceteris paribus, the only difference in thermal performance were due to the materials properties, rather than surface area, thickness, or distribution. Due to the low thermal conductivity of the aerogel granules, a significant reduction of the U_f was achieved (1.23 $W/m^2{\cdot}K$, using EPS). The value found after the simulation process was 0.66 $W/m^2{\cdot}K$ for the frame and divider conductance. This further comparison demonstrates that one of the most interesting advantages of using aerogel for thermal insulation consists in reducing the thickness of the materials adopted, in the face of significant increases in thermal resistance.

The temperature field of the frame embodying granular aerogel shows the effectiveness of the aerogel "thermal break". In fact, the large number of isothermal lines concentrated within the aerogel "thermal break", in which a thermal variation of about 10 °C is detected in a few centimeters of thickness, demonstrates that the thermal performance of the insulating material used is very good (Figure 5a). This confirms that aerogel can act as an effective thermal barrier at the interface between the two materials that constitute the cross-section of the designed frame (wood and aluminium). As a comparison, the use of the polystyrene insulation (Figure 5b) clearly shows that the thermal variation between the faces and the distribution of the isothermal lines are much less effective.

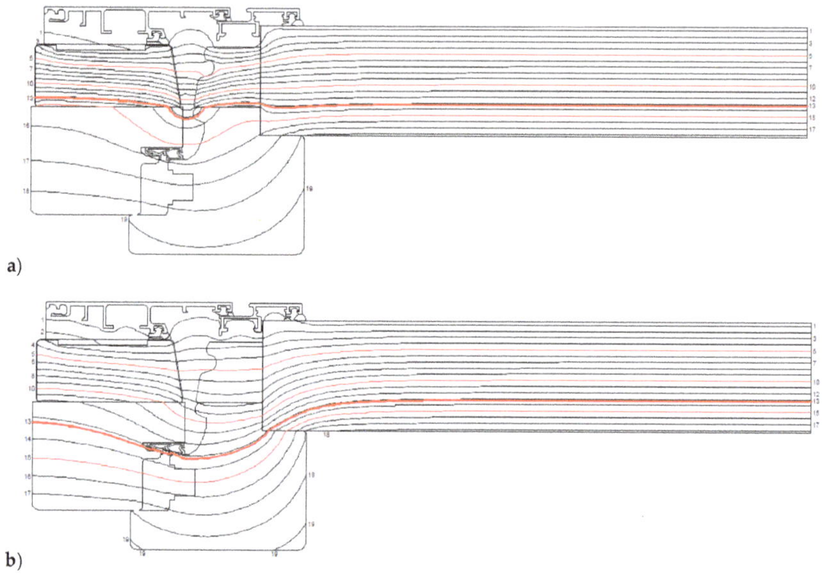

Figure 5. Isothermal lines within the cross section of the newly designed frame with aerogel insulation (**a**) and with expanded polystyrene insulation (**b**). Numbers reported in the figure represent temperatures corresponding to isothermal lines, expressed in °C.

3.2. Energy Consumption

In the subsequent analysis, the energy consumption for heating and cooling was presented in terms of the normalized value per unit floor surface and per year, in order to provide easily comparable results. With reference to heating (Figure 6), in both the cities, maximum heating energy use was observed for the North-facing façade, as expected. The minimum results were found to occur on the Southern façade. On the other hand, intermediate consumptions were observed for the East and West orientations. Predictably, the energy required for heating in winter was higher in the city of Toronto than in Bari. In Toronto, during the heating season, the innovative window, embodying aerogel, allowed an energy saving of 6.9 $kWh/m^2 \cdot yr$ on the South exposure, compared to the Refurbishment scenario. This figure reached ~8 $kWh/m^2 \cdot yr$ on the East and West exposures. The maximum amount of energy saving reached 12 $kWh/m^2 \cdot yr$ on the Northern facade. Nevertheless, the difference in energy saving between the innovative window and the one compliant to regulations was much lower. In fact, the comparison between the newly designed window and the Standard scenario gave the observation that the amount of energy saving strictly due to the superior aerogel thermal performance was ~2 $kWh/m^2 \cdot yr$ on the Northern orientation. The relevant change in thermal conductance, among the frame technologies adopted in this study, affects the output of dynamic simulations, generating an offset in energy consumption that may be explained in terms of thermal performance of the window frame, when all the other parameters are kept constant. Such an effect was amplified in Toronto, rather than in Bari, as thermal power is proportional to the temperature difference between the external environment and indoor air.

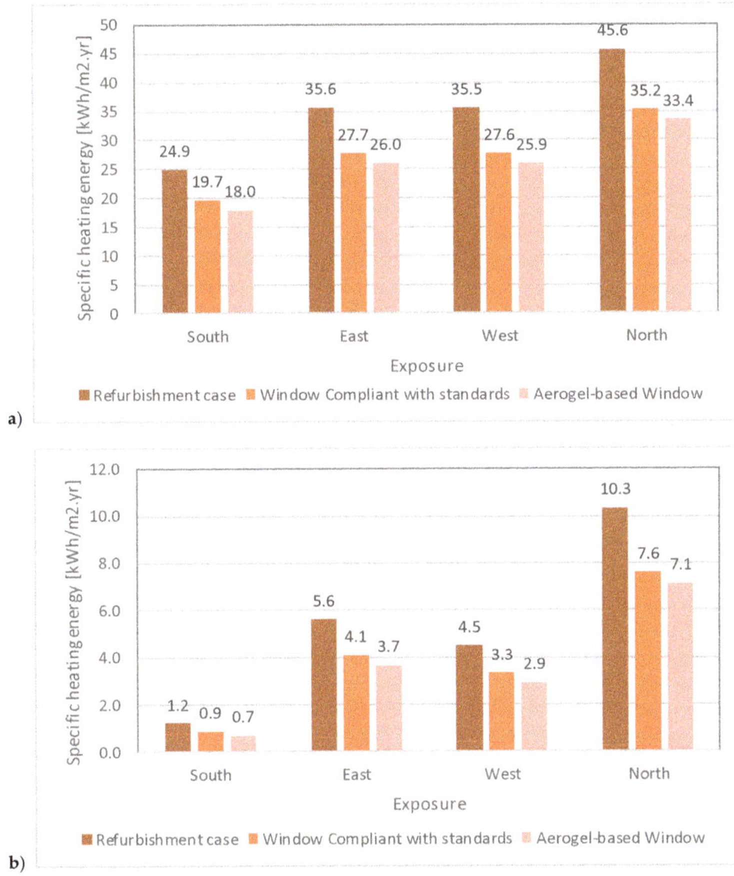

Figure 6. Energy use (kWh/m^2·yr) in Toronto (**a**) and in Bari (**b**), according to exposures and window type.

For this reason, energy saving was lower in Bari, mainly due to its climate conditions, though showing similar tends compared to Toronto: the highest saving during the heating season in Bari was observed assuming the North orientation with reference to the Refurbishment case (3.2 kWh/m^2·yr) and ~2 kWh/m^2·yr for the East and West facades. In Bari, the saving achieved by switching from the compliant window to the aerogel-insulated one was lower than in Toronto (below 1 kWh/m^2·yr); these results are consistent with the considerations exposed above.

The situation was completely reversed between the two locations when the consumption for summer cooling was taken into consideration (Figure 7). In this case, the consumptions of energy for cooling prevailed in Bari, where they were approximately double compared to those observed in Toronto. The façade orientations with higher consumptions were South, East, and West, both in Toronto and Bari. In all cases, the saving achievable by adopting the innovative frame was negligible, whereas the differences observed with respect to the reference window (Refurbishment scenario) were mainly due to the reduction in the incoming radiation through the triple glass with low transmittance. The maximum savings were 4 kWh/m^2·yr in Toronto on the South façade and ~6 kWh/m^2·yr for the same orientation in Bari. In both locations, cooling energy was slightly higher in the aerogel-base window, compared to the one compliant with standards. As the method used in dynamic simulations using the Energyplus platform is deterministic in nature, the observed differences can only be explained in terms of differences in the thermal conductance of the window frame. Consequently, it may be deduced that the lower thermal transmittance of the window containing aerogel, during the summer season, reduces heat exchanges during the night time, thus limiting the beneficial removal of cooling loads associated with sensible thermal power due to equipment, lighting density, persons, and heat transfer through the opaque and transparent envelopes.

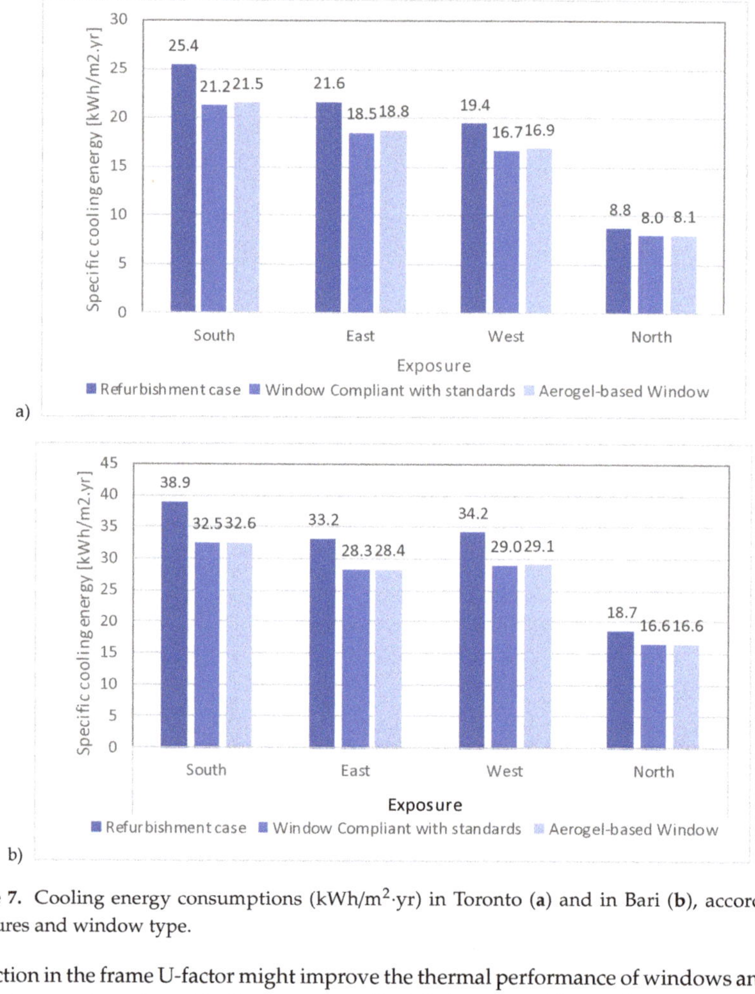

Figure 7. Cooling energy consumptions (kWh/m^2·yr) in Toronto (**a**) and in Bari (**b**), according to exposures and window type.

Reduction in the frame U-factor might improve the thermal performance of windows and influence the indoor temperatures in cold climates, as verified in Toronto. The analyses carried out clearly showed the extent to which the location may influence the achievable energy performance, especially in terms of energy saving. The highest saving (~27%) for heating consumption was attained in Toronto on the North-facing exposure, comparing the performance obtained using the innovative window with those reported for the Refurbishment scenario, and 6% comparing it with a Standard compliant scenario. In Bari, the results were equally high in percentage terms, reaching 27% and 7%, respectively, with reference to the winter season and energy use for heating, although these results were not so significant in absolute terms.

3.3. Frame Temperatures

The hourly surface frame temperature inside the room was reported in Figure 8 on an annual basis, taking into account only the facades exposed to North, as they were those in which the innovative frame showed the best performance. It is clear that during the entire winter season, in Toronto, the strongly insulating properties of the frame of the innovative window might guarantee a rather sharp increase in temperature (up to 5 °C) compared to the other two frame technologies. During the summer season, the frame temperature was lower than in the frames adopted for the other scenarios. In Bari, the performance difference was barely noticeable as well, because the different climatic conditions did not emphasize the difference in thermal performances. Regardless, also in this case, frame temperature was higher in winter and lower in summer, confirming the behavior observed in Toronto. In Bari, the maximum temperature differences observed in January were 4.1 °C and 6.19 °C, respectively, compared to the Standard compliant scenario frame and the Refurbishment scenario. In June, the highest temperature differences were 3.35 °C and 7.44 °C, in the cooling season, taking into account the same reference scenarios as above. As shown in Table 1, the value of the frame conductance is halved when switching from the Refurbishment scenario to the Standard Compliant scenario, but is further reduced by a factor of 5.5, by switching the latter to the innovative, aerogel-based frame. This justifies the different data of the internal temperature in the months from December to April (Figure 8a), in which we take better advantage of the greater level of thermal insulation offered by the best performing frame in the cold seasons, in the climate conditions of Toronto.

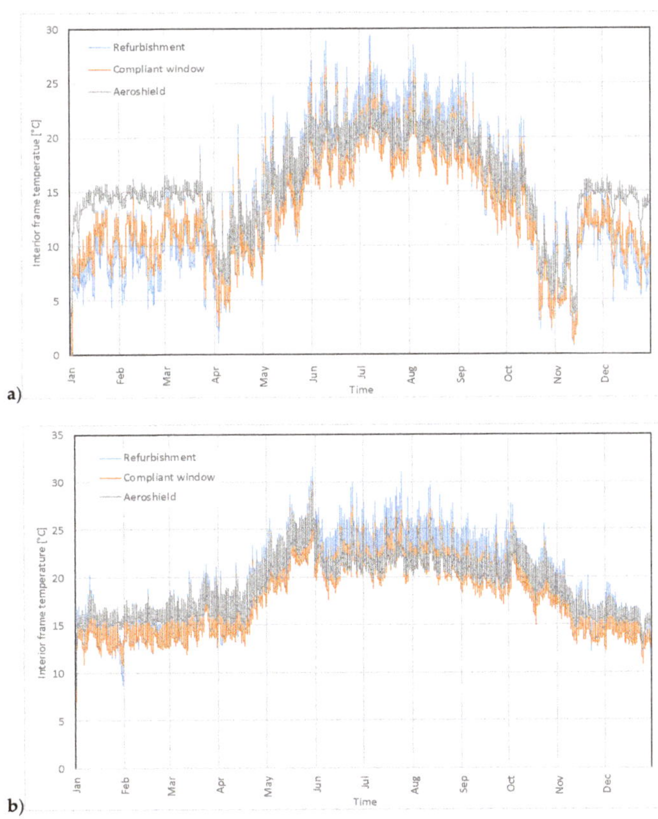

Figure 8. Frame surface temperature for North exposure, plotted with reference to one-year time, in Toronto (**a**) and Bari (**b**).

3.4. North-Facades in Toronto

Starting from these remarks, further analyses were carried out to compare performances due to different values of Window-to-Wall Ratios (WWRs) of North-facing walls in Toronto. Three different values of Window-to-Wall Ratio were considered to take into account the effect of window size on the North-facing exposure as well as the effect of the frame surface, in each case. The window used in previous simulations was then assumed as the "intermediate" case window. Two more sizes were added: a "small" and a "large" window, as suitable terms of comparison. For each case, the Frame-to-Glazing Ratio (FGR) was also calculated, expressing the ratio between the frame surface and the glazing surface, in percentage terms (Figure 9).

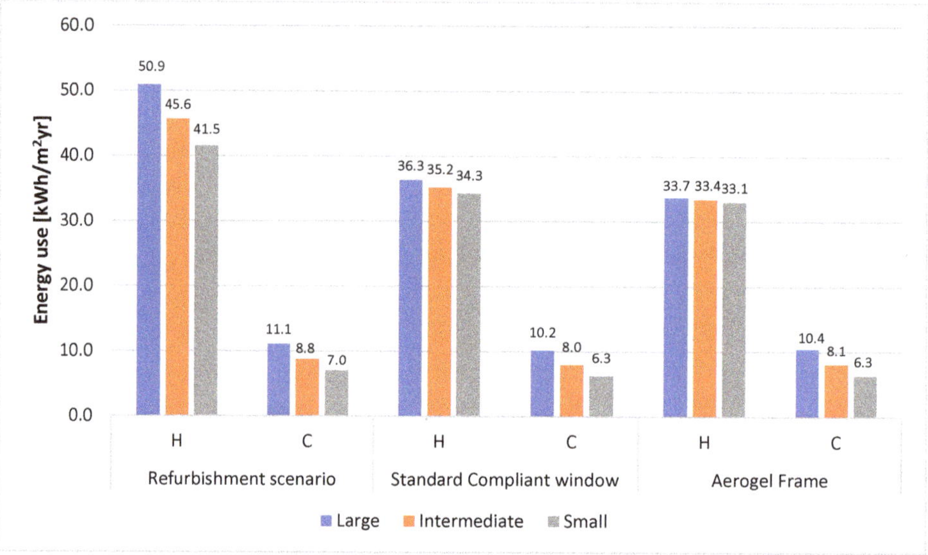

Figure 9. Energy use for heating and cooling using three different window sizes on the North-facing exposure, in Toronto. (H= Heating, C= Cooling).

When the aerogel window was compared with the Refurbishment scenario, the "large" window (WWR = 100%) showed the highest saving (17 kWh/m^2·yr), passing from 51 to 34 kWh/m^2·yr yearly consumption for heating. This figure dropped to 12 kWh/m^2·yr when the intermediate window was considered (WWR = 60%) and to 8.4 kWh/m^2·yr in the small window (WWR = 43%).

When the aerogel window was compared with the Standard compliant scenario, the highest contribution to energy saving due to the aerogel frame only was observed in the large window (3 kWh/m^2·yr). The better performance of this case was due to the greater contribution of the frame on the overall façade surface, expressed by the higher FGR (46%), compared to the "intermediate" and "small" window, with lower FGRs (41% and 39%, respectively).

The effect of FGR on energy consumption, for all the window sizes and frame technologies, was taken into account by normalizing the overall yearly energy consumption (for cooling and heating) by the actual frame surface area (Table 3). The table shows that the increase of glazed area has a larger impact than the frame technology adopted, implying larger consumption, in all the cases investigated. Anyway, lower energy consumption occurred in the aerogel-insulated windows, although the standard compliant window and aerogel-insulated window show almost comparable results. The lowest normalized consumption (about 5%), for the aerogel-insulated window, compared to the standard compliant window, was obtained when the WWR was set to 43%.

Table 3. Yearly energy consumption normalized to frame surface area.

Window Size	WWR (%)	FGR (%)	Window Surface Area (m^2)	Frame Surface Area (m^2)	Normalized Yearly Energy Consumption ($kWh/m^2 \cdot yr$)		
					Refurbishment	Standard-Compliant	Aerogel-Insulated
Small	43	39	5.3	2.0	221.6	166.3	157.5
Intermediate	60	42	7.3	3.0	362.6	287.8	276.5
Large	100	46	12.2	5.6	484.7	405.7	393.8

4. Discussion

The results obtained in this study showed that even when reducing U_f from 3.5 to 0.66 $W/m^2 \cdot K$, by means of the best available technologies, the achievable benefits in terms of energy saving were significant only on North-facing walls, and in more extreme cold climates. Clearly, the transparent component of the window plays a major role in determining the final result, as the comparison between the "refurbishment" and the "standard compliant" cases showed, but FGR values proved that frame may be equally significant when smaller glass panes are used. Thus, proper economic considerations might be useful to improve understanding of the practical advantages of such technologies. According to Koebel et al. (2012), the price of aerogel could drop below 1500 USD/m^3 by 2020. In line with this, other studies reveal that the cost of a meter cube of aerogel will achieve 50% cost reduction in production within the next few years and will decrease to 660 USD/m^3 by 2050 [13,35]. Assuming the latter value for the volume cost of granular aerogel, less than 20 USD would be required to fabricate the super-insulated aerogel-based window at the basis of this study, considering that the aerogel volume in the frame would be ~0.011 m^3. Furthermore, the real challenge for a significant cost reduction for the aerogel does not seem to come from mass production using the supercritical drying process (the process that is currently used by all the manufactures of aerogel products), but from the definition of new methods for evacuating the liquid component of the gel. In particular, the ambient drying process has already shown some preliminary advantages depicted over a decade ago by Bhagat et al. in 2007 [36]. More recently, Koebel et al. [37] and Wu et al. [38] explored successful ways to fabricate large quantities of aerogel using ambient-dried silica aerogel. These methods have also been applied by Berardi and Zaidi [6], who in 2019 presented a new ambient pressure drying aerogel blanket.

5. Conclusions

The results presented in this paper constitute an interesting premise for further experimental investigations on the here proposed innovative window. The aerogel-based window outperformed both the obsolete window adopted for a "Refurbishment scenario", and the highly performing "Standard compliant" one, showing reduction of energy uses by 27% and 6%, respectively.

The overall effect on energy use due to the improvement of the frame thermal resistance is limited superiorly by its surface, compared to that of the glazed component, generally lower than 50%. In the present work, it has been proven that the conductance of the frame can be much lower compared with that of a highly performing triple glass.

This work has shown the effectiveness of significantly lowering the frame conductance, opening the way to further investigations taking into account both energy saving and the increase of cost due to the use of super-insulating materials, inside the frame.

Author Contributions: Conceptualization, A.C.; methodology, A.C.; writing—original draft preparation, A.C.; writing—review and editing, U.A., U.B., A.C., V.D.C., S.L., F.M., and C.R. All authors have read and agreed to the published version of the manuscript.

Funding: This research was partially funded by the Action Co-founded by Cohesion and Development Fund 2007–2013 – APQ Research Puglia Region "Regional programme supporting smart specialization and social and environmental sustainability – Future in Research".

Acknowledgments: Antonio Ferrara and the technical team from DCS Group s.r.l. are kindly acknowledged for their support during the fabrication of the aerogel-based window; Cabot Company is gratefully acknowledged for providing the granular aerogel used for the fabrication of the new window frame prototype.

Conflicts of Interest: The authors declare no conflict of interest.

References

1. Parliament, E. Directive (EU) 2018/844 of the European Parliament and of the Council of 30 May 2018 Amending Directive 2010/31/EU on the Energy Performance of Buildings and Directive 2012/27/EU on Energy Efficiency. Available online: https://eur-lex.europa.eu/eli/dir/2018/844/oj (accessed on 1 February 2020).
2. Wen, R.T.; Arvizu, M.A.; Niklasson, G.A.; Granqvist, C.G. Electrochromics for energy efficient buildings: Towards long-term durability and materials rejuvenation. *Surf. Coat. Technol.* **2015**, *278*, 121–125. [CrossRef]
3. Eperon, G.E.; Burlakov, V.M.; Goriely, A.; Snaith, H.J. Neutral color semitransparent microstructured perovskite solar cells. *ACS Nano* **2014**, *8*, 591–598. [CrossRef] [PubMed]
4. Cannavale, A.; Ierardi, L.; Hörantner, M.; Eperon, G.E.; Snaith, H.J.; Ayr, U.; Martellotta, F. Improving energy and visual performance in offices using building integrated perovskite-based solar cells: A case study in Southern Italy. *Appl. Energy* **2017**, *205*, 834–846. [CrossRef]
5. Jelle, B.P. Traditional, state-of-the-art and future thermal building insulation materials and solutions - Properties, requirements and possibilities. *Energy Build.* **2011**, *43*, 2549–2563. [CrossRef]
6. Berardi, U. Characterization of commercial aerogel-enhanced blankets obtained with supercritical drying and of a new ambient pressure drying blanket. *Energy Build.* **2019**, *198*, 542–552. [CrossRef]
7. De Matteis, V.; Cannavale, A.; Martellotta, F.; Rinaldi, R.; Calcagnile, P.; Ferrari, F.; Ayr, U.; Fiorito, F. Nano-encapsulation of phase change materials: From design to thermal performance, simulations and toxicological assessment. *Energy Build.* **2019**, *188*, 1–11. [CrossRef]
8. Ascione, F.; Bianco, N.; De Masi, R.F.; de' Rossi, F.; Vanoli, G.P. Energy refurbishment of existing buildings through the use of phase change materials: Energy savings and indoor comfort in the cooling season. *Appl. Energy* **2014**, *113*, 990–1007. [CrossRef]
9. Buratti, C.; Belloni, E.; Palladino, D. Evolutive Housing System: Refurbishment with new technologies and unsteady simulations of energy performance. *Energy Build.* **2014**, *74*, 173–181. [CrossRef]
10. Bahaj, A.S.; James, P.A.B.; Jentsch, M.F. Potential of emerging glazing technologies for highly glazed buildings in hot arid climates. *Energy Build.* **2008**, *40*, 720–731. [CrossRef]
11. Jelle, B.P.; Baetens, R.; Gustavsen, A. Aerogel Insulation for Building Applications. In *Sol-Gel Handbook*; Wiley-VCH Verlag GmbH & Co. KGaA: Weinheim, Germany, 2015; Volume 3, pp. 1385–1412.
12. Walker, R.; Pavía, S. Thermal performance of a selection of insulation materials suitable for historic buildings. *Build. Environ.* **2015**, *94*, 155–165. [CrossRef]
13. Cuce, E.; Cuce, P.M.; Wood, C.J.; Riffat, S.B. Toward aerogel based thermal superinsulation in buildings: A comprehensive review. *Renew. Sustain. Energy Rev.* **2014**, *34*, 273–299. [CrossRef]
14. Berardi, U. Aerogel-enhanced insulation for building applications. In *Nanotechnology in Eco-efficient Construction (Second Edition), Materials, Processes and Applications*; Woodhead Publishing: Cambridge, UK, 2018; pp. 395–416.
15. Parameshwaran, R.; Kalaiselvam, S. *Nano and Biotech Based Materials for Energy Building Efficiency*; Pacheco Torgal, F., Buratti, C., Kalaiselvam, S., Granqvist, C.-G., Ivanov, V., Eds.; Springer International Publishing: Cham, Switzerland, 2016; pp. 215–243; ISBN 978-3-319-27505-5.
16. Buratti, C.; Moretti, E.; Belloni, E. Aerogel plasters for building energy efficiency. In *Nano and Biotech Based Materials for Energy Building Efficiency*; Springer Nature: Cham, Switzerland, 2016; pp. 17–40.
17. Ebert, H.-P. Thermal Properties of Aerogels. In *Aerogels Handbook*; Springer Nature: Cham, Switzerland, 2011; pp. 537–564.
18. Berardi, U. The development of a monolithic aerogel glazed window for an energy retrofitting project. *Appl. Energy* **2015**, *154*, 603–615. [CrossRef]
19. Neugebauer, A.; Chen, K.; Tang, A.; Allgeier, A.; Glicksman, L.R.; Gibson, L.J. Thermal conductivity and characterization of compacted, granular silica aerogel. *Energy Build.* **2014**, *79*, 47–57. [CrossRef]
20. Ihara, T.; Jelle, B.P.; Gao, T.; Gustavsen, A. Aerogel granule aging driven by moisture and solar radiation. *Energy Build.* **2015**, *103*, 238–248. [CrossRef]
21. Hoseini, A.; McCague, C.; Andisheh-Tadbir, M.; Bahrami, M. Aerogel blankets: From mathematical modeling to material characterization and experimental analysis. *Int. J. Heat Mass Transf.* **2016**, *93*, 1124–1131. [CrossRef]

22. Cuce, E.; Cuce, P.M. The impact of internal aerogel retrofitting on the thermal bridges of residential buildings: An experimental and statistical research. *Energy Build.* **2016**, *116*, 449–454. [CrossRef]
23. Ng, S.; Jelle, B.P.; Sandberg, L.I.C.; Gao, T.; Wallevik, Ó.H. Experimental investigations of aerogel-incorporated ultra-high performance concrete. *Constr. Build. Mater.* **2015**, *77*, 307–316. [CrossRef]
24. Gao, T.; Jelle, B.P.; Gustavsen, A.; He, J. Lightweight and thermally insulating aerogel glass materials. *Appl. Phys. A Mater. Sci. Process.* **2014**, *117*, 799–808. [CrossRef]
25. Berardi, U. Development of glazing systems with silica aerogel. *Energy Procedia* **2015**, *78*, 394–399. [CrossRef]
26. Jelle, B.P.; Hynd, A.; Gustavsen, A.; Arasteh, D.; Goudey, H.; Hart, R. Fenestration of today and tomorrow: A state-of-the-art review and future research opportunities. *Sol. Energy Mater. Sol. Cells* **2012**, *96*, 1–28. [CrossRef]
27. Dowson, M.; Harrison, D.; Craig, S.; Gill, Z. Improving the Thermal Performance of Single Glazed Windows using Translucent Granular Aerogel. *Int. J. Sustain. Eng.* **2011**, *4*, 266–280. [CrossRef]
28. Huang, Y.; Niu, J. Application of super-insulating translucent silica aerogel glazing system on commercial building envelope of humid subtropical climates e Impact on space cooling load. *Energy* **2015**, *83*, 316–325. [CrossRef]
29. Gao, T.; Ihara, T.; Grynning, S.; Petter, B.; Gunnarshaug, A. Perspective of aerogel glazings in energy efficient buildings. *Build. Environ.* **2016**, *95*, 405–413. [CrossRef]
30. Buratti, C.; Moretti, E. Experimental performance evaluation of aerogel glazing systems. *Appl. Energy* **2012**, *97*, 430–437. [CrossRef]
31. Buratti, C.; Moretti, E. Glazing systems with silica aerogel for energy savings in buildings. *Appl. Energy* **2012**, *98*, 396–403. [CrossRef]
32. Valachova, D.; Zdrazilova, N.; Panovec, V.; Skotnicova, I. Using of aerogel to improve thermal insulating properties of windows. *Civ. Environ. Eng.* **2018**, *14*, 2–11. [CrossRef]
33. EnergyPlus 8.9, Building Technologies Program, National Renewable Energy Laboratory (NREL). Available online: https://energyplus.net/ (accessed on 1 July 2019).
34. ISO 10077-2:2017 - Thermal Performance of Windows, Doors and Shutters—Calculation of Thermal Transmittance Numerical Method for Frames; Standards Norway. 2017. Available online: https://www.iso.org/standard/64995.html (accessed on 1 July 2019).
35. Koebel, M.; Rigacci, A.; Achard, P. Aerogel-based thermal superinsulation: An overview. *J. Sol-Gel Sci. Technol.* **2012**, *3*, 315–339. [CrossRef]
36. Bhagat, S.D.; Kim, Y.-H.; Moon, M.-J.; Ahn, Y.-S.; Yeo, J.-G. A cost-effective and fast synthesis of nanoporous SiO_2 aerogel powders using water-glass via ambient pressure drying route. *Solid State Sci.* **2007**, *9*, 628–635. [CrossRef]
37. Koebel, M.M.; Huber, L.; Zhao, S.; Malfait, W.J. Breakthroughs in cost-effective, scalable production of superinsulating, ambient-dried silica aerogel and sili-ca-biopolymer hybrid aerogels: From laboratory to pilot scale. *J. Sol-Gel Sci. Technol.* **2016**, *79*, 308–318. [CrossRef]
38. Wu, X.; Fan, M.; Mclaughlin, J.F.; Shen, X.; Tan, G. A novel low-cost method of silica aerogel fabrication using fly ash and trona ore with ambient pressure drying technique. *Powder Technol.* **2018**, *323*, 310–322. [CrossRef]

© 2020 by the authors. Licensee MDPI, Basel, Switzerland. This article is an open access article distributed under the terms and conditions of the Creative Commons Attribution (CC BY) license (http://creativecommons.org/licenses/by/4.0/).

buildings

MDPI

Review

Perovskite Solar Cells for BIPV Application: A Review

Anurag Roy, Aritra Ghosh *, Shubhranshu Bhandari, Senthilarasu Sundaram and Tapas K. Mallick

Environment and Sustainability Institute, University of Exeter, Penryn Campus, Cornwall TR10 9FE, UK; a.roy30@exeter.ac.uk (A.R.); sb964@exeter.ac.uk (S.B.); s.sundaram@exeter.ac.uk (S.S.); t.k.mallick@exeter.ac.uk (T.K.M.)

* Correspondence: a.ghosh@exeter.ac.uk

Received: 14 May 2020; Accepted: 10 July 2020; Published: 13 July 2020

Abstract: The rapid efficiency enhancement of perovskite solar cells (PSCs) make it a promising photovoltaic (PV) research, which has now drawn attention from industries and government organizations to invest for further development of PSC technology. PSC technology continuously develops into new and improved results. However, stability, toxicity, cost, material production and fabrication become the significant factors, which limits the expansion of PSCs. PSCs integration into a building in the form of building-integrated photovoltaic (BIPV) is one of the most holistic approaches to exploit it as a next-generation PV technology. Integration of high efficiency and semi-transparent PSC in BIPV is still not a well-established area. The purpose of this review is to get an overview of the relative scope of PSCs integration in the BIPV sector. This review demonstrates the benevolence of PSCs by stimulating energy conversion and its perspective and gradual evolution in terms of photovoltaic applications to address the challenge of increasing energy demand and their environmental impacts for BIPV adaptation. Understanding the critical impact regarding the materials and devices established portfolio for PSC integration BIPV are also discussed. In addition to highlighting the apparent advantages of using PSCs in terms of their demand, perspective and the limitations, challenges, new strategies of modification and relative scopes are also addressed in this review.

Keywords: renewable energy; perovskite solar cells; BIPV; semi-transparent; challenges

1. Introduction

Currently, building sector consumes 40% energy globally, which is expected to reach double or triple by 2050 because of population growth, changes of household size with improved electrical and cooking appliances, increasing levels of wealth and lifestyle changes at the global level [1–6]. In addition, this consumed building energy is responsible for emitting 40% of total carbon dioxide. The global emission is further expected to be 50% by 2050 if the current trend of energy consumption occurs [7–11]. Hence, replacement of building energy generation from fossil fuel, coal-based power plant to renewable energy sources is highly demanding. Photovoltaic (PV) power generation can displace fossil fuel-generated energy. The installed PV capacity surpassed 500 GW already in worldwide, and another 500 GW installation is expected by 2023 [12]. However, PV generated power has low-density power supply compared to other renewable sources. Moreover, for large scale PV plant needs a large land area while transmission and distribution power losses are also very high. Hence, PV technology manifests potential opportunism in building architecture. The integration of PV in a building is known as building-integrated PV (BIPV), where PV replaces the traditional building envelopes such as window, roof, wall and offset building construction cost. At the same time, these new BIPV envelopes generate power and also protect the building interior from harsh external ambient [6,13–16]. Semitransparency or transparent nature is the precondition for BIPV envelopes as they are not only responsible for producing the benevolent power but also controls the net building energy consumption by offering suitable indoor ambient. Semi-transparent or transparent BIPV allow sufficient daylight, which reduces lighting energy demand and also controls heat loss and gain, which, in turn, reduce the heating and cooling load demand [17]. Although buildings have a strong potential of reducing their energy consumption and greenhouse gas emissions; the main challenge is to achieve this objective without compromising thermal comfort needs [18–20].

First-generation cost-effective, crystalline silicon for BIPV application needs space between cells to create semitransparency [21,22]. While second-generation thin film [23–25] and third-generation [26] emerging perovskite solar cells (PSCs) [27], dye-sensitized solar cells (DSSCs) [28], quantum dot solar cells (QDSSC) and organic–polymer solar cells [29] types possess semitransparency by tuning their corresponding cell thickness.

The photosynthesis process inspired the fundamental novelty of the third-generation PV concepts in the way of assembling the cell architecture, which allows easy and cost-effective processing to generate power from effective sunlight absorption [30]. Among the third generation PV system, PSCs are the most efficient technology that promises a cheaper and accessible fabrication route to produce more efficient photo-conversion efficiency (PCE). The perovskite materials satisfy almost all the requirements to achieve the best performance, as listed in Table 1.

Table 1. Summarized requirements for the best-performance photovoltaic (PV) applications in terms of materials and properties categories.

Categories		Requirement
Properties	1	Suitable bandgap matching the solar spectrum
	2	Strong absorption coefficient
	3	Excellent and balanced carrier mobility
	4	Defect tolerant
	5	Ambipolar dopability
	6	Long carrier lifetime and diffusion length
	7	Reasonably low exciton binding
Materials	1	Earth-abundant/low cost
	2	Nontoxic
	3	Long-term stability
	4	High performance

In a typical PSC, the perovskite absorber layer is sandwiched between the electron and hole transport layer (ETL and HTL). Generally, the ETL deposits on a fluorine-based tin oxide (FTO)-based glass followed by the perovskite layer and HTL. Finally, a back contact introduces on the top of the HTL and thus a complete cell form. The sequential layers of a typical PSC are schematically described in Figure 1. Briefly, PSCs show a photovoltaic phenomenon when exposed to light, consequently producing photovoltage and photocurrent.

This kind of cell structure benefits for a high collection efficiency and low recombination of carriers, which are indispensable to realize a high conversion efficiency. Compared to other solar technology, PSC reacts to different wavelengths of light. This is due to the structure of the cells, which enables electrons to travel through various interfacial layers [31,32]. As a result, they can convert a proportionately higher amount of sunlight into electricity. Thus, it is required to mobilize the material quality of the perovskite absorber by enhancing their charge carrier and reduce the defect density. Despite the material quality improvement, the layer thickness adjustment is also a possible way to enhance cell performance. Maximum light trapping can provide mutual benefits for both the optical and electrical properties of the PSCs.

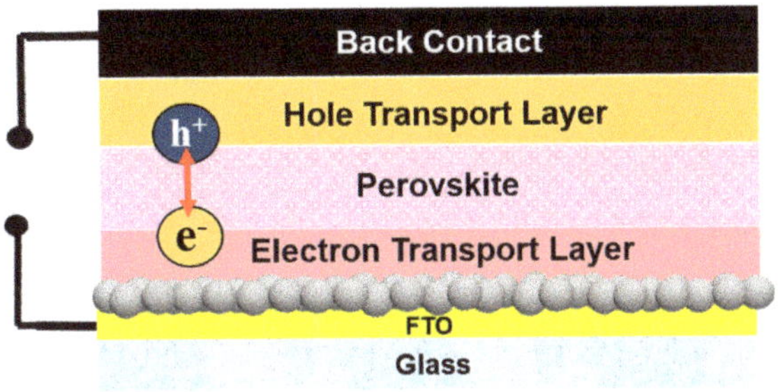

Figure 1. Schematic configuration of a typical perovskite solar cell.

To date, the number of publications on PSCs exceeded almost 17,726 according to the web of science while searching the keyword of "perovskite solar." Among them, 1143 publications were related to review articles [33]. Not only that, but most of the article also comes under the category of materials science multidisciplinary (10,865), dealing with (1) emergence of perovskite solar cells, (2) fundamental relation with optoelectronics, ion mobility properties, (3) recombination effect, (4) theoretical studies to achieve high efficiency, (5) defect and interfacial related studies, (6) device architecture such as mesoscopic, planar, inverted, hole-transport material free tandem, etc., (7) stability and scalability related gripping topics [34–37]. Around 2554 publications are available including 865 proceeding papers and 150 review articles according to the web of science by searching the keyword "building-integrated photovoltaic". Nevertheless, interestingly, searching by the keywords "building-integrated photovoltaic perovskite solar" indicates only 37 articles, 9 reviews and 3 proceeding papers as per web of science. The status of the publications lists as gathered from the web of science thus highlights the adequate requirement of more research-based PSCs integration in BIPV (Figure 2).

Figure 2. Schematic representation of the theme of this review represents perovskite solar cells integration to building-integrated photovoltaic (BIPV).

In this review, we have put great emphasis on the potentiality of PSCs integration for BIPV application through its appearance as an emerging PV technology to futuristic employment in BIPV application. A general phenomenon and the importance of the review are recapitulated in Section 1. Next, Section 2 illustrates the efficiency evaluation of PSCs as a breakthrough trendsetter in PV Afterwards; Section 3 highlights the difficulties associated with PSCs, mainly stability and toxicity.

Then, Section 4 addresses on the PV technological aspects of PSCs includes transparency and efficiency tradeoff and up-scaling status. Then Section 5 discusses the advancement of PSCs integration in BIPV. This includes recent trends of PSCs in BIPV includes lightweight, solar paint techniques and zero-energy building conceptualization for sustainable BIPV adaptation. Next, Section 6 describes the future scope of PSCs integration in BIPV, followed by a constructive discussion in Section 7. Finally, the review is summarized with the concluding remarks in Section 8.

2. Performance Trends Evaluation of PSC

PSCs has become one of the hot spots owing to its dramatic development in efficiency levels and thus extensively studied in recent years [38–40]. The organic-inorganic perovskite material such as $CH_3NH_3PbI_3$ ($MAPbI_3$) and $CH_3NH_3PbBr_3$ was successfully applied to first DSSCs as a promising alternative to the conventional molecular dyes. However, its instability in the liquid electrolyte seemed to put an obstacle for further development of higher efficiency [41]. Research activities on PSC were swiftly increased since the first reported solid-state PSC in 2012, having PCE of 9.7% [42]. After this breakthrough, in the following years, the investigation of PSCs became gradually dominant in PV research. Eventually, the efficiency of PSCs was enhanced to 22.1% in early 2016 [43]. PSCs have achieved an efficiency of >23% in just a few years. While writing this review article, the highest certified PCE reached 25.2%, confirmed by the National Renewable Energy Laboratory [44]. However, the solution processing fabrication routes of perovskite resulted in its crystallization, defect passivation, which influences the interfacial optoelectronic properties. Therefore, the employment of various additive-assisted strategies for interface optimization can escalate the performance of a PSC. Table 2 highlights the recorded significant PCE of various PSCs in consecutive years in terms of their perovskite absorber modification and implementation of various device fabrication strategies in order to achieve the best performance.

Table 2. Significant performance achievement of experimental planner structure-based perovskite solar cells (PSCs) in consecutive years *.

Perovskite Sensitizer	Device Assembly	PCE (%)	Year	Reference
$MAPbI_3$	$TiO_2/CH_3NH_3PbI_3$/Electrolyte solution/Pt-FTO	3.8	2009	[42]
$MAPbI_3$	FTO/m-TiO_2/Spriro-OMeTAD/Au	9.7	2012	[45]
$MAPbI_3$	FTO/m-TiO_2/c-TiO_2/spiro-MeOTAD/FTO Au	17.0	2014	[45]
$FAPbI_3$	FTO/bl-TiO_2/mp-TiO_2/PTAA/Au	20.2	2015	[46]
$MA_{0.6}FA_{0.4}PbI_3$	ITO/PTAA/ICBA/C_{60}/BCP/Cu	18.3	2016	[47]
$Cs_{0.2}FA_{0.8}PbI_3$	FTO/SnO_2/C_{60}-SAM/Spriro-OMeTAD/Au	19.6	2016	[48]
$Cs_{0.15}FA_{0.85}Pb(I_{0.83}Br_{0.17})_3$	FTO/bl-TiO_2/mp-TiO_2/Spriro-OMeTAD/Au	20.0	2016	[49]
$MAPbI_3$	ITO/PTAA/C_{60}/BCP/Cu	20.7	2016	[50]
$FA_{0.81}MA_{0.15}PbI_{2.51}Br_{0.45}$	FTO/bl-TiO_2/mp-TiO_2/spiro-OMeTAD/Au	20.8	2016	[51]
$Cs_{0.05}(MA_{0.17}FA_{0.83})_{0.95}Pb(I_{0.83}Br_{0.17})_3$	FTO/bl-TiO_2/mp-TiO_2/PTAA/Au	21.1	2016	[49]
$Rb_{0.05}[Cs_{0.05}(MA_{0.17}FA_{0.83})_{0.95}]_{0.95}Pb(I_{0.83}Br_{0.17})_3$	FTO/bl-TiO_2/mp-TiO_2/Spriro-OMeTAD/Au	21.6	2016	[46]
$MAPbBr_3$ into $FAPbI_3$	FTO/thin-barrier TiO_2/m-TiO_2/PTAA/Au	22.1	2017	[52]
$FAPbI_3$- $MAPbBr_3$	FTO/d-TiO_2/mp-TiO_2/NBH/P3HT/Au	22.7	2019	[53]
$Cs_{0.17}FA_{0.83}Pb(I_{0.97-x}Br_xCl_{0.03})_3$	FTO/bl-TiO_2/mp-TiO_2/Spriro-OMeTAD/Au	20.5	2020	[54]

* MA—CH_3NH_3; PTTA—Poly-$HC(NH_2)_2PbI_3$; BCP—bathocuproine; ICBA—indene-C_{60} bisadduct; SpiroOMeTAD—2,2′,7,7′-tetrakis[*N*,*N*-di(4-methoxyphenyl)amino]-9,9′-spirobifluorene; NBH—narrow-bandgap-halide; P3HT—poly(3-hexylthiophene)

Recently, the Oxford photovoltaics has achieved a certified PCE of 27.3% for 1 cm^2 perovskite-silicon tandem solar cell [55]. This result defeats the highest recorded PCE of the single-junction silicon solar cell (26.7%). This also indicates the emerging performance and rapid development of the PSCs in comparison to silicon solar cells to achieve a similar efficiency [56]. The prevalent efficiency trend of various PV cells is further exhibited in Figure 3. Impressive progress on the PCE is being achieved in case of PSC, which is highly dominating over other PV cells. However, solving the problem of transferring high efficiency from laboratory small-area devices to large-area perovskite modules is the pivotal challenge. The maximum theoretical PCE (Shockley–Queisser limit) of the PSCs, employing $MAPbI_{3-x}Cl_x$ is predicted to 31.4%. Therefore, there is still enough space for development [57].

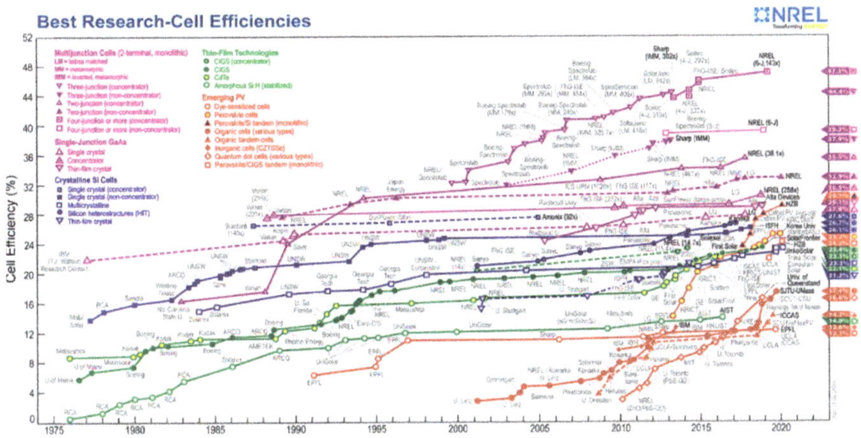

Figure 3. Cell efficiency progress trend of various PV devices (This plot is courtesy of the National Renewable Energy Laboratory, Golden, CO) [44].

Recently, Nishigaki et al. (2020) show simulations study based on light absorption characteristics of chalcogenide alloy-based perovskites such as $Ba(Zr,Ti)S_3$, indicating a maximum potential of 38.7% PCE in a perovskite/crystalline Si tandem structure [58]. Despite intrinsically excellent optoelectronic properties of the perovskite absorber, a gap between the theoretical efficiency and the experimental one opens up an ample scope of further investigation and detailed scrutiny to unveil the hidden information of achieving high efficiency and stable PSC. All other kinds of non-Si solar cells suffer significant barriers in terms of their materialization, fabrication and cost and thus became almost saturated to develop next-generation PV devices. In this scenario, the rapid improvement of PSCs thus is expected to be a prominent choice than the traditional silicon solar cells [59].

Efficiency and operation stability both are the crucial factors to determine the practical applications of these devices. Therefore, to execute the best performance of a PSC, there are several fundamental concerns needs to be addressed as mentioned:

- Excellent improvement of photo physics-chemistry and dynamics of exciton/charge-carrier;
- Well-established interfacial energy alignment, interface electronic structure and charge-transfer (carrier-collection) processes;
- Structural and composition characterization;
- Improved stability by developing material and compositional engineering;
- High-efficiency solar cell fabrication with device performance and stability testing;
- Scale-up, printing, slot-die coating and roll-to-roll manufacturing;
- Understanding of solar parameter behaviour.

3. Device Challenges: Stability and Toxicity Affairs of PSCs

To develop highly efficient and environmentally stable, benign perovskite devices is critical and challenging. Particularly for BIPV application, semi-transparent or transparent PSCs are required as they not only generate power for the building but also allows daylight and improve the indoor environment. However, for opaque façade application, the requirement for the transparency can be eliminated. Long-term instability, use of toxic lead (Pb) and semitransparency, while having higher power generating capability are the significant challenges with PSC.

As mentioned, stability stands out as the main challenge of a PSC device. The perovskite layer degrades rapidly in the presence of high temperature, ultraviolet (UV) light and moisture (Figure 4). Moreover, Pb-based perovskite ($MAPbI_3$) causes severe toxicological implications on the environment with a dramatic impact on the environment and human health. Rapid degradation is the most common phenomenon for PSC while they are exposed to moisture and UV radiation [60,61]. This is quite a realistic point to be considered for outdoor application. Consequently, effective strategies are required to overcome the stability issue and employed for long-term running [62]. As a result, besides the implementation of high-efficiency devices, it is also essential to consider on environment and safety issues for an optimum PSC device.

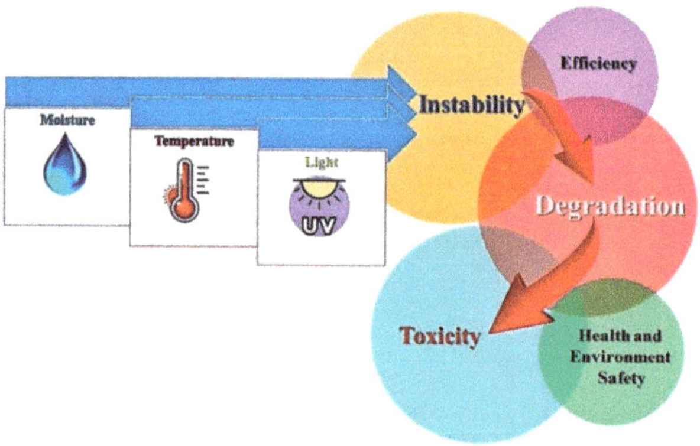

Figure 4. Major challenges of PSCs according to instability and toxicity.

3.1. Stability Issue

$MAPbI_3$-based perovskite is the most efficient candidate for high PCE generation, but they suffer from material degradation in the presence of high temperature, intense UV light and the high percentage of moisture content [63,64]. The hygroscopic character of amine salt present in MA^+ and highly hydrophilic nature of perovskite lead the hydrate products formation, which eventually broadens the degradation time [65–67]. In the presence of UV-light photogenerated electron ease the formation of superoxide [68,69], which are moderately reactive with perovskite. The degradation leads to the formation of PbI_2 and HI, which further produce I_2 in the presence of oxygen, as shown in the equation (i-iii). The UV-degradation can also happen for encapsulated devices [70,71]. Organic-inorganic perovskites have various thermal stabilities depending on the time of exposure. Grain boundaries play a crucial role in temperature-dependent stability [72,73]. Depends on the organic part of the material, degradation can occur at 60 °C or can start at 100 °C [74–76]. Not only the perovskite materials; there are other layers like HTL, ETL and metal-based electrode materials, which also have their fundamental issues [77–80]. The metal electrode has the tendency to be corroded by the degradation of halide perovskite and tends to produce metal halide due to temperature stress. Stability issues can also occur from a structural point of view. The crystal structure of the perovskite and tolerance factor plays a pivotal role as they dictate the performance of the perovskite [81].

$$CH_3\,NH_3\,PbI_3\,(s) \rightarrow PbI_2\,(s) + CH_3\,NH_3\,I\,(aq.) \tag{i}$$

$$CH_3\,NH_3\,I\,(aq.) \rightarrow CH_3\,NH_2\,(aq.) + HI\,(aq.) \tag{ii}$$

$$HI\,(aq.) + O_2\,(g) \rightarrow 2I_2\,(s) + 2H_2O\,(L) \tag{iii}$$

To solve the moisture instability, two-dimensional (2D) perovskites and mixed dimensional perovskites are the most promising ones [82,83]. Incorporation of 2D perovskite in the grain boundary of three-dimensional (3D) perovskite has also proven to be effective in passivating the defect states of grain boundary [84,85]. Protection of PSCs from UV-light can also be achieved by using mixed dimensional (2D-3D) perovskite materials. At the same time, incorporation of inorganic charge transport materials like CuSCN, NiO, WO_3 also provides some promising light stability in ambient conditions [86–88]. In spite of that, encapsulation is another way out to protect the devices from external factors [85]. Encapsulating materials play a huge role in commercializing PSCs, although UV instability is a matter of concern even with encapsulated devices. Thermal stability can be managed using the perovskite materials having high tolerance factor and quasi-2D structures. Using inorganic charge transport layers via sol-gel and pyrolysis methods effectively reduce the chances of degradation [89–91]. Presently, a considerable amount of study is going on by replacing the metal electrodes with carbon-based materials [92], which are non-degradable by corrosion. Hence, incorporation of carbon nanostructures as the charge transport layer take new initiatives for stable and economically viable PCE [93]. Few examples are given in Table 3, showing the developments of PSCs to overcome different instability factors. Perovskite-silicon tandem cells are coming in the market to make it more feasible for energy application [94].

Table 3. Types of PSCs along with their efficiency and stability.

Device Structure	PCE (%)	Stability Status	Reference
ITO/ZnO-JTCA/$MAPbI_3$/spiro-MeOTAD/Au	18.8	70% left after 840 h in air	[95]
ITO/PEDOT:PSS/$MAPbI_{3-x}Cl_x$/PCBM/ZnO/Al	16.8	Excellent stability	[96]
ITO/Cu:NiOx/$MAPb(I_{1-x}Br_x)_3$/$PC_{61}BM$/BCP/Al	15.4	244 h stability	[97]
ITO/NiO/$MAPbI_3$/ZnO/Al	16.1	>60 d stability	[98]
FTO/SnO_2/$FA_{1-x}(MACs)_xPbI_3$/spiro-MeOTAD/Au	20.7	83% left after 60 h	[96]
FTO/TiO_2/$MAPbI_{3-x}Clx$/sputtered NiO_x/Ni	7.3	>2 months stability	[99]
FTO/NiO_x/$FA_{1-x}MA_xPbI_3$/PCBM/TiO_x/Ag	20.7	90% left after 500 h under 85 °C	[100]
ITO/PTAA/$(FA_{0.83}MA_{0.17})_{0.95}Cs_{0.05}$ $Pb(I_{0.6}Br_{0.4})_3$/ICBA/C_{60}/BCP/Cu	18.3	90% left after 720 h in N	[101]
FTO/BI-TiO_2/Mp-TiO_2/FA-perovskite/spiro-OMeTAD/Au	20.0	800 h	[102]
FTO/bl-TiO_2/mp-TiO_2/RuCsFAMAPbI1-$_x$Br$_x$/spiro-MeOTAD/Au	21.8	95% left after 500 h under 85 °C	[103]
FTO/bl-TiO_2/mp-TiO_2/$CsFAMAPbI_{3-x}Br_x$/CuSCN/RGO/Au	20.4	95% left after 1000 h under 60 °C	[87]
FTO/c-TiO_2/m-TiO_2/m-ZrO_2/Co_3O_4/carbon/$MAPbI_3$	11.7	~2500 h in ambient condition in presence of light	[104]
FTO/m-TiO_2/m-ZrO_2/NiO/carbon/$MAPbI_3$	13.7	PCE decreased to 80% of initial after ~150 h in presence of light	[105]
FTO/c-TiO_2/m-TiO_2/$CH_3NH_3PbI_3$/C-CuS	10.2	Over 600 h in ambient condition with 30%–50% humidity in dark	[106]
FTO/c-TiO_2/m-TiO_2/m-ZrO_2/carbon/$MAPbI_3$	6.5	~850 h in dry air condition at room temperature in dark	[107]
FTO/c-TiO_2/m-TiO_2/m-Al_2O_3/carbon/$MAPbI_3$	12.3	PCE decreased to 1% of initial after ~480 h under light at room temperature	[107]
FTO/c-TiO_2/m-TiO_2/m-Al_2O_3/SWCNT-NiO/$MAPbI_3$	12.7	~300 h in ambient condition	[108]
FTO/c-TiO_2/m-TiO_2/m-Al_2O_3/carbon-WO_3/$MAPbI_3$	10.3	85% of initial PCE retains after ~500 h in the ambient condition in the presence of light	[109]

Degradation of perovskite is the main reason for instability as mentioned above. Poor thermal stability over a long period and also the humidity related adverse effects contribute towards the degradation of the Pb-based perovskite [63,64,92]. Resolving the stability issues of perovskite material is the censorious strategy to upgrade the long-term performances.

The growth of PSC confirmed that trying to leap directly into the marketplace BIPV installations would require substantial upfront capital investment. However, the standard test condition, STC (using the air–mass 1.5 spectrum, an intensity of 100 mW·cm^{-2} also known as 1 sun of illumination and a cell temperature of 25 °C) is vastly different from real-world conditions, where temperature and the solar light intensity level and its spectrum always change with time [110,111]. Tress et al. (2019) recently varied both the light intensity and temperature of a PSC to monitor the PCE performing outdoor experiments at Lausanne, Switzerland. They have realized that temperature and solar light intensity closely depend on the associate weather condition, which determines the PV parameters associated with the PSCs [111]. As an example, based on weather station data from Bangor University in the UK from 2018, there were just 51.4 h over the whole year where the solar irradiance was equal to or greater than 1 sun and the mean irradiance during daylight hours was only 0.226 sun [110].

3.2. Toxicity Issue

As mentioned in Section 2, the most efficient PSCs are fabricated using Pb-based halide perovskites ($MAPbI_3$). However, they suffer from the toxicity issues during device fabrication, deployment and disposal [112,113], which also retards the pace of commercialization. Pb is a carcinogenic element and has no safe threshold limit of exposure. Degradation of $MAPbI_3$ due to pH from rainwater on $MAPbI_3$ films were characterized and found that it degrades completely in water [114]. Moreover, a recent report by Li et al. (2020) highlights Pb- leakage into the ground, which can further provide serious impact on plants and human's food cycle [115]. In addition, Su et al. (2020) have examined the impact of Pb-leakage on the environment. They have observed that the total organic carbon and chemical oxygen demand analysis signify discarded PSCs could increase the oxygen consumption and may release CO_2 into the environment [116].

The toxic effect of Pb and polymers pushed researchers to think about greener alternatives [60,112,117,118]. Tin, germanium, bismuth can replace the toxic Pb [119,120]. Moreover, some new series of materials as chalcogenide perovskites, double perovskites, etc. are also worth investigating [121,122]. Various alternative Pb-free perovskite materials can be employed for light absorbers in PSCs. Table 4 indicates about the performance of some notable Pb-free perovskite absorber development. However, the efficiency is not as much like Pb-based perovskites due to high minority and low mobility of carrier-effective masses. Still, they are in the investigation because of their better stability over Pb-based PSCs at ambient condition.

Table 4. Significant performance of various lead-free perovskite-based materials in PSCs.

Pb-Free Absorber	Features	PCE (%)	Reference
$CsSnI_3$	$CsSnX_3$ (X = Cl, Br and I) quantum rods with tunable emission wavelength ranging from 625 to 709 nm has been prepared via solvothermal synthesis using different halide salt conditions	13.0	[123]
$CsSnI_3$	Addition of SnF_2 reduces the background charge carrier density by lowering the defect concentrations	2.0	[124]
Cs_2TiBr_6	First-ever solar cells using Cs_2TiBr_6 thin films show a stable efficiency of up to 3.3	3.3	[125]
Sn-based	Tin perovskite solar cells by using high energy level Indene-C_{60} bisadduct	12.4	[126]
$MASnI_3$	Redshifted and absorption up to 950 nm, compared to $MAPbI_3$ counterpart (1.55 eV).	5.7	[127]
$FASnI_3$	Lewis acid-base adduct formed crystallization with trimethylamine	7.1	[128]

Table 4. *Cont.*

Pb-Free Absorber	Features	PCE (%)	Reference
$MAGeI_{2.7}Br_{0.3}$	Flash-photolysis time-resolved microwave conductivity and photoelectron yield spectroscopy techniques	0.6	[129]
$Cs_2AgBiBr_6$	Fabricated $Cs_2AgBiBr_6$ films and incorporation to working devices	2.5	[130]
$MA_3Bi_2I_9$	Pinhole-free, large-grained films fabrication using two steps approach. The absorption coefficient, trap density of states characterization and charge diffusion length comparison to lead perovskite.	1.6	[131]
$MASbSI_2$	methylammonium antimony sulfur diiodide was fabricated the first time for solar cell	3.1	[132]
$MA_3Sb_2Cl_XI_{9-X}$	A joint experimental and theoretical study. Inclusion of methylammonium chloride into the Precursor solution; low-cost fabrication	2.2	[132]
MA_2KBiCl_6	Indirect bandgap, density functional calculations	3.0	[133]
$Ba(Zr,Ti)S_3$ and $BaZr(S,Se)_3$	Perovskite/crystalline Si tandem architecture, indicating the highest band-edge α among all known solar cell materials; Theoretical study	38.0	[58]

4. Photovoltaic Consequences of PSCs

4.1. Transparency and Efficiency Tradeoff of PSCs

Transparency is a crucial factor to determine the PCE of PSCs. Therefore, it is imperative to study the semi-transparent PSCs towards its useful energy harvesting through the windows and roofs of buildings. The transmitted sunlight through the transparent or semi-transparent window of a building significantly influences on the visual comfort of building occupant. Besides PCE, there are some other crucial parameters such as average visible transmittance (AVT), transparency colour perception [134–136], corresponding colour temperature (CCT) [137,138] and colour rendering index (CRI) [139] should be investigated for optimizing the semi-transparent PSCs. Usually, AVT range between 20%–30% is required for window applications. Bandgap tuning of the perovskite absorber resolves the tradeoff between transparency and efficiency and thus becoming as a smart choice as a window. Transparency is associated with the thickness of the cell. Sometimes, highly transparent cells suffer low efficiency due to less absorption of sunlight. In contrast, less transparent cells are sometimes insufficient to generate the required power [140,141]. Jung et al. (2015) have described transparency enhancement by reducing the Perovskite absorber coverage. The PCE significantly boosts up by 128%, while the thickness increased from 40 to 280 nm, followed by the AVT decreased from 35% to 10% [142].

The modern architects prefer to build exteriors designed with artwork on the glass while building engineers consider low-cost energy-efficient construction [143,144]. Transparent or semi-transparent solar cell devices can replace the transparent glass for building. However, the fully transparent solar cell possesses low solar light absorption capability. Hence it is a critical task to fabricate a fully transparent solar cell, which will also generate high power [145]. Thus, optimization between the efficiency and transparency at acceptable levels by varying the thickness of the different layers are required [146–148]. The light transmission with opaque crystalline Si PV is achieved by spacing the cells in the module [21,22,149] whereas PSCs are semi-transparent elements. An effort has been executed to find a suitable medium between high electrical efficiency and a high AVT, as shown in Figure 5. The thickness and structure of the mesoporous layer create an impact on PSC performance [150]. The optical transparency of the other interlayers and the semi-transparent electrode of PSC influence to archive high PCE [151]. Therefore, to achieve a highly transparent PSC device, the perovskite layer should be thin enough, which inevitably leads to obtain brownish cells [152]. In addition, reducing the thickness of the perovskite layer cannot wholly solve the tradeoff between transparency and efficiency as improvement of the transparency decrease the efficiency [153]. Mesoscopic cells that use TiO_2 photoanodes are partially circumvented. By providing a large $MAPbI_3/TiO_2$ interfacial surface area, electrons are therefore extracted from the perovskite throughout the entire thickness of the layer [71]. Thin mesoporous TiO_2 layer has the potential to fill pore, which results in an improved PCE [154].

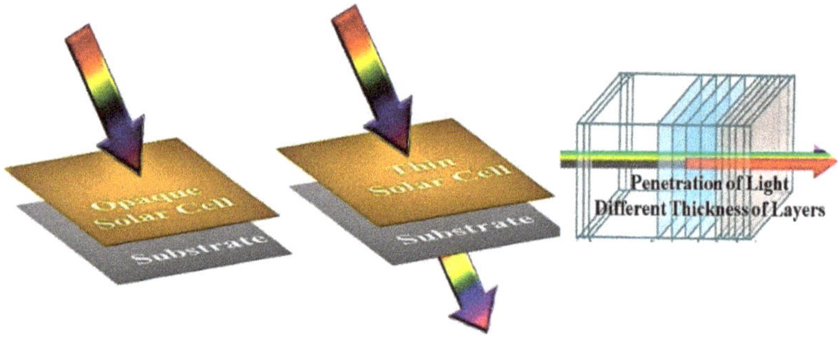

Figure 5. Schematic representation of the transparency of light, indicating the thickness of the different layer of a PSC.

Various reports claimed that a thin film of $MAPbI_3$ could be developed, consisting of a thickness of 100–350 nm [155–157]. This resulted in the substantial differences in carrier diffusion lengths of the compositions, a much deeper understanding of the role of film thickness and morphology on the efficiency of $MAPbI_3$-based planar heterojunctions is required. The influence of the TiO_2 layer during the growth of the perovskite "pores", determined mainly by the thickness of the film by using their non-stoichiometric mixture of MAI and $PbCl_2$ precursors was also studied. Employing perovskite de-wetting technique, neutral colour perovskite was synthesized, which has an array of perovskite microstructured island. Each island could absorb visible light while the transparent region was colour neutral. However, this structure offers high transmission, but the gap between perovskite island create high shunt resistance [158,159]. This shunt resistance lowers the power generation, which can be further improved by -post-treatment of perovskite using alkyl-siloxane molecular layers [160]. The morphologic control of the perovskite layer is much determined by their precursors, deposition method and preparation temperature [161]. Without hole transport material, PSC having photoanode of 800 nm shows 12.22% cell efficiency [162]. It is observed that thinner n-type layers would improve the conductivity and accordingly lead to better device performance. Consequently, an appropriate thickness of n-type layers is also paramount to achieve superior device performance [163,164].

Yuan et al. (2018) investigated systematically by varying the I: Br ratio for $MAPb(Br, I)_3$ -based perovskites and thickness of the corresponding film, which led to an increase in the AVT as well as PCE [141]. They have realized that >20% AVT and ~10% PCE, even when the device thickness of wide-bandgap perovskite was >200 nm. A sandwich-type $MAPbI_3$ to obtain semi-transparent PSCs with high efficiency by modification of morphology and structure, which can increase transparency to some extent, as reported by Heo et al. (2016) [163]. Wei et al. (2019) reported semi-transparent PSC-based on MoO_x/ITO/Ag/ITO (MoO_x/IAI) with $MAPbI_3$, which exhibited a PCE of 12.85% with an AVT of ~80%. Ag nanowire is deposited on the spiro-OMeTAD or perovskite to attain a semi-transparent PSC, which achieved a peak transmission of 77% approximately 800 nm and an efficiency of 12.7% as reported by Bailie et al. (2015) [165]. The PCE of 14.2% along with 72% AVT in the near-infrared region has been reported Fu et al. (2015) for an FTO/ZnO/PCMB/$MAPbI_3$/Au based device architecture [166].

On the other hand, the strategy for achieving high transparency relies on uniform microstructure within the perovskite film, such as inserting a passivating material to confine the growth of perovskite grains. Mesh assisted grid deposition technique for mixed cation mixed halide perovskite on mesoporous TiO_2 was employed to fabricate PSC, which resulted in an average PCE ~10% with an AVT of 28%. In this case, islands of perovskite well-connected across the walls of the mesh are formed, which are capable of efficient light conversion and provide intermittent surface coverage for average transparency [167]. Table 5 highlights the performance of the investigated semi-transparent PSCs. Notably, both electrically conductive and optically transparent perovskite materials are still challenging to develop.

Table 5. The efficiency of various semi-transparent PSCs at their different transmission wavelength.

Device Structure	PCE (%)	Transmittance (%)	Wavelength (nm)	Reference
ITO/PTAA/$Cs_{0.05}FA_{0.8}MA_{0.15}PbI_{2.55}Br_{0.45}/C_{60}/SnO_2$/ZTO/IZO	18.2	75	760–1200	[168]
ITO/PEDOT:PSS/$MAPbI_xCl_{3-x}$-PVP/CYTOP/$PC_{61}BM$/PEIE/Au	5.4	34	400–800	[169]
FTO/c-$TiO_2/MAPbI_3$/Spiro-OMeTAD/MoO_3/ Au/MoO_3	5.3–13.6	7–31	370–740	[170]
FTO/c-$TiO_2/MAPbI_XCl_{3-X}$/Spiro-OMeTAD/Ni	6.1	38	300–850	[160]
FTO/c-TiO_2/AAO + $MAPbI_{3-X}Cl_X$/Spiro-OMeTAD/MoO_X/ITO	8.5–13.3	26.3–45.4	400–800	[171]
In_2O_3:H/PTAA/$MAPbI_3/PC_{61}BM$/ZnO ZnO:Al/Ni/Al	16.1	80.4	800–1200	[146]
ITO/CuSCN/$MAPbI_3$/PCBM/Bis-C_{60}/Ag	7.5–10.7	13–37.5	300–850	[142]
FTO/c-$TiO_2/MAPbI_3$/PTAA/PEDOT:PSS/ITO/glass	12.6–15.8	6.3–17.3	300–800	[172]
ITO/PEDOT:PSS/$MAPbI_3$/PCBM/AZO SnO_X/Ag/SnO_X	11.8	29	400–900	[173]
FTO/c-TiO_2/mp-TiO_2/Al_2O_3/(5-AVA)$_X$(MA)$_{1-X}PbI_3$-Carbon grid/MWCNT	8.2	24	300–800	[174]
ITO/ZnO/$CsPbBr_3$/Spiro-OMeTAD/PH1000	5.9	59.8	300–800	[175]
FTO/$TiO_2/MAIPbI_3$-CH_3NH_2- $MAPbI_3$/SWCNT/PEDOT:PSS/Ni microgrid	11.3	68	380–740	[176]
ITO/c-TiO_2/mp-TiO_2/Rb-$FA_{0.75}MA_{0.15}Cs_{0.1}PbI_2Br$/PTAA/$MoO_X$/ITO	16	84	720–1100	[177]
ITO/$NiO_X/FA_{0.83}Cs_{0.17}PbBr_{0.5}I_{0.25}$/LiF/$PC_{60}BM/SnO_2$/ZTO:ITO/LiF/Ag grid	14.5	74	765–1200	[178]
ITO/PEDOT:PSS/$MAPbI_3/PC_{61}BM$/Au/LiF	3.4–7.7	10–35.4	400–800	[179]
ITO/NiO_X/MA_3PbI_3/PCBM/BCP/Ag/SiO_2/ZnS/Ag/ZnS	10.5, 10.7, 11.2	28.8, 12.4, 3.4	Red, Green, Blue	[180]
ITO/ZnO/PCBM/$MAPbI_3/NiO_X$/ITO/dielectric mirror	10.12	28.1, 22.4, 17.4	Red, Green, Blue	[181]
FTO/c-TiO_2/m-$TiO_2/MAPbCl_XI_{3-X}$/Spiro-OMeTAD/PEDOT:PSS	11.6, 13.8, 11.6	11.6, 10.5, 9.1	400–750	[182]
ITO/c-$TiO_2/MAPbI_3$/Spiro-OMeTAD/MoO_3/ Ag/ZnS	11.9–13.3	7.42–10.9	380–750	[183]
FTO/c-TiO_2/mp-$TiO_2/Al_2O_3/MAPbI_3$-Carbon grid/WO_3	8.1	20	380–780	[184]
FTO/c-TiO_2/mp-TiO_2/Al_2O_3/graphene-PEDOT:PSS/$MAPbI_3$-carbon grid	7.2, 11.6	26, 20	380–780	[185]

4.2. Upscaling Stature of PSCs

Scaled up of the PSC is essential to implement them for their on-site application. However, scaling up PSC exerts as a hugely challenging issue specifically for the fabrication technology, large-scale material development and sequential right tracklayer deposition. Scaling up often reduce the efficiency of the overall system. Therefore, the development of large-scale PSC is a big challenge in terms of both the cost-effectiveness and necessary longevity. In this regard, sheet-to-sheet and roll-to-roll deposition methods stand out as a suitable scaling up methods for large-scale PSC fabrication. Reduction of bulk defect recombination and electric leakage followed by adjusting its bandgap and increasing the charge generation are the crucial factors that require to be perfectly attained in large-scale PSCs, which is undoubtedly a challenging task. There are many attempts that have been made for upscaling of PSCs, providing importance on high efficiency and excellent stability too. However, increasing the cell dimension also drops the power conversion. For example, cell dimension increment from 0.12 to 1.1 cm^2 resulted 11.5% reduction of efficiency [186]. PSC 10.1 cm^2 modules exhibited a PCE of 10.4%, which was enhanced to 13.3% for reducing the area [187]. A 198 cm^2 area based PSC offered PCE of 6.6% [188] while Gardner et al. (2016) [189] achieved only 4.3% PCE for a cell having an active area of 100 cm^2. However, Hu et al. (2017) achieved a mesoscopic 100 cm^2 area PSC having PCE of 10.4% [190].

Another work showed fully printable large-area 7 m^2 perovskite panels made by 10% efficient 10 × 10 cm^2 perovskite modules [190]. Weihua Solar manufactured large-area 45 × 65 cm^2 modules by using 10.6% efficient 5 × 5 cm^2 cells [191]. Recently, one 703 cm^2 (24.15 × 29.10 cm^2) large size with excellent efficiency Perovskite was fabricated where crystal properties were controlled during the fabrication of the crystal growth process [192]. Microquanta successfully fabricated a 200 × 800 cm^2 module with a PCE of 14.24% [193]. Very recently, the Panasonic electronics manufacturer has produced a lightweight 30 × 30 cm^2 perovskite with an efficiency of 16.09% [194].

5. Advancement of PSCs for BIPV

Building integration of PSCs is one of the most holistic approaches as transparent and semi-transparent PSCs are possible by tuning the material thickness and bandgap. BIPV may hold the potential to increase PSC technology suitable space on buildings. Uniform transparency over the entire device area makes PSC a very engaging for BIPV applications, which combine the production of green energy and esthetic architectural features of elements that are part of the green environment for human habitat. Integration of PSC in BIPV is still a less-explored work compared to the other PVs. The transparency-efficiency tradeoff can be considered for a proper design of BIPV applications where transparency and efficiency both represent design parameters. Figure 6 highlights the major required characteristics for BIPV integration in the less energy-hungry building using PSCs.

- Outdoor stability;
- Lightweight: the total mass of the cell does not strain the construction of the building;
- Semitransparency and;
- Low-scattered light: able to harvest solar energy also in shadowed areas or cloudy conditions that can bring up the PSC for a suitable BIPV employment.

Figure 6. Schematic representation of the suitable PSC features for BIPV.

5.1. Recent Progress of PSCs Integration in BIPV

BIPV systems are promising for energy conscious building design techniques. Design considerations for BIPV systems must include the building's architecture, its location, any appropriate building and safety codes. Integration of PSC involves effective design for the local climate and environment and considering consolidated daylighting and photovoltaic collection. In this regard, various attempts have been promoted to employ PSC in BIPV application.

Bifacial colourful ST-PSCs in an n–i–p architecture using CuSCN as a hole transporting material, was developed by Wang et al. (2020), exhibits wide colourful tunability and excellent bifacial photovoltaic behaviour and emerging as a promising candidate for BIPV applications [195]. Martellotta et al. (2017) pointed out with particular reference to semi-transparent PV cells using amorphous silica (a-Si), and perovskite may have a significant impact on BIPV application. PSC with 42.4% transmission exhibited higher PCE (6.64%) compared to lesser transparent (30.1%) amorphous Si cell, which resulted in maximum efficiency of 4.80%. Moreover, the authors claimed that the PSC offered a more balanced visual comfort behaviour compared to the a-Si device and produced electricity with different behaviours concerning solar radiation. This study offers the futuristic scope of PSCs integration as a potential substitute candidate for silicon solar cells for BIPV. Despite that, Aernouts et al. (2019) [196] demonstrates about the efficient structures and processes for reliable perovskite solar modules in BIPV facade application, with support of outdoor test results, which opens then the door for technology validation for a fully integrated product (Figure 7). The authors demonstrated a cost-effective, highly efficient (>17%) perovskite-based 35 × 35 cm^2 module architecture having over 20 years of reliability.

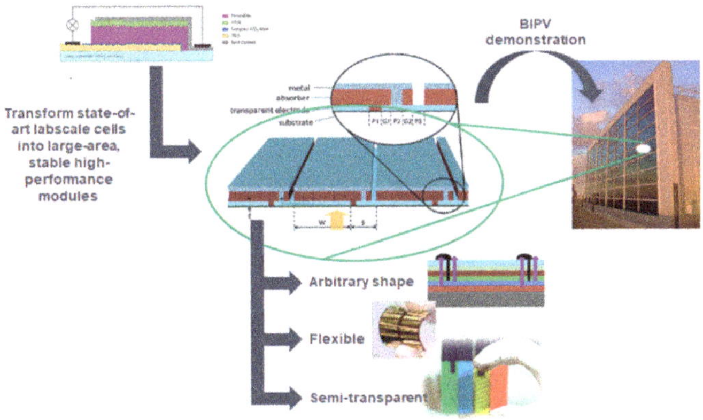

Figure 7. Schematic represents of efficient structures and processes for reliable PSC modules for BIPV application. Reproduced with permission [196].

Ghosh et al. (2019) developed carbon-based PSC having which had PCE of 8.13%, 30% average solar transmission and 20% average visible transmission, at the laboratory environment for BIPV application. In addition, daylighting and glare control analysis employing the subjective rating methods was investigated. For midday period of a typical bright sunny day at the location of the University of Exeter, Penryn (50.16° N, 5.10° W), UK, high glare was observed [184]. Improved visual comfort is attainable while Perovskite PV windows are employed in a building by replacing solar control window [7]. Coloured PSC reflects light which is also potential for BIPV application as lowering the thermal gain comfortable indoor environment is achievable. Furthermore, reflected light reduces the possibility of overheating of the solar cell. The PSC developed by Zhang et al. (2015) was capable of tuning colour across the visible spectrum while maintained the high efficiency. This product is suitable for sustainable coloured BIPV architecture [197]. PSCs also changes its transmission in the presence of temperature, which is known as thermochromism. Transmission variation in the visible range for thermochromic (TC) perovskite is suitable for retrofit or newly glazed façade application [198]. Figure 8 shows a colour neutral PSC-BIPV, while amorphous Si cells absorb most of the blue-green radiation and appear as brown-orange [199].

Tandem structure of PSCs is attractive candidates where at the top [118] or the bottom silicon are stacked mechanically. Mechanically stacked silicon bottom cell PSC achieved a PCE of 23.9% [200,201]. This is the highest achieved efficiency from the four-terminal tandem device.

Figure 8. Simulation of the appearance of the different PV technologies applied to the glazing such as (**a**) baseline condition; (**b**) a-Si cell and (**c**) PSC. Reprinted with permission [199].

However, stability was encountered as the most challenging for PSC efforts, were given to characterize it at the outdoor condition. Of course, the low-cost and flexible manufacturing prospects of PSCs offers a desirable integration in the BIPV market. The flexible PSC modules laminated into a glass façade element, which was manufactured by Saule Technologies and installed in Skanska's Spark building in Warsaw, Poland [202]. The BIPV market expects tremendous growth from €1 billion in 2015 to €6.3 billion in 2022.

5.2. Lightweight PSCs for BIPV

Lightweight BIPV technology is essential for building integration especially retrofit the building. Hence flexible and wearable PV systems are gaining importance because of their potential application in building. Flexible PV indicates a reduced weight PV which can be quantified as specific power (watts/kilogram). Fabrication of flexible transparent PSC requires flexible transparent electrodes which should have high conductivity and excellent mechanical stability. Hence, the replacement of the conventional brittle electrodes is required [203,204]. In terms of flexibility consideration, the polymer substrates are considerable candidate due to cheaper and low-temperature solution processing. The inexpensive polymer substrates such as polyethene terephthalate (PET) and polyethylene naphthalate (PEN) have been found suitable to fabricate flexible PSCs [101].

Along with, the mechanical stability of the device is also essential to generate adequate power per weight for their utilization. The potential of reaching extremely high specific powers in PSC reporting ~29.4 $W{\cdot}g^{-1}$ with only 12% efficient device as demonstrated by Kang et al. (2019) [205]. This value is orders of magnitude higher than what typically characterizes traditional crystalline Si cells lies in the range of 0.1–1 $W{\cdot}g^{-1}$ [206]. Notable performances of lightweight-based PSCs are summarized in Table 6.

Table 6. Significant reports on the performance of flexible PSCs *.

Substrate	PCE (%)	Power-per-Weight	Reference
PAN foils-ITO (1.3 μm)	15.2	29.4 W/g	[205]
PET foil (1.4 μm)	12	23 W/g	[206]
PET/Graphene (20 μm)	11.5	5 W/g	[207]
Ultrathin PET/Silver mesh (57 μm)	14	1.96 kW/kg	[208]
ITO-coated PET	14.2	23.26 W/g	[209]
PET/Ag Nanowire (17 μm)	12.9	4.16 W/g	[210]

* PEN: Polyethylene naphthalate; PET: Polyethylene terephthalate;

5.3. Solar Ink-Based PSCs

Solar paint can replace the high-temperature annealing processed fabrication for high-efficiency PSCs [211]. This process can further be improvised by addition of chemical steps to enhance the electronic inter-grain connectivity, which will enhance the conversion efficiency. The roll-to-roll fabrication process excels as a compatible technique in order to make large-scale PSCs. The roll-to-roll deposition techniques such as ink-jet printing, slot die coating, blade coating, spray coating, screen-printing, flexography, etc. are one of the most convenient processes to use the solar paints to fabricate PSCs in a large scale (Figure 9a,b) [201,212]. Having the advantage of large wafer substrates allowance, faster deposition technique, a highly scalable and cost-effectiveness of the overall process further provides the roll-on roll fabrication approach as one of the best solutions for large scale production [191,212]. Implementation of such solar paint-based protocol on a roll-to-roll cell manufacturing is now seeking attention. The solar paint can be applied in BIPV, where the photoanode can be directly painted on the walls or windows of a building. Solar paints are a class of thin-film technology in which it can be fabricated in the form of solution. This solution is applied on glass or plastic substrate by spraying or brushing to make a complete solar cell. Some of the most attractive features of PV paints are thin layer deposition, substrate availability, less waste during processing and non-fragile. The solar paint can be further directly applied in BIPV, where the ETL can be directly painted on the walls or windows of a building.

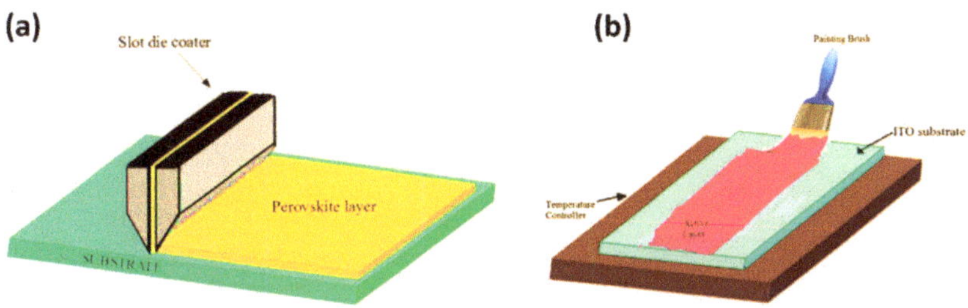

Figure 9. Perovskite layer deposition using (**a**) slot die coating, (**b**) brush painting techniques, for solar paint-based PSCs, respectively.

So far, different approaches such as additive inclusion, thermal annealing and different kind of precursors, were explored to control morphology, crystalline structure and moisture stability of perovskite. In combination with these strategies, the use of anti-solvent dripping during the deposition of perovskite precursors is widely exploited to obtain better film morphology and properties. Beside solvent toxicity, anti-solvent dripping is hardly compatible with scalable productions technologies; therefore, strategies to control the perovskite film formation in one-step and solvent dripping free is a technological challenge. The spinning in the typical perovskite process flings away most of the liquid, so a better approach is therefore required. The acetonitrile/methylamine system is unusual for a few reasons: First, it utilizes methylamine as the Pb-alkylamine former, thereby reducing the chance of impurities and over passivation [213]. The acetonitrile system is shelf-stable for long periods. Acetonitrile as a solvent also has many desirable properties for high-speed manufacturing such as low boiling point, low viscosity and low surface tension relative to the commonly used dimethylformamide (DMF) [214], dimethyl sulfoxide (DMSO) [154] and N-Methyl-2-pyrrolidone (NMP) [215] systems. Instead of using the toxic solvents, starch biopolymer exploits as a rheological modifier to tailor the viscosity of perovskite precursor solutions to obtain a stable ink for uniform perovskite thin films [216]. TCs have attracted broad interest due to their simple reversible mechanism of using thermal energy for switching transmittance. Typically, commercially available TCs are liquid crystals and leuco dyes and recently vanadium dioxide. The crystallization temperature in inverse temperature crystallization depends on the halogen constituent of the perovskite. This observation inspired to combine the tunable optical properties and unusual crystallization processes of perovskites to design a facile synthesis of TC inks that realize smart windows with more extensive chromatic variation than thought possible in hybrid perovskites [217,218].

5.4. Zero-Energy Building Approach of PSCs in BIPV

Significant insights were gained while theoretical approaches based on $MAPbI_3$ perovskite was employed [57,219,220]. The modelling studies of such PSCs include formability of the perovskite structure, effect of reduced structural dimensionality and Pb-ion insertion, which can pave way further improvement of perovskite development. On the other hand, features like Rashba effect [221], the ferroelectric effect [222], carrier localization effect [223], ion transport and halogen atom inclusion [224], provides high optoelectronics property modulation for PSCs. Alternatively, achieving zero energy is an ambitious yet increasingly achievable goal, i.e., gaining momentum across geographic regions and markets [225]. Reduced energy consumption of PSCs, therefore, makes it simpler and less expensive to meet the building's energy needs for BIPV. Figure 10 highlights about a scheme of all these features, which need to be co-inside to develop a zero-energy building using PSC.

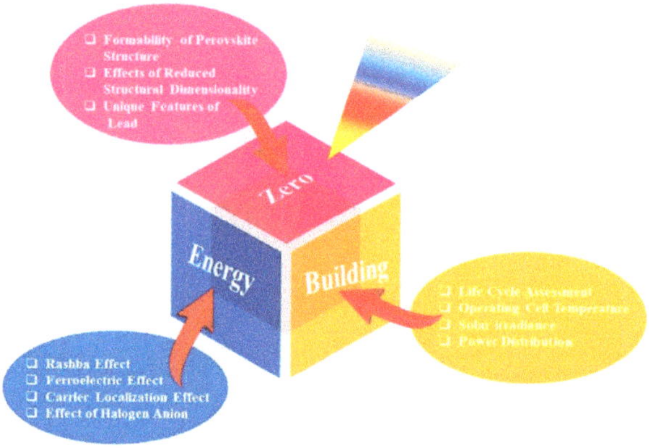

Figure 10. Scheme of the PSC-integrated BIPV modelling study for zero-energy building development.

Reports are available on the top floor of the apartment, which has been investigated by using the $4 \times 4 \times 3\ m^3$ single box model covered by PSCs [226–230]. The hourly energy gains from the rooftops and façades are calculated based on the measured angle-dependent performance of the solar cells and the hourly solar intensities. Assoa et al. (2017) attempted different approaches, including a linear model, lumped elements models and models that make use of commercial software solvers for rooftop BIPV modelling study [227]. The forecasted solar irradiance 7 kW solar PV system for net-zero building was evaluated in India climate [228]. One the other hand, Poulek et al. (2015) rationalized an annual estimation of temperature difference 3.2 K and 3.5% in energy production between roof integrated and free-standing PV panels through thermal models [229]. Recently, Walker et al. (2019) reported a thin-film CIGS module with longitudinal cell direction to outperform modules with orthogonal cell direction by up to 8%. The thin-film CIGS modules on the scale of one module under partial shading conditions, one curved module and two connected modules [226]. The model was validated using recorded experimental data over an entire year. Recently, Gong et al. (2019) used the single box model of $4 \times 4 \times 3\ m^3$ which converted 35% solar radiation to electricity is a promising solution for distributed power generation and the zero-energy residential building in the Wuhan urban area in China [231].

6. Future of PSCs Based BIPV

Concerning the present status of PV technologies, improvements in several areas are required, such as conversion efficiency, cost, durability and sustainability. With the emergence and rapid growth of nanotechnology, harnessing of the novel nanomaterials for PSCs has increased over the past two decades, and the process of discovery and invention is continuing.

The calculation of LCOE, i.e., Levelized cost of electricity signifies that the total cost per unit of electricity generated over the lifetime from the PSC system is competitive compared to other available PV technology [207,232,233]. The "Golden Triangle" model indicates about the performance, durability and cost of the PV system, as shown in Figure 11a. As an example, Liu et al. (2018) developed the PSC having a PCE of 19.1% ± 0.4% with good reproducibility. The un-encapsulated PSCs exhibit above 1600 h lifetime under continuous dry nitrogen environment. This study recommends a suitable example of the "golden triangle" report for a PSC in terms of their conversion rate of sunlight into electricity must be high, inexpensive to produce, and possess a long lifespan [207]. In the case of BIPV implementation, the functionality depends on building's design, durability, geographical position, safety and maintenance as shown in Figure 11b, which can control the effective justice to the golden triangle theory.

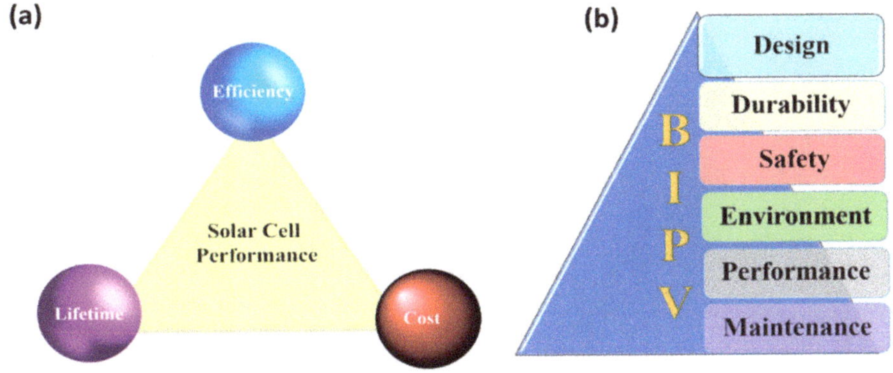

Figure 11. (**a**) Golden triangle of solar cells bonded with the cost, efficiency and lifetime parameters and (**b**) parameters for PV building integration.

6.1. Scope of PSCs in Building Energy Storage Application

It is advantageous to design PSC-electrochemical storage hybrid devices to store the power as achieved form PSC. In this regard, PSC introduces as a window frame around the electrochromic glass. This can make solid-state electrochromic batteries in smart windows. In such a configuration, the solar-cell window frame harvests electricity and provide power to the electrochromic glass [234–237]. Table 7 summarizes performances of various significant PSCs for electrochemical storage hybrid devices. Xia et al. (2016) demonstrated a solid-state electrochromic battery made with $MAPbI_3$-based PSC and WO_3-based electrochromic device, which exhibits reasonable cycling stability with a capacity of 65 $mA·h·g^{-1}$ at 1 $A·g^{-1}$ after 2500 cycles with retention of 86.7% [45]. This result indicates the successful implantation of PSC in electrochromic devices.

Table 7. Significant performance of various PSCs for electrochemical storage hybrid devices *.

Perovskite	Storage Active Layer	Device Performance	Storage Performance	Reference
$MAPbI_3$	Bacterial cellulose membrane/Polypyrrole/MWCNT	Efficiency: 13.6% V_{OC}: 0.98 V	Capacitance: 572 mF·cm^{-2} (1.0 mA·cm^{-2}), SE: 49%	[238]
$MAPbI_3$	Copper hydroxide nanotubes/MnO_2	Efficiency: 10.41% V_{OC}: 0.96 V	Capacitance: 37.93 mF·cm^{-2} (1.0 mA·cm^{-2}), Energy Density: 3.36 mW·h·cm^{-2}, Power Density 74.02 mW·cm^{-2}, SE, 67%	[239]
$MAPbI_3$	Carbon-supported graphene/PEDOT	Efficiency: 8.65% V_{OC}: 1 V	Capacitance: 388 mF·cm^{-2} (1.0 mA·cm^{-2}), Energy Density: 0.18 W·h·cm^{-2}	[240]
$CH_3NH_3PbI_{3-x}Cl_x$	WO_3	Efficiency: 8.25% V_{OC}: 0.98 V	Co-anode (co-cathode) Energy Density: 13.4 (24.5) mW·h·cm^{-2}, Power Density: 187.6 (377.0) mW.m^{-2}, Capacitance: 286.8 (430.7) F.m^{-2}	[241]
$MAPbI_3$	Asymmetric: TiO_2 Nanotubes and MnO_2/Co_6S_8	Efficiency: 5.6% V_{OC}: 0.81 V	Capacitance: 262.5 mF·cm^{-2} (0.5 mA·cm^{-2}), SE: 80%	[242]
$MA_{1-y}FA_yPbI_{3-x}Cl_x$	Anode: $Li_4Ti_5O_{12}$/rGO, Cathode: Carbon	Efficiency: 14.01% V_{OC}: 1.05 V	Energy Density: 60 W·h·kg^{-1}, SE: 80%	[243]
$(C_6H_9C_2H_4NH_3)_2PbI_4$	$(C_6H_9C_2H_4NH_3)_2PbI_4$/rGO/PVDF	Efficiency †: 0.003%	Capacity: 410 mA·h·g^{-1}	[244]

* MWCNT: Multi-walled carbon nanotube; rGO: Reduced graphene oxide; S.E.: Storage efficiency; PEDOT: poly(3,4-ethylene dioxythiophene) polystyrene sulfonate; MA: Methylammonium; F.A.: Formamidinium; PVDF: Polyvinylidene; Fluoride; V_{OC}: Open circuit voltage. † After device integration

On the other hand, $MAPbI_3$-based series-connected PSCs directly charge lithium-ion batteries with a storage efficiency of 7.80% [245,246]. Recently, Kin et al. (2020) have combined a high open-circuit voltage (V_{OC} > 1 V) PSC single cells with single-cell batteries via a direct-current–direct-current (DC–DC) boost converter to explore the feasibility of single-cell-to-single-cell integration of a single PSC with a Li-ion battery. They developed the PSCs charged Li-ion battery, in which, PSCs with a boost converter resulted in maximum overall efficiency of 9.9% and a high 14.9% solar to battery charging efficiency [247].

6.2. Commercial and Building Integration Confronts of PSCs

The outdoor stability of PSC is a crucial parameter for commercial application besides their high efficiency [248]. Despite the structural modification and various perovskite absorbers employment, the availability of sufficient quantities of precursors, cost-effective deposition and synthesis and easy process for a large-scale device with long-term stability are still the main challenges for PSCs (Figure 12) [249]. The crucial challenges that need to be overcome before PSCs can be considered a viable solution for extensive PV market deployment, primarily related to the stability of low bandgap perovskite. Nevertheless, the potential is unprecedented. However, there are ample of scope and opportunity to develop new and economically viable fabrication methods to develop an adequate PSC.

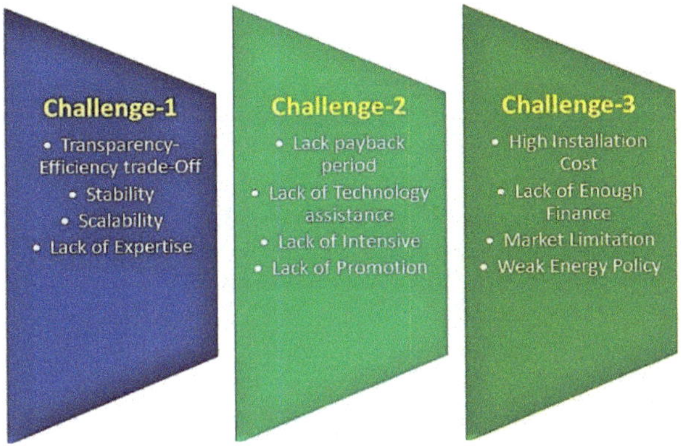

Figure 12. Challenges of a semi-transparent PSC for BIPV application.

Currently, the laboratory-scale record of PSC is at >20% level, and commercial cells are <9% PCE [250]. Stability of PSC is the main obstacle for commercialization as currently, best PSC shows one-year lifetime, which is much shorter than silicon-based 25 years stable PV [251]. Although, the current trending of PSC research signifies that the perovskite technology holds a great promise towards cost-effective manufacturing among other PV technologies. This can be achievable by variation in device architectures, incorporation of novel materials and synthesis approaches [252]. Use of such variations allows high charge transfer mobility and fewer grain boundaries between the layers to improve the performance. Despite fundamental research, synergies between environmental, socioeconomic and regional approaches are needed for perovskite research.

For the BIPV industry, the major challenge is the lack of standards [253]. There are no international standard rules for integration of PV technology into a building. Most often, the engaging parameters are PV inclination, orientation and occupant comfort. However, there are electrical cables involved for BIPV and the other electronic components [254]. Besides, Perovskite integrated into a building in northern latitude place where snow and rain are the most common phenomenon that should be considered before integration [255,256]. In the case of a hot climate and dust environment, how this technology will perform can be a significant investigating area before building integration [257–263] that must be a vital fact when considering PSC integration in BIPV.

Moreover, the fire safety norms for PSC-based BIPV should also be evaluated [264,265]. Therefore, PSC emerges as a possible candidate for BIPV application. In contrast, lack of proper building adaptive policy and stability challenges retard the PSC integration for BIPV.

7. Perspective and Discussion

Plenty of successful demonstrations of PSCs at low temperature and ambient condition processes on flexible substrates are available that can ease the fabrication process. The $MAPbI_3$-based perovskite interplays various inter-structural physicochemical interactions with the different components of the perovskites device and generate excellent optoelectronic properties for PV applications [266]. For commercialization, use of toxic Pb element should be monitored by following the Restriction of Hazardous Substances (RoHS) criteria made by the European Union [267]. There is a drive to replace Pb with Sn/Cs, which is less toxic. Nevertheless, its natural oxidation creates Sn^{4+} and low electron mobility of Cs^{2+} that originates a metal-like behaviour in the semiconductor, which lowers the photovoltaic performance. The high-efficiency evaluation of PSC is now giving robust and stiff competition to first- and second-generation solar cells. The positive inherent features of PSCs like conformability, flexibility, workability under low-light conditions, and more straightforward integration possibility in buildings as PV windows will strengthen their market entry.

A proper forecast of future energy yield is also required from the investors to make a successful integration of PSC in BIPV. This cannot be accurately achieved with only obtained STC data. Furthermore, as electricity infrastructure transforms towards smart grids, understanding variability in PV energy production will become a significant component in the design and optimization of energy exchange and storage at all temporal and spatial scales. Within this framework, the findings on the different energy yields obtained under STC and real-life temperature and irradiance data are of crucial importance [268–270]. Various PV companies such as Oxford PV, GreatCell Energy, Solar Cube, Toshiba, Frontier Energy Solution, Microquanta Semiconductor, Hunt Perovskite Technologies, Swift Solar, Sekisui Chemical, Irish Photovoltaics, Panasonic, Saule Technologies, etc. are already manufacturing large-scale PVs module [27]. Constant research activities are going on, and combined efforts from both industry and research labs are in progress to scaling-up of PSCs. It is expected that in near future performance gap will be minimal between laboratory and industrial modules. Cost per kilowatt-hour is also an essential parameter, which is dominated by the efficiency and lifetime of PV systems. Simple fabrication and material formulation anticipate less production cost of PSC compare to conventional PV technologies. The lower production cost and highly durable is more sustainable and makes the PSCs commercially competitive. The crucial challenges that need to be overcome before all-perovskite multi-junctions can be considered a viable solution ready for high-value market deployment, primarily related to the stability of low bandgap perovskite. Many scientific articles and academic studies describe the cost analysis to manufacture the PSC in BIPV technology once it is at scale. Unfortunately, there is still a lack of research to address the actual cost figure and factors affecting economies of scale for the PSC integration in BIPV. Additionally, other routes more towards customization and facilitation of integration are to be opened. Further, to besides the research and development challenges, it is necessary to address other technological aspects for PSCs integration in BIPV, which are:

- Improvement of the Perovskite technology, by increasing the stability and efficiency and reliability;
- Reduction of negative life cycle environment cost, operation and maintenance cost;
- Reduce the technological risks and develop new capabilities for improved components and systems

A reliable simulation tool is required to predict the manufacturing cost of these modules under real situation is essential. We need to have a plan as to how we get the material and the technology to scale. Thus, are we going to have commercialization of PSC in BIPV on tomorrow? The answer is no, not yet. However, we are getting closure to reach our target every day.

8. Conclusions

In summary, understanding of the underlying photo-physical, interfacial energy alignment and electronic interface structure and charge-transfer phenomenon will further help improving device structures and better selection of materials. PSC holds great promise for addressing our energy concerns. Long-term poor stability is the major issue with PSCs; however, progress has been made in the manufacturing of more giant cells as well as modules, which is indispensable for commercialization of the technology. Stability improvement of the PSC requires an interdisciplinary research approach to explore new stable materials by selecting the choice of electrodes, barrier layers, charge transport layers and encapsulation strategies. Several approaches are underway in understanding many critical issues related to the fundamental physiochemical characteristics of each component in PSC and to enhance the stability. Besides, there is a lack of proper resources to perceive the value of esthetic designs and a suitable module for building adaptive PSC. Still, the discussed characteristics may help mitigate the disadvantages of higher prices and excels more opportunity that is significant for PSCs. Tandem perovskite can be a solution for a highly stable system. The continued research and development within PSCs materials and BIPV technologies will yield a better solution for in future. In this present study, we present the opportunities for PSCs towards considerable variation in the available BIPV research. We have made an effort on the current situation of PSCs research and the crucial factor associated with it to integrate them for BIPV application.

Author Contributions: Conceptualization, A.G., methodology, A.R.; A.G., investigation, A.G., A.R., S.B., resources, A.R., A.G., S.B., writing—original draft preparation, A.R.; A.G., S.B., writing—review and editing, A.R., A.G., S.B., supervision, A.G., S.S., T.K.M., project administration, A.G., S.S., T.K.M., funding acquisition, A.G., S.S., T.K.M. All authors have read and agreed to the published version of the manuscript.

Funding: This research received no external funding.

Acknowledgments: S.B. acknowledges the College of Engineering, Mathematics and Physical Sciences, the University of Exeter for his PhD fellowship.

Conflicts of Interest: The authors declare no conflicts of interest.

References

1. Selvaraj, P.; Ghosh, A.; Mallick, T.K.; Sundaram, S. Investigation of semi-transparent dye-sensitized solar cells for fenestration integration. *Renew. Energy* **2019**, *141*, 516–525. [CrossRef]
2. Ghosh, A.; Norton, B.; Duffy, A. First outdoor characterisation of a PV powered suspended particle device switchable glazing. *Sol. Energy Mater. Sol. Cells* **2016**, *157*, 1–9. [CrossRef]
3. Ghosh, A.; Norton, B. Optimization of PV powered SPD switchable glazing to minimise probability of loss of power supply. *Renew. Energy* **2019**, *131*, 993–1001. [CrossRef]
4. Ghosh, A.; Norton, B. Durability of switching behaviour after outdoor exposure for a suspended particle device switchable glazing. *Sol. Energy Mater. Sol. Cells* **2017**, *163*, 178–184. [CrossRef]
5. Ghosh, A.; Norton, B.; Duffy, A. Measured thermal & daylight performance of an evacuated glazing using an outdoor test cell. *Appl. Energy* **2016**, *177*, 196–203. [CrossRef]
6. Ghosh, A.; Norton, B.; Duffy, A. Behaviour of a SPD switchable glazing in an outdoor test cell with heat removal under varying weather conditions. *Appl. Energy* **2016**, *180*, 695–706. [CrossRef]
7. Cannavale, A.; Ayr, U.; Martellotta, F. Energetic and visual comfort implications of using perovskite-based building-integrated photovoltaic glazings. *Energy Procedia* **2017**, *126*, 636–643. [CrossRef]
8. Jelle, B. Building Integrated Photovoltaics: A Concise Description of the Current State of the Art and Possible Research Pathways. *Energies* **2015**, *9*, 21. [CrossRef]
9. Osseweijer, F.J.; Hurk, L.B.V.D.; Teunissen, E.J.; Van Sark, W. A comparative review of building integrated photovoltaics ecosystems in selected European countries. *Renew. Sustain. Energy Rev.* **2018**, *90*, 1027–1040. [CrossRef]
10. Vasiliev, M.; Nur-E-Alam, M.; Alameh, K. Recent Developments in Solar Energy-Harvesting Technologies for Building Integration and Distributed Energy Generation. *Energies* **2019**, *12*, 1080. [CrossRef]
11. Heinstein, P.; Ballif, C.; Perret-Aebi, L.-E. Building Integrated Photovoltaics (BIPV): Review, Potentials, Barriers and Myths. *Green* **2013**, *3*, 125–156. [CrossRef]
12. Haegel, N.M.; Atwater, H.; Barnes, T.; Breyer, C.; Burrell, A.; Chiang, Y.-M.; De Wolf, S.; Dimmler, B.; Feldman, D.; Glunz, S.; et al. Terawatt-scale photovoltaics: Transform global energy. *Science* **2019**, *364*, 836–838. [CrossRef]
13. Karthick, A.; Murugavel, K.K.; Ghosh, A.; Sudhakar, K.; Ramanan, P. Investigation of a binary eutectic mixture of phase change material for building integrated photovoltaic (BIPV) system. *Sol. Energy Mater. Sol. Cells* **2020**, *207*, 110360. [CrossRef]
14. Karthick, A.; Ramanan, P.; Ghosh, A.; Stalin, B.; Kumar, R.V.; Baranilingesan, I. Performance enhancement of copper indium diselenide photovoltaic module using inorganic phase change material. *Asia-Pac. J. Chem. Eng.* **2020**, e2480. [CrossRef]
15. K, P.R.; Gupta, M.V.N.S.; Nundy, S.; Karthick, A.; Ghosh, A. Status of BIPV and BAPV System for Less Energy-Hungry Building in India—A Review. *Appl. Sci.* **2020**, *10*, 2337. [CrossRef]
16. Ghosh, A.; Norton, B.; Duffy, A. Measured overall heat transfer coefficient of a suspended particle device switchable glazing. *Appl. Energy* **2015**, *159*, 362–369. [CrossRef]
17. Ghosh, A.; Norton, B. Advances in switchable and highly insulating autonomous (self-powered) glazing systems for adaptive low energy buildings. *Renew. Energy* **2018**, *126*, 1003–1031. [CrossRef]

18. Soebarto, V.; Zhang, H.; Schiavon, S. A thermal comfort environmental chamber study of older and younger people. *Build. Environ.* **2019**, *155*, 1–14. [CrossRef]
19. Zhang, F.; De Dear, R.; Hancock, P. Effects of moderate thermal environments on cognitive performance: A multidisciplinary review. *Appl. Energy* **2019**, *236*, 760–777. [CrossRef]
20. López-Pérez, L.; Flores-Prieto, J.; Ríos-Rojas, C. Adaptive thermal comfort model for educational buildings in a hot-humid climate. *Build. Environ.* **2019**, *150*, 181–194. [CrossRef]
21. Ghosh, A.; Sundaram, S.; Mallick, T.K. Investigation of thermal and electrical performances of a combined semi-transparent PV-vacuum glazing. *Appl. Energy* **2018**, *228*, 1591–1600. [CrossRef]
22. Ghosh, A.; Sundaram, S.; Mallick, T.K. Colour properties and glazing factors evaluation of multicrystalline based semi-transparent Photovoltaic-vacuum glazing for BIPV application. *Renew. Energy* **2019**, *131*, 730–736. [CrossRef]
23. AlRashidi, H.; Ghosh, A.; Issa, W.; Sellami, N.; Mallick, T.; Sundaram, S. Thermal performance of semitransparent CdTe BIPV window at temperate climate. *Sol. Energy* **2020**, *195*, 536–543. [CrossRef]
24. AlRashidi, H.; Issa, W.; Sellami, N.; Ghosh, A.; Mallick, T.K.; Sundaram, S. Performance assessment of cadmium telluride-based semi-transparent glazing for power saving in façade buildings. *Energy Build.* **2020**, *215*, 109585. [CrossRef]
25. AlRashidi, H.; Ghosh, A.; Issa, W.; Sellami, N.; Mallick, T.; Sundaram, S. Evaluation of solar factor using spectral analysis for CdTe photovoltaic glazing. *Mater. Lett.* **2019**, *237*, 332–335. [CrossRef]
26. Roy, A.; Ghosh, A.; Bhandari, S.; Sundaram, S.; Mallick, T.K. Realization of Poly(methyl methacrylate)-Encapsulated Solution-Processed Carbon-Based Solar Cells: An Emerging Candidate for Buildings' Comfort. *Ind. Eng. Chem. Res.* **2020**, *59*, 11063–11071. [CrossRef] [PubMed]
27. Roy, P.; Sinha, N.K.; Tiwari, S.; Khare, A. A review on perovskite solar cells: Evolution of architecture, fabrication techniques, commercialization issues and status. *Sol. Energy* **2020**, *198*, 665–688. [CrossRef]
28. Roy, A.; Ghosh, A.; Bhandari, S.; Selvaraj, P.; Sundaram, S.; Mallick, T.K. Color Comfort Evaluation of Dye-Sensitized Solar Cell (DSSC) Based Building-Integrated Photovoltaic (BIPV) Glazing after 2 Years of Ambient Exposure. *J. Phys. Chem. C* **2019**, *123*, 23834–23837. [CrossRef]
29. Lucera, L.; Machui, F.; Schmidt, H.; Ahmad, T.; Kubis, P.; Strohm, S.; Hepp, J.; Vetter, A.; Egelhaaf, H.-J.; Brabec, C. Printed semi-transparent large area organic photovoltaic modules with power conversion efficiencies of close to 5 %. *Org. Electron.* **2017**, *45*, 209–214. [CrossRef]
30. ORegan, B.; Gratzel, M. A Low-Cost, High-Efficiency Solar Cell Based on Dye-Sensitized Colloidal TiO_2 Films. *Nature* **1991**, *353*, 737–739. [CrossRef]
31. Duan, J.; Xu, H.; Sha, W.E.I.; Zhao, Y.; Wang, Y.; Yang, X.; Tang, Q. Inorganic perovskite solar cells: An emerging member of the photovoltaic community. *J. Mater. Chem. A* **2019**, *7*, 21036–21068. [CrossRef]
32. Xu, Q.; Yang, D.; Lv, J.; Sun, Y.-Y.; Zhang, L. Perovskite Solar Absorbers: Materials by Design. *Small Methods* **2018**, *2*, 1700316. [CrossRef]
33. Park, N. Research Direction toward Scalable, Stable, and High Efficiency Perovskite Solar Cells. *Adv. Energy Mater.* **2020**, *10*, 1–14. [CrossRef]
34. Stoumpos, C.C.; Kanatzidis, M.G. The Renaissance of Halide Perovskites and Their Evolution as Emerging Semiconductors. *Acc. Chem. Res.* **2015**, *48*, 2791–2802. [CrossRef]
35. Niu, G.; Guo, X.; Wang, L. Review of recent progress in chemical stability of perovskite solar cells. *J. Mater. Chem. A* **2015**, *3*, 8970–8980. [CrossRef]
36. Gao, P.; Grätzel, M.; Nazeeruddin, M.K. Organohalide lead perovskites for photovoltaic applications. *Energy Environ. Sci.* **2014**, *7*, 2448–2463. [CrossRef]
37. Green, M.A.; Ho-Baillie, A.W.Y.; Snaith, H.J. The emergence of perovskite solar cells. *Nat. Photonics* **2014**, *8*, 506–514. [CrossRef]
38. A decade of perovskite photovoltaics. *Nat. Energy* **2019**, *4*, 1. Available online: https://www.nature.com/articles/s41560-018-0323-9 (accessed on 6 May 2020). [CrossRef]
39. Ansari, M.I.H.; Qurashi, A.; Nazeeruddin, M.K. Frontiers, opportunities, and challenges in perovskite solar cells: A critical review. *J. Photochem. Photobiol. C Photochem. Rev.* **2018**, *35*, 1–24. [CrossRef]
40. Jena, A.K.; Kulkarni, A.; Miyasaka, T. Halide Perovskite Photovoltaics: Background, Status, and Future Prospects. *Chem. Rev.* **2019**, *119*, 3036–3103. [CrossRef]
41. Seok, S.I.; Grätzel, M.; Park, N.-G. Methodologies toward Highly Efficient Perovskite Solar Cells. *Small* **2018**, *14*, e1704177. [CrossRef] [PubMed]

42. Kojima, A.; Teshima, K.; Shirai, Y.; Miyasaka, T. Organometal halide perovskites as visible-light sensitizers for photovoltaic cells. *J. Am. Chem. Soc.* **2009**, *131*, 6050–6051. [CrossRef]
43. Tang, H.; He, S.; Peng, C. A Short Progress Report on High-Efficiency Perovskite Solar Cells. *Nanosi. Res. Lett.* **2017**, *12*, 410. [CrossRef] [PubMed]
44. NERL Best Research-Cell Efficiency Chart. Available online: https://www.nrel.gov/pv/cell-efficiency.html (accessed on 5 May 2020).
45. Minemoto, T.; Murata, M. Device modeling of perovskite solar cells based on structural similarity with thin film inorganic semiconductor solar cells. *J. Appl. Phys.* **2014**, *116*, 54505. [CrossRef]
46. Yang, W.S.; Noh, J.H.; Jeon, N.J.; Kim, Y.C.; Ryu, S.; Seo, J.; Seok, S.I. High-performance photovoltaic perovskite layers fabricated through intramolecular exchange. *Science* **2015**, *348*, 1234–1237. [CrossRef] [PubMed]
47. Deng, Y.; Dong, Q.; Bi, C.; Yuan, Y.; Huang, J. Air-Stable, Efficient Mixed-Cation Perovskite Solar Cells with Cu Electrode by Scalable Fabrication of Active Layer. *Adv. Energy Mater.* **2016**, *6*, e1600372. [CrossRef]
48. Yu, Y.; Wang, C.; Grice, C.R.; Shrestha, N.; Chen, J.; Zhao, D.; Liao, W.; Cimaroli, A.J.; Roland, P.J.; Ellingson, R.J.; et al. Improving the Performance of Formamidinium and Cesium Lead Triiodide Perovskite Solar Cells using Lead Thiocyanate Additives. *ChemSusChem* **2016**, *9*, 3288–3297. [CrossRef]
49. Saliba, M.; Matsui, T.; Seo, J.-Y.; Domanski, K.; Correa-Baena, J.-P.; Nazeeruddin, M.K.; Zakeeruddin, S.M.; Tress, W.; Abate, A.; Hagfeldt, A.; et al. Cesium-containing triple cation perovskite solar cells: Improved stability, reproducibility and high efficiency. *Energy Environ. Sci.* **2016**, *9*, 1989–1997. [CrossRef]
50. Zhao, J.; Zheng, X.; Deng, Y.; Li, T.; Shao, Y.; Gruverman, A.; Shield, J.; Huang, J. Is Cu a stable electrode material in hybrid perovskite solar cells for a 30-year lifetime? *Energy Environ. Sci.* **2016**, *9*, 3650–3656. [CrossRef]
51. Li, X.; Bi, D.; Yi, C.; Décoppet, J.-D.; Luo, J.; Zakeeruddin, S.M.; Hagfeldt, A.; Grätzel, M. A vacuum flash–assisted solution process for high-efficiency large-area perovskite solar cells. *Science* **2016**, *353*, 58–62. [CrossRef]
52. Yang, W.S.; Park, B.-W.; Jung, E.H.; Jeon, N.J.; Kim, Y.C.; Lee, D.U.; Shin, S.S.; Seo, J.; Kim, E.K.; Noh, J.H.; et al. Iodide management in formamidinium-lead-halide–based perovskite layers for efficient solar cells. *Science* **2017**, *356*, 1376–1379. [CrossRef] [PubMed]
53. Jung, E.H.; Jeon, N.J.; Park, E.Y.; Moon, C.S.; Shin, T.J.; Yang, T.-Y.; Noh, J.H.; Seo, J. Efficient, stable and scalable perovskite solar cells using poly(3-hexylthiophene). *Natural* **2019**, *567*, 511–515. [CrossRef] [PubMed]
54. Gao, X.-X.; Luo, W.; Zhang, Y.; Hu, R.; Zhang, B.; Züttel, A.; Feng, Y.; Nazeeruddin, M.K. Stable and High-Efficiency Methylammonium-Free Perovskite Solar Cells. *Adv. Mater.* **2020**, *32*, e1905502. [CrossRef]
55. OXFORDPV. Oxford PV Perovskite Solar Cell Achieves 28% Efficiency. Available online: https://www.oxfordpv.com/news/oxford-pv-perovskite-solar-cell-achieves-28-efficiency (accessed on 5 May 2020).
56. Fujiwara, H. Almost 40% Conversion Efficiency Predicted in New Perovskite Solar Cell. 2020. Available online: https://www.advancedsciencenews.com/almost-40-conversion-efficiency-predicted-in-new-perovskite-solar-cell/ (accessed on 31 January 2020).
57. Yin, W.-J.; Yang, J.-H.; Kang, J.; Yan, Y.; Wei, S.-H. Halide perovskite materials for solar cells: A theoretical review. *J. Mater. Chem. A* **2015**, *3*, 8926–8942. [CrossRef]
58. Nishigaki, Y.; Nagai, T.; Nishiwaki, M.; Aizawa, T.; Kozawa, M.; Hanzawa, K.; Kato, Y.; Sai, H.; Hiramatsu, H.; Hosono, H.; et al. Extraordinary Strong Band-Edge Absorption in Distorted Chalcogenide Perovskites. *Sol. RRL* **2020**, *4*. [CrossRef]
59. Dupre, O.; Niesen, B.; De Wolf, S.; Ballif, C. Field Performance versus Standard Test Condition Efficiency of Tandem Solar Cells and the Singular Case of Perovskites/Silicon Devices. *J. Phys. Chem. Lett.* **2018**, *9*, 446–458. [CrossRef]
60. Lee, S.-W.; Kim, S.; Bae, S.; Cho, K.; Chung, T.; Mundt, L.E.; Lee, S.; Park, S.; Park, H.; Schubert, M.C.; et al. UV Degradation and Recovery of Perovskite Solar Cells. *Sci. Rep.* **2016**, *6*, 38150. [CrossRef]
61. Fu, Q.; Tang, X.; Huang, B.; Hu, T.; Tan, L.; Chen, L.; Chen, Y. Recent Progress on the Long-Term Stability of Perovskite Solar Cells. *Adv. Sci.* **2018**, *5*, 1700387. [CrossRef]
62. Yi, Z.; Ladi, N.H.; Shai, X.; Li, H.; Shen, Y.; Wang, M.-K. Will organic–inorganic hybrid halide lead perovskites be eliminated from optoelectronic applications? *Nanoscale Adv.* **2019**, *1*, 1276–1289. [CrossRef]
63. Yang, J.; Siempelkamp, B.D.; Mosconi, E.; De Angelis, F.; Kelly, T.L. Origin of the Thermal Instability in CH3NH3PbI3 Thin Films Deposited on ZnO. *Chem. Mater.* **2015**, *27*, 4229–4236. [CrossRef]

64. Conings, B.; Drijkoningen, J.; Gauquelin, N.; Babayigit, A.; D'Haen, J.; D'Olieslaeger, L.; Ethirajan, A.; Verbeeck, J.; Manca, J.V.; Mosconi, E.; et al. Intrinsic Thermal Instability of Methylammonium Lead Trihalide Perovskite. *Adv. Energy Mater.* **2015**, *5*, 1500477. [CrossRef]
65. Kwon, Y.S.; Lim, J.; Yun, H.-J.; Kim, Y.-H.; Park, T. A diketopyrrolopyrrole-containing hole transporting conjugated polymer for use in efficient stable organic–inorganic hybrid solar cells based on a perovskite. *Energy Environ. Sci.* **2014**, *7*, 1454. [CrossRef]
66. Philippe, B.; Park, B.-W.; Lindblad, R.; Oscarsson, J.; Ahmadi, S.; Johansson, E.M.J.; Rensmo, H. Chemical and Electronic Structure Characterization of Lead Halide Perovskites and Stability Behavior under Different Exposures—A Photoelectron Spectroscopy Investigation. *Chem. Mater.* **2015**, *27*, 1720–1731. [CrossRef]
67. Christians, J.A.; Herrera, P.A.M.; Kamat, P.V. Transformation of the Excited State and Photovoltaic Efficiency of $CH_3NH_3PbI_3$ Perovskite upon Controlled Exposure to Humidified Air. *J. Am. Chem. Soc.* **2015**, *137*, 1530–1538. [CrossRef]
68. Aristidou, N.; Sanchez-Molina, I.; Chotchuangchutchaval, T.; Brown, M.; Martínez, L.; Rath, T.; Haque, S.A. The Role of Oxygen in the Degradation of Methylammonium Lead Trihalide Perovskite Photoactive Layers. *Angew. Chem. Int. Ed.* **2015**, *54*, 8208–8212. [CrossRef]
69. Wang, R.; Mujahid, M.; Duan, Y.; Wang, Z.-K.; Xue, J.; Yang, Y. A Review of Perovskites Solar Cell Stability. *Adv. Funct. Mater.* **2019**, *29*, 1–25. [CrossRef]
70. Leijtens, T.; Eperon, G.E.; Pathak, S.; Abate, A.; Lee, M.M.; Snaith, H.J. Overcoming ultraviolet light instability of sensitized TiO_2 with meso-superstructured organometal tri-halide perovskite solar cells. *Nat. Commun.* **2013**, *4*, 2885. [CrossRef]
71. Liu, D.; Gangishetty, M.K.; Kelly, T.L. Effect of $CH_3NH_3PbI_3$ thickness on device efficiency in planar heterojunction perovskite solar cells. *J. Mater. Chem. A* **2014**, *2*, 19873–19881. [CrossRef]
72. Wang, Q.; Chen, B.; Liu, Y.; Deng, Y.; Bai, Y.; Dong, Q.; Huang, J. Scaling behavior of moisture-induced grain degradation in polycrystalline hybrid perovskite thin films. *Energy Environ. Sci.* **2017**, *10*, 516–522. [CrossRef]
73. Ran, C.; Xua, J.; Gao, W.; Huang, C.; Dou, S.X. Defects in metal triiodide perovskite materials towards high-performance solar cells: Origin, impact, characterization, and engineering. *Chem. Soc. Rev.* **2018**, *47*, 4581–4610. [CrossRef]
74. Rong, Y.; Liu, L.; Mei, A.; Li, X.; Han, H. Beyond Efficiency: The Challenge of Stability in Mesoscopic Perovskite Solar Cells. *Adv. Energy Mater.* **2015**, *5*, 1501066. [CrossRef]
75. Supasai, T.; Rujisamphan, N.; Ullrich, K.; Chemseddine, A.; Dittrich, T. Formation of a passivating $CH_3NH_3PbI_3/PbI_2$ interface during moderate heating of $CH_3NH_3PbI_3$ layers. *Appl. Phys. Lett.* **2013**, *103*, 183906. [CrossRef]
76. Kim, H.-S.; Seo, J.-Y.; Park, N.-G. Material and Device Stability in Perovskite Solar Cells. *ChemSusChem* **2016**, *9*, 2528–2540. [CrossRef] [PubMed]
77. Lian, J.; Lu, B.; Niu, F.; Zeng, P.; Zhan, X. Electron-Transport Materials in Perovskite Solar Cells. *Small Methods* **2018**, *2*, e1800082. [CrossRef]
78. Krishna, A.; Grimsdale, A.C. Hole transporting materials for mesoscopic perovskite solar cells—towards a rational design? *J. Mater. Chem. A* **2017**, *5*, 16446–16466. [CrossRef]
79. Seo, S.; Jeong, S.; Bae, C.; Park, N.-G.; Shin, H. Perovskite Solar Cells with Inorganic Electron- and Hole-Transport Layers Exhibiting Long-Term (≈500 h) Stability at 85 °C under Continuous 1 Sun Illumination in Ambient Air. *Adv. Mater.* **2018**, *30*, e1801010. [CrossRef]
80. Kato, Y.; Ono, L.K.; Lee, M.; Wang, S.; Raga, S.R.; Qi, Y.B. Silver Iodide Formation in Methyl Ammonium Lead Iodide Perovskite Solar Cells with Silver Top Electrodes. *Adv. Mater. Interfaces* **2015**, *2*, 1500195. [CrossRef]
81. Amat, A.; Mosconi, E.; Ronca, E.; Quarti, C.; Umari, P.; Nazeeruddin, M.K.; Grätzel, M.; De Angelis, F. Cation-Induced Band-Gap Tuning in Organohalide Perovskites: Interplay of Spin–Orbit Coupling and Octahedra Tilting. *Nano Lett.* **2014**, *14*, 3608–3616. [CrossRef] [PubMed]
82. Koh, T.M.; Shanmugam, V.; Schlipf, J.; Oesinghaus, L.; Müller-Buschbaum, P.; Ramakrishnan, N.; Swamy, V.; Mathews, N.; Boix, P.P.; Mhaisalkar, S.G. Nanostructuring Mixed-Dimensional Perovskites: A Route Toward Tunable, Efficient Photovoltaics. *Adv. Mater.* **2016**, *28*, 3653–3661. [CrossRef] [PubMed]
83. Cao, D.H.; Stoumpos, C.C.; Farha, O.K.; Hupp, J.T.; Kanatzidis, M.G. 2D Homologous Perovskites as Light-Absorbing Materials for Solar Cell Applications. *J. Am. Chem. Soc.* **2015**, *137*, 7843–7850. [CrossRef]

84. Shao, S.; Liu, J.; Portale, G.; Fang, H.-H.; Blake, G.R.; Brink, G.H.T.; Koster, L.J.A.; Loi, M.A. Highly Reproducible Sn-Based Hybrid Perovskite Solar Cells with 9% Efficiency. *Adv. Energy Mater.* **2017**, *8*, 1702019. [CrossRef]
85. Tsai, H.; Nie, W.; Blancon, J.-C.; Stoumpos, C.C.; Asadpour, R.; Harutyunyan, B.; Neukirch, A.J.; Verduzco, R.; Crochet, J.J.; Tretiak, S.; et al. High-efficiency two-dimensional Ruddlesden–Popper perovskite solar cells. *Natural* **2016**, *536*, 312–316. [CrossRef] [PubMed]
86. Cruz, S.H.T.; Hagfeldt, A.; Saliba, M. Methylammonium-free, high-performance, and stable perovskite solar cells on a planar architecture. *Science* **2018**, *362*, 449–453. [CrossRef] [PubMed]
87. Arora, N.; Dar, M.I.; Hinderhofer, A.; Pellet, N.; Schreiber, F.; Zakeeruddin, S.M.; Grätzel, M. Perovskite solar cells with CuSCN hole extraction layers yield stabilized efficiencies greater than 20%. *Science* **2017**, *358*, 768–771. [CrossRef]
88. Hou, Y.; Du, X.; Scheiner, S.; McMeekin, D.P.; Wang, Z.; Li, N.; Killian, M.S.; Chen, H.; Richter, M.; Levchuk, I.; et al. A generic interface to reduce the efficiency-stability-cost gap of perovskite solar cells. *Science* **2017**, *358*, 1192–1197. [CrossRef] [PubMed]
89. Chen, C.; Cheng, Y.; Dai, Q.; Song, H. Radio Frequency Magnetron Sputtering Deposition of TiO2 Thin Films and Their Perovskite Solar Cell Applications. *Sci. Rep.* **2015**, *5*, 17684. [CrossRef] [PubMed]
90. Fakharuddin, A.; Di Giacomo, F.; Palma, A.L.; Matteocci, F.; Razza, S.; Epifanio, A.D.; Licoccia, S.; Ismail, J.; Carlo, A.D.; Brown, T.M.; et al. Vertical TiO_2 Nanorods as a Medium for Durable and High Efficiency Perovskite Solar Modules. *Chem. Sci.* **2015**, *9*, 8420–8429. [CrossRef]
91. Qiu, W.; Paetzold, U.W.; Gehlhaar, R.; Smirnov, V.; Boyen, H.-G.; Tait, J.G.; Conings, B.; Zhang, W.; Nielsen, C.B.; McCulloch, I.; et al. An electron beam evaporated TiO_2 layer for high efficiency planar perovskite solar cells on flexible polyethylene terephthalate substrates. *J. Mater. Chem. A* **2015**, *3*, 22824–22829. [CrossRef]
92. Chen, H.; Yang, S. Carbon-Based Perovskite Solar Cells without Hole Transport Materials: The Front Runner to the Market? *Adv. Mater.* **2017**, *29*, 1603994. [CrossRef]
93. Collavini, S.; Delgado, J.L. Carbon Nanoforms in Perovskite-Based Solar Cells. *Adv. Energy Mater.* **2016**, *7*, 1601000. [CrossRef]
94. Torabi, N.; Behjat, A.; Zhou, Y.; Docampo, P.; Stoddard, R.J.; Hillhouse, H.W.; Ameri, T. Progress and challenges in perovskite photovoltaics from single- to multi-junction cells. *Mater. Today Energy* **2019**, *12*, 70–94. [CrossRef]
95. Barbe, J.M.; Tietze, M.L.; Neophytou, M.; Banavoth, M.; Alarousu, E.; El Labban, A.; Abulikemu, M.; Yue, W.; Mohammed, O.F.; McCulloch, I.; et al. Amorphous Tin Oxide as a Low-Temperature-Processed Electron-Transport Layer for Organic and Hybrid Perovskite Solar Cells. *ACS Appl. Mater. Interfaces* **2017**, *9*, 11828–11836. [CrossRef] [PubMed]
96. Anaraki, E.H.; Kermanpur, A.; Steier, L.; Domanski, K.; Matsui, T.; Tress, W.; Saliba, M.; Abate, A.; Grätzel, M.; Hagfeldt, A.; et al. Highly efficient and stable planar perovskite solar cells by solution-processed tin oxide. *Energy Environ. Sci.* **2016**, *9*, 3128–3134. [CrossRef]
97. Yu, W.; Li, F.; Wang, H.; Alarousu, E.; Chen, Y.; Ling, Y.; Wang, L.; Hedhili, M.N.; Li, Y.; Wu, K.; et al. Ultrathin Cu_2O as an efficient inorganic hole transporting material for perovskite solar cells. *Nanoscale* **2016**, *8*, 6173–6179. [CrossRef] [PubMed]
98. You, J.; Meng, L.; Song, T.-B.; Guo, T.-F.; Yang, Y.; Chang, W.-H.; Hong, Z.; Chen, H.; Zhou, H.; Chen, Q.; et al. Improved air stability of perovskite solar cells via solution-processed metal oxide transport layers. *Nat. Nanotechnol.* **2015**, *11*, 75–81. [CrossRef] [PubMed]
99. Nejand, B.A.; Ahmadi, V.; Shahverdi, H.R. New Physical Deposition Approach for Low Cost Inorganic Hole Transport Layer in Normal Architecture of Durable Perovskite Solar Cells. *ACS Appl. Mater. Interfaces* **2015**, *7*, 21807–21818. [CrossRef] [PubMed]
100. Higgins, M.; Ely, F.; Nome, R.C.; Nome, R.A.; Dos Santos, D.P.; Choi, H.; Nam, S.; Quevedo-Lopez, M. Enhanced reproducibility of planar perovskite solar cells by fullerene doping with silver nanoparticles. *J. Appl. Phys.* **2018**, *124*, e065306. [CrossRef]
101. Li, S.; Zhang, H.; Zhao, W.; Ye, L.; Yao, H.; Yang, B.; Zhang, S.; Hou, J. Green-Solvent-Processed All-Polymer Solar Cells Containing a Perylene Diimide-Based Acceptor with an Efficiency over 6.5%. *Adv. Energy Mater.* **2015**, *6*, e1501991. [CrossRef]

102. Zheng, H.; Liu, G.; Zhang, C.-N.; Zhu, L.; Alsaedi, A.; Hayat, T.; Pan, X.; Dai, S. The influence of perovskite layer and hole transport material on the temperature stability about perovskite solar cells. *Sol. Energy* **2018**, *159*, 914–919. [CrossRef]
103. Ponseca, J.C.S.; Chábera, P.; Uhlig, J.; Persson, P.; Sundström, V. Ultrafast Electron Dynamics in Solar Energy Conversion. *Chem. Rev.* **2017**, *117*, 10940–11024. [CrossRef]
104. Bashir, A.; Shukla, S.; Lew, J.H.; Shukla, S.; Bruno, A.; Gupta, D.; Baikie, T.; Patidar, R.; Akhter, Z.; Priyadarshi, A.; et al. Spinel Co_3O_4 nanomaterials for efficient and stable large area carbon-based printed perovskite solar cells. *Nanoscale* **2018**, *10*, 2341–2350. [CrossRef] [PubMed]
105. Xu, X.; Liu, Z.; Zuo, Z.; Zhang, M.; Zhao, Z.; Shen, Y.; Zhou, H.; Chen, Q.; Yang, Y.; Wang, M.-K. Hole Selective NiO Contact for Efficient Perovskite Solar Cells with Carbon Electrode. *Nano Lett.* **2015**, *15*, 2402–2408. [CrossRef] [PubMed]
106. Hu, R.; Zhang, R.; Ma, Y.; Liu, W.; Chu, L.; Mao, W.; Zhang, J.; Yang, J.; Pu, Y.; Li, X. Enhanced hole transfer in hole-conductor-free perovskite solar cells via incorporating CuS into carbon electrodes. *Appl. Surf. Sci.* **2018**, *462*, 840–846. [CrossRef]
107. Ku, Z.; Rong, Y.; Xu, M.; Liu, T.; Han, H. Full Printable Processed Mesoscopic $CH_3NH_3PbI_3/TiO_2$ Heterojunction Solar Cells with Carbon Counter Electrode. *Sci. Rep.* **2013**, *3*, 3132. [CrossRef]
108. Liu, S.; Cao, K.; Li, H.; Song, J.; Han, J.; Shen, Y.; Wang, M.-K. Full printable perovskite solar cells based on mesoscopic TiO_2/Al_2O_3/NiO (carbon nanotubes) architecture. *Sol. Energy* **2017**, *144*, 158–165. [CrossRef]
109. Bhandari, S.; Roy, A.; Ghosh, A.; Mallick, T.K.; Sundaram, S. Performance of WO_3-Incorporated Carbon Electrodes for Ambient Mesoscopic Perovskite Solar Cells. *ACS Omega* **2019**, *5*, 422–429. [CrossRef]
110. Kettle, J. Ready cells for large-scale systems. *Nat. Energy* **2019**, *4*, 536–537. [CrossRef]
111. Tress, W.; Domanski, K.; Carlsen, B.; Agarwalla, A.; Alharbi, E.A.; Graetzel, M.; Hagfeldt, A. Performance of perovskite solar cells under simulated temperature-illumination real-world operating conditions. *Nat. Energy* **2019**, *4*, 568–574. [CrossRef]
112. Babayigit, A.; Ethirajan, A.; Muller, M.; Conings, B. Toxicity of organometal halide perovskite solar cells. *Nat. Mater.* **2016**, *15*, 247–251. [CrossRef]
113. Luceño, J.A.; Pascual, A.M.D.; Capilla, R.P. Materials for Photovoltaics: State of Art and Recent Developments. *Int. J. Mol. Sci.* **2019**, *20*, 976. [CrossRef]
114. Hailegnaw, B.; Kirmayer, S.; Edri, E.; Hodes, G.; Cahen, D. Rain on Methylammonium Lead Iodide Based Perovskites: Possible Environmental Effects of Perovskite Solar Cells. *J. Phys. Chem. Lett.* **2015**, *6*, 1543–1547. [CrossRef] [PubMed]
115. Li, J.; Cao, H.-L.; Jiao, W.-B.; Wang, Q.; Wei, M.; Cantone, I.; Lü, J.; Abate, A. Biological impact of lead from halide perovskites reveals the risk of introducing a safe threshold. *Nat. Commun.* **2020**, *11*, 1–5. [CrossRef]
116. Su, P.; Liu, Y.; Zhang, J.; Chen, C.; Yang, B.; Zhang, C.; Zhao, X. Pb-Based Perovskite Solar Cells and the Underlying Pollution behind Clean Energy: Dynamic Leaching of Toxic Substances from Discarded Perovskite Solar Cells. *J. Phys. Chem. Lett.* **2020**, *11*, 2812–2817. [CrossRef]
117. Babayigit, A.; Duy-Thanh, D.; Ethirajan, A.; Manca, J.; Muller, M.; Boyen, H.-G.; Conings, B. Assessing the toxicity of Pb- and Sn-based perovskite solar cells in model organism Danio rerio. *Sci. Rep.* **2016**, *6*, 18721. [CrossRef]
118. Wang, Z.; Fang, J.; Mi, Y.; Zhu, X.; Ren, H.; Liu, X.; Yan, Y. Enhanced performance of perovskite solar cells by ultraviolet-ozone treatment of mesoporous TiO. *Appl. Surf. Sci.* **2018**, *436*, 596–602. [CrossRef]
119. Hasan, S.A.U.; Lee, D.S.; Im, S.H.; Hong, K.-H. Present Status and Research Prospects of Tin-based Perovskite Solar Cells. *Sol. RRL* **2019**, *4*, 4. [CrossRef]
120. Parida, B.; Yoon, S.; Jeong, S.M.; Cho, J.S.; Kim, J.-K.; Kang, D.-W. Recent progress on cesium lead/tin halide-based inorganic perovskites for stable and efficient solar cells: A review. *Sol. Energy Mater. Sol. Cells* **2020**, *204*, 110212. [CrossRef]
121. Sun, Y.-Y.; Agiorgousis, M.L.; Zhang, P.; Zhang, S. Chalcogenide Perovskites for Photovoltaics. *Nano Lett.* **2015**, *15*, 581–585. [CrossRef]
122. Khalfin, S.; Bekenstein, Y. Advances in lead-free double perovskite nanocrystals, engineering band-gaps and enhancing stability through composition tunability. *Nanoscale* **2019**, *11*, 8665–8679. [CrossRef]
123. Chen, L.-J.; Lee, C.-R.; Chuang, Y.-J.; Wu, Z.-H.; Chen, C. Synthesis and Optical Properties of Lead-Free Cesium Tin Halide Perovskite Quantum Rods with High-Performance Solar Cell Application. *J. Phys. Chem. Lett.* **2016**, *7*, 5028–5035. [CrossRef]

124. Kumar, M.H.; Dharani, S.; Leong, W.L.; Boix, P.P.; Prabhakar, R.R.; Baikie, T.; Shi, C.; Ding, H.; Ramesh, R.; Asta, M.; et al. Lead-Free Halide Perovskite Solar Cells with High Photocurrents Realized Through Vacancy Modulation. *Adv. Mater.* **2014**, *26*, 7122–7127. [CrossRef]
125. Chen, M.; Ju, M.-G.; Carl, A.D.; Zong, Y.; Grimm, R.L.; Gu, J.; Zeng, X.C.; Zhou, Y.; Padture, N.P. Cesium Titanium(IV) Bromide Thin Films Based Stable Lead-free Perovskite Solar Cells. *Joule* **2018**, *2*, 558–570. [CrossRef]
126. Jiang, X.; Wang, F.; Wei, Q.; Li, H.; Shang, Y.; Zhou, W.; Wang, C.; Cheng, P.; Chen, Q.; Chen, L.; et al. Ultra-high open-circuit voltage of tin perovskite solar cells via an electron transporting layer design. *Nat. Commun.* **2020**, *11*, 1–7. [CrossRef]
127. Hao, F.; Stoumpos, C.C.; Cao, D.H.; Chang, R.P.H.; Kanatzidis, M.G. Lead-free solid-state organic–inorganic halide perovskite solar cells. *Nat. Photonics* **2014**, *8*, 489–494. [CrossRef]
128. Zhu, Z.; Chueh, C.-C.; Li, N.; Mao, C.; Jen, A.K.-Y. Realizing Efficient Lead-Free Formamidinium Tin Triiodide Perovskite Solar Cells via a Sequential Deposition Route. *Adv. Mater.* **2017**, *30*, 1703800. [CrossRef]
129. Nishikubo, R.; Ishida, N.; Katsuki, Y.; Wakamiya, A.; Saeki, A. Minute-Scale Degradation and Shift of Valence-Band Maxima of $(CH_3NH_3)SnI_3$ and $HC(NH_2)_2SnI_3$ Perovskites upon Air Exposure. *J. Phys. Chem. C* **2017**, *121*, 19650–19656. [CrossRef]
130. Greul, E.; Petrus, M.L.; Binek, A.; Docampo, P.; Bein, T. Highly stable, phase pure $Cs_2AgBiBr_6$ double perovskite thin films for optoelectronic applications. *J. Mater. Chem. A* **2017**, *5*, 19972–19981. [CrossRef]
131. Zhang, Z.; Li, X.; Xia, X.; Wang, Z.; Huang, Z.; Lei, B.; Gao, Y. High-Quality $(CH_3NH_3)_3Bi_2I_9$ Film-Based Solar Cells: Pushing Efficiency up to 1.64%. *J. Phys. Chem. Lett.* **2017**, *8*, 4300–4307. [CrossRef]
132. Nie, R.; Mehta, A.; Park, B.-W.; Kwon, H.-W.; Im, J.; Seok, S.I. Mixed Sulfur and Iodide-Based Lead-Free Perovskite Solar Cells. *J. Am. Chem. Soc.* **2018**, *140*, 872–875. [CrossRef]
133. Wei, F.; Deng, Z.; Sun, S.-J.; Xie, F.; Kieslich, G.; Evans, D.M.; A Carpenter, M.; Bristowe, P.D.; Cheetham, A.K. The synthesis, structure and electronic properties of a lead-free hybrid inorganic–organic double perovskite $(MA)_2KBiCl_6$ (MA = methylammonium). *Mater. Horiz.* **2016**, *3*, 328–332. [CrossRef]
134. Ghosh, A.; Norton, B. Interior colour rendering of daylight transmitted through a suspended particle device switchable glazing. *Sol. Energy Mater. Sol. Cells* **2017**, *163*, 218–223. [CrossRef]
135. Ghosh, A.; Norton, B.; Mallick, T.K. Daylight characteristics of a polymer dispersed liquid crystal switchable glazing. *Sol. Energy Mater. Sol. Cells* **2018**, *174*, 572–576. [CrossRef]
136. Ghosh, A.; Norton, B.; Mallick, T.K. Influence of atmospheric clearness on PDLC switchable glazing transmission. *Energy Build.* **2018**, *172*, 257–264. [CrossRef]
137. Ghosh, A.; Mallick, T. Evaluation of colour properties due to switching behaviour of a PDLC glazing for adaptive building integration. *Renew. Energy* **2018**, *120*, 126–133. [CrossRef]
138. Nundy, S.; Ghosh, A. Thermal and visual comfort analysis of adaptive vacuum integrated switchable suspended particle device window for temperate climate. Renew. *Energy* **2019**. [CrossRef]
139. Ghosh, A.; Selvaraj, P.; Sundaram, S.; Mallick, T.K. The colour rendering index and correlated colour temperature of dye-sensitized solar cell for adaptive glazing application. *Sol. Energy* **2018**, *163*, 537–544. [CrossRef]
140. Traverse, C.J.; Pandey, R.; Barr, M.C.; Lunt, R.R. Correction: Emergence of highly transparent photovoltaics for distributed applications (Nature Energy (2017) doi:10.1038/s41560-017-0016-9). *Nat. Energy* **2018**, *3*, 157. [CrossRef]
141. Yuan, L.; Wang, Z.; Duan, R.; Huang, P.; Zhang, K.; Chen, Q.; Allam, N.K.; Zhou, Y.; Song, B.; Li, Y. Semi-transparent perovskite solar cells: Unveiling the trade-off between transparency and efficiency. *J. Mater. Chem. A* **2018**, *6*, 19696–19702. [CrossRef]
142. Jung, J.W.; Chueh, C.-C.; Jen, A.K.-Y. High-Performance Semitransparent Perovskite Solar Cells with 10% Power Conversion Efficiency and 25% Average Visible Transmittance Based on Transparent CuSCN as the Hole-Transporting Material. *Adv. Energy Mater.* **2015**, *5*, 1500486. [CrossRef]
143. Ghosh, A.; Norton, B.; Duffy, A. Effect of sky conditions on light transmission through a suspended particle device switchable glazing. *Sol. Energy Mater. Sol. Cells* **2017**, *160*, 134–140. [CrossRef]
144. Ghosh, A.; Norton, B.; Duffy, A. Effect of atmospheric transmittance on performance of adaptive SPD-vacuum switchable glazing. *Sol. Energy Mater. Sol. Cells* **2017**, *161*, 424–431. [CrossRef]
145. Sun, J.; Jasieniak, J. Semi-transparent solar cells. *J. Phys. D: Appl. Phys.* **2017**, *50*, 093001. [CrossRef]

146. Fu, F.; Feurer, T.; Weiss, T.P.; Pisoni, S.; Avancini, E.; Andres, C.; Buecheler, S.; Tiwari, A.N. High-efficiency inverted semi-transparent planar perovskite solar cells in substrate configuration. *Nat. Energy* **2016**, *2*, 16190. [CrossRef]
147. Novak, S.; Hillebrandt, M. Analysing the trade-off between transparency and efficiency in the Council of the European Union. *J. Eur. Public Policy* **2019**, *27*, 141–159. [CrossRef]
148. Shin, D.; Choi, S.-H. Recent Studies of Semitransparent Solar Cells. *Coatings* **2018**, *8*, 329. [CrossRef]
149. Ghosh, A.; Sarmah, N.; Sundaram, S.; Mallick, T.K. Numerical studies of thermal comfort for semi-transparent building integrated photovoltaic (BIPV)-vacuum glazing system. *Sol. Energy* **2019**, *190*, 608–616. [CrossRef]
150. Roy, A.; Bhandari, S.; Ghosh, A.; Sundaram, S.; Mallick, T.K. Incorporating Solution-Processed Mesoporous WO3 as an Interfacial Cathode Buffer Layer for Photovoltaic Applications. *J. Phys. Chem. A* **2020**. [CrossRef]
151. Kwon, H.-C.; Moon, J. Recent advances in high-performance semitransparent perovskite solar cells. *Curr. Opin. Electrochem.* **2018**, *11*, 114–121. [CrossRef]
152. Husain, A.A.; Hasan, W.Z.W.; Shafie, S.; Hamidon, M.N.; Pandey, S.S. A review of transparent solar photovoltaic technologies. *Renew. Sustain. Energy Rev.* **2018**, *94*, 779–791. [CrossRef]
153. Shi, B.; Duan, L.; Zhao, Y.; Luo, J.; Zhang, X. Semitransparent Perovskite Solar Cells: From Materials and Devices to Applications. *Adv. Mater.* **2019**, *32*, e1806474. [CrossRef]
154. Leijtens, T.; Lauber, B.; Eperon, G.E.; Stranks, S.D.; Snaith, H.J. The Importance of Perovskite Pore Filling in Organometal Mixed Halide Sensitized TiO_2-Based Solar Cells. *J. Phys. Chem. Lett.* **2014**, *5*, 1096–1102. [CrossRef]
155. Stranks, S.D.; Eperon, G.E.; Grancini, G.; Menelaou, C.; Alcocer, M.; Leijtens, T.; Herz, L.M.; Petrozza, A.; Snaith, H.J. Electron-Hole Diffusion Lengths Exceeding 1 Micrometer in an Organometal Trihalide Perovskite Absorber. *Science* **2013**, *342*, 341–344. [CrossRef] [PubMed]
156. Malinkiewicz, O.; Yella, A.; Lee, Y.H.; Espallargas, G.M.; Graetzel, M.; Nazeeruddin, M.K.; Bolink, H.J. Perovskite solar cells employing organic charge-transport layers. *Nat. Photonics* **2013**, *8*, 128–132. [CrossRef]
157. Liu, D.; Kelly, T.L. Perovskite solar cells with a planar heterojunction structure prepared using room-temperature solution processing techniques. *Nat. Photonics* **2013**, *8*, 133–138. [CrossRef]
158. Eperon, G.E.; Burlakov, V.M.; Docampo, P.; Goriely, A.; Snaith, H.J. Morphological Control for High Performance, Solution-Processed Planar Heterojunction Perovskite Solar Cells. *Adv. Funct. Mater.* **2013**, *24*, 151–157. [CrossRef]
159. Eperon, G.E.; Burlakov, V.M.; Goriely, A.; Snaith, H.J. Neutral Color Semitransparent Microstructured Perovskite Solar Cells. *ACS Nano* **2013**, *8*, 591–598. [CrossRef]
160. Hörantner, M.T.; Nayak, P.K.; Mukhopadhyay, S.; Wojciechowski, K.; Beck, C.; McMeekin, D.; Kamino, B.; Eperon, G.E.; Snaith, H.J. Shunt-Blocking Layers for Semitransparent Perovskite Solar Cells. *Adv. Mater. Interfaces* **2016**, *3*, 1500837. [CrossRef]
161. Burlakov, V.M.; Eperon, G.E.; Snaith, H.J.; Chapman, S.J.; Goriely, A. Controlling coverage of solution cast materials with unfavourable surface interactions. *Appl. Phys. Lett.* **2014**, *104*, 091602. [CrossRef]
162. Xiao, Y.; Han, G.; Chang, Y.; Zhang, Y.; Li, Y.; Li, M. Investigation of perovskite-sensitized nanoporous titanium dioxide photoanodes with different thicknesses in perovskite solar cells. *J. Power Sources* **2015**, *286*, 118–123. [CrossRef]
163. Heo, J.H.; Jang, M.H.; Lee, M.H.; Han, H.J.; Kang, M.G.; Lee, M.L.; Im, S.H. Efficiency enhancement of semi-transparent sandwich type $CH_3NH_3PbI_3$ perovskite solar cells with island morphology perovskite film by introduction of polystyrene passivation layer. *J. Mater. Chem. A* **2016**, *4*, 16324–16329. [CrossRef]
164. Dou, L.; You, J.; Hong, Z.; Xu, Z.; Li, G.; Street, R.A.; Yang, Y. 25th Anniversary Article: A Decade of Organic/Polymeric Photovoltaic Research. *Adv. Mater.* **2013**, *25*, 6642–6671. [CrossRef]
165. Bailie, C.; Christoforo, G.; Mailoa, J.P.; Bowring, A.R.; Unger, E.L.; Nguyen, W.H.; Burschka, J.; Pellet, N.; Lee, J.Z.; Grätzel, M.; et al. Semi-transparent perovskite solar cells for tandems with silicon and CIGS. *Energy Environ. Sci.* **2015**, *8*, 956–963. [CrossRef]
166. Fu, F.; Feurer, T.; Jäger, T.; Avancini, E.; Bissig, B.; Yoon, S.; Buecheler, S.; Tiwari, A.N. Low-temperature-processed efficient semi-transparent planar perovskite solar cells for bifacial and tandem applications. *Nat. Commun.* **2015**, *6*, 8932. [CrossRef] [PubMed]
167. Rai, M.; Rahmany, S.; Lim, S.S.; Magdassi, S.; Wong, L.H.; Etgar, L. Hot dipping post treatment for improved efficiency in micro patterned semi-transparent perovskite solar cells. *J. Mater. Chem. A* **2018**, *6*, 23787–23796. [CrossRef]

168. Tong, J.; Song, Z.; Kim, D.H.; Chen, X.; Chen, C.; Palmstrom, A.F.; Ndione, P.F.; Reese, M.O.; Dunfield, S.P.; Reid, O.G.; et al. Carrier lifetimes of >1 μs in Sn-Pb perovskites enable efficient all-perovskite tandem solar cells. *Science* **2019**, *364*, 475–479. [CrossRef] [PubMed]
169. Guo, Y.; Shoyama, K.; Sato, W.; Nakamura, E. Polymer Stabilization of Lead(II) Perovskite Cubic Nanocrystals for Semitransparent Solar Cells. *Adv. Energy Mater.* **2016**, *6*, 1502317. [CrossRef]
170. Della Gaspera, E.; Peng, Y.; Hou, Q.; Spiccia, L.; Bach, U.; Jasieniak, J.; Cheng, Y.-B. Ultra-thin high efficiency semitransparent perovskite solar cells. *Nano Energy* **2015**, *13*, 249–257. [CrossRef]
171. Kwon, H.-C.; Kim, A.; Lee, H.; Lee, D.; Jeong, S.; Moon, J. Parallelized Nanopillar Perovskites for Semitransparent Solar Cells Using an Anodized Aluminum Oxide Scaffold. *Adv. Energy Mater.* **2016**, *6*, 1601055. [CrossRef]
172. Heo, J.H.; Han, H.J.; Lee, M.; Song, M.; Kim, D.-H.; Im, S.H. Stable semi-transparent CH3NH3PbI3planar sandwich solar cells. *Energy Environ. Sci.* **2015**, *8*, 2922–2927. [CrossRef]
173. Zhao, J.; Brinkmann, K.O.; Hu, T.; Pourdavoud, N.; Becker, T.; Gahlmann, T.; Heiderhoff, R.; Polywka, A.; Görrn, P.; Chen, Y.; et al. Self-Encapsulating Thermostable and Air-Resilient Semitransparent Perovskite Solar Cells. *Adv. Energy Mater.* **2017**, *7*, 1602599. [CrossRef]
174. Li, F.R.; Xu, Y.; Chen, W.; Xie, S.; Li, J. Nanotube enhanced carbon grids as top electrodes for fully printable mesoscopic semitransparent perovskite solar cells. *J. Mater. Chem. A* **2017**, *5*, 10374–10379. [CrossRef]
175. Chen, W.; Zhang, J.; Xu, G.; Xue, R.; Li, Y.; Zhou, Y.; Hou, J.; Li, Y. A Semitransparent Inorganic Perovskite Film for Overcoming Ultraviolet Light Instability of Organic Solar Cells and Achieving 14.03% Efficiency. *Adv. Mater.* **2018**, *30*, 1–10. [CrossRef] [PubMed]
176. Wheeler, L.M.; Moore, D.T.; Ihly, R.; Stanton, N.J.; Miller, E.M.; Tenent, R.C.; Blackburn, J.L.; Neale, N.R. Switchable photovoltaic windows enabled by reversible photothermal complex dissociation from methylammonium lead iodide. *Nat. Commun.* **2017**, *8*, 1722. [CrossRef] [PubMed]
177. Duong, T.; Wu, Y.; Shen, H.; Peng, J.; Fu, X.; Jacobs, D.; Wang, E.-C.; Kho, T.C.; Fong, K.C.; Stocks, M.; et al. Rubidium Multication Perovskite with Optimized Bandgap for Perovskite-Silicon Tandem with over 26% Efficiency. *Adv. Energy Mater.* **2017**, *7*, 1700228. [CrossRef]
178. Bush, K.; Palmstrom, A.F.; Yu, Z.J.; Boccard, M.; Cheacharoen, R.; Mailoa, J.P.; McMeekin, D.P.; Hoye, R.L.; Bailie, C.D.; Leijtens, T.; et al. 23.6%-efficient monolithic perovskite/silicon tandem solar cells with improved stability. *Nat. Energy* **2017**, *2*, 17009. [CrossRef]
179. Roldán-Carmona, C.; Malinkiewicz, O.; Betancur, R.; Longo, G.; Momblona, C.; Jaramillo, F.; Camacho, L.; Bolink, H.J. High efficiency single-junction semitransparent perovskite solar cells. *Energy Environ. Sci.* **2014**, *7*, 2968–2973. [CrossRef]
180. Lu, J.-H.; Yu, Y.-L.; Chuang, S.-R.; Yeh, C.-H.; Chen, C.P. High-Performance, Semitransparent, Easily Tunable Vivid Colorful Perovskite Photovoltaics Featuring Ag/ITO/Ag Microcavity Structures. *J. Phys. Chem. C* **2016**, *120*, 4233–4239. [CrossRef]
181. Lee, K.T.; Jang, J.-Y.; Park, S.J.; Ok, S.A.; Park, H.J. Incident-angle-controlled semitransparent colored perovskite solar cells with improved efficiency exploiting a multilayer dielectric mirror. *Nanoscale* **2017**, *9*, 13983–13989. [CrossRef]
182. Jiang, Y.; Luo, B.; Jiang, F.; Jiang, F.; Fuentes-Hernandez, C.; Liu, T.; Mao, L.; Xiong, S.; Li, Z.; Wang, T.; et al. Efficient Colorful Perovskite Solar Cells Using a Top Polymer Electrode Simultaneously as Spectrally Selective Antireflection Coating. *Nano Lett.* **2016**, *16*, 7829–7835. [CrossRef]
183. Kim, H.; Kim, H.-S.; Ha, J.; Yoo, S.; Park, N.-G. Empowering Semi-Transparent Solar Cells with Thermal-Mirror Functionality. *Adv. Energy Mater.* **2016**, *6*, 1502466. [CrossRef]
184. Ghosh, A.; Bhandari, S.; Sundaram, S.; Mallick, T.K. Carbon counter electrode mesoscopic ambient processed & characterised perovskite for adaptive BIPV fenestration. *Renew. Energy* **2020**, *145*, 2151–2158. [CrossRef]
185. Bhandari, S.; Roy, A.; Mallick, T.K.; Sundaram, S. Impact of different light induced effect on organic hole-transporting layer in perovskite solar cells. *Mater. Lett.* **2020**, *268*, 127568. [CrossRef]
186. Yang, M.; Zhou, Y.; Zeng, Y.; Jiang, C.-S.; Padture, N.P.; Zhu, K. Square-Centimeter Solution-Processed Planar $CH_3NH_3PbI_3$ Perovskite Solar Cells with Efficiency Exceeding 15%. *Adv. Mater.* **2015**, *27*, 6363–6370. [CrossRef] [PubMed]
187. Razza, S.; Hermosa, S.A.C.; Di Carlo, A.; Brown, T.M. Research Update: Large-area deposition, coating, printing, and processing techniques for the upscaling of perovskite solar cell technology. *APL Mater.* **2016**, *4*, 091508. [CrossRef]

188. Rossi, L.M.; Fiorio, J.L.; Garcia, M.A.S.; Ferraz, C.P. The role and fate of capping ligands in colloidally prepared metal nanoparticle catalysts. *Dalton Trans.* **2018**, *47*, 5889–5915. [CrossRef] [PubMed]
189. Gardner, K.L.; Tait, J.G.; Merckx, T.; Qiu, W.; Paetzold, U.W.; Kootstra, L.; Jaysankar, M.; Gehlhaar, R.; Cheyns, D.; Heremans, P.; et al. Nonhazardous Solvent Systems for Processing Perovskite Photovoltaics. *Adv. Energy Mater.* **2016**, *6*, 1600386. [CrossRef]
190. Hu, Y.; Si, S.; Mei, A.; Rong, Y.; Liu, H.; Li, X.; Han, H. Stable Large-Area (10 × 10 cm^2) Printable Mesoscopic Perovskite Module Exceeding 10% Efficiency. *Sol. RRL* **2017**, *1*, 1600019. [CrossRef]
191. Cai, L.; Liang, L.; Wu, J.; Ding, B.; Gao, L.; Fan, B. Large area perovskite solar cell module. *J. Semicond.* **2017**, *38*, 14006. [CrossRef]
192. Corporation, N.T. NEDO and Toshiba Develops World's Largest Film-Based Perovskite Photovoltaic Module—703 cm^2 Module Achieves 11.7% Power Conversion Efficiency. Available online: https://www.nedo.go.jp/english/news/AA5en_100391.html (accessed on 4 May 2020).
193. Microquanta. *Microquanta Semiconductor*; Microquanta: Hangzhou, China, 2019.
194. Bellini, E. PANASONIC Claims 16.09% Efficiency for Lightweight Perovskite Solar Module. Available online: https://www.pv-magazine.com/2020/01/21/panasonic-claims-16-1-efficiency-for-lightweight-perovskite-solar-module/ (accessed on 3 May 2020).
195. Wang, H.; Dewi, H.A.; Koh, T.M.; Bruno, A.; Mhaisalkar, S.G.; Mathews, N. Bifacial, Color-Tunable Semitransparent Perovskite Solar Cells for Building-Integrated Photovoltaics. *ACS Appl. Mater. Interfaces* **2019**, *12*, 484–493. [CrossRef] [PubMed]
196. Aernouts, T. Efficient Structures and Processes for Reliable Perovskite Solar Modules. In Proceedings of the International Conference on Hybrid and Organic Photovoltaics, Roma, Italy, 12–15 May 2019.
197. Zhang, W.; Anaya, M.; Lozano, G.; Calvo, M.E.; Johnston, M.B.; Míguez, H.; Snaith, H.J. Highly Efficient Perovskite Solar Cells with Tunable Structural Color. *Nano Lett.* **2015**, *15*, 1698–1702. [CrossRef]
198. Lin, J.; Lai, M.; Dou, L.; Kley, C.S.; Chen, H.; Peng, F.; Sun, J.; Lu, D.; Hawks, S.A.; Xie, C.; et al. Thermochromic halide perovskite solar cells. *Nat. Mater.* **2018**, *17*, 261–267. [CrossRef] [PubMed]
199. Martellotta, F.; Cannavale, A.; Ayr, U. Comparing energy performance of different semi-transparent, building-integrated photovoltaic cells applied to "reference" buildings. *Energy Procedia* **2017**, *126*, 219–226. [CrossRef]
200. Shen, H.; Duong, T.; Peng, J.; Jacobs, D.; Wu, N.; Gong, J.; Wu, Y.; Karuturi, S.K.; Fu, X.; Weber, K.; et al. Mechanically-stacked perovskite/CIGS tandem solar cells with efficiency of 23.9% and reduced oxygen sensitivity. *Energy Environ. Sci.* **2018**, *11*, 394–406. [CrossRef]
201. Shen, G.; Du, Z.; Pan, Z.; Du, J.; Zhong, X. Solar Paint from TiO_2 Particles Supported Quantum Dots for Photoanodes in Quantum Dot–Sensitized Solar Cells. *ACS Omega* **2018**, *3*, 1102–1109. [CrossRef] [PubMed]
202. Skanska. Skanska Launches First Perovskite Solar Cell Application In Office Buildings Together with Saule Technologies. Available online: https://www.linkedin.com/feed/update/urn:li:activity:6537688750716182528 (accessed on 1 May 2020).
203. Jung, H.S.; Han, G.S.; Park, N.-G.; Ko, M.J. Flexible Perovskite Solar Cells. *Joule* **2019**, *3*, 1850–1880. [CrossRef]
204. Heo, J.H.; Lee, D.S.; Shin, D.H.; Im, S.H. Recent advancements in and perspectives on flexible hybrid perovskite solar cells. *J. Mater. Chem. A* **2019**, *7*, 888–900. [CrossRef]
205. Kang, S.; Jeong, J.; Cho, S.; Yoon, Y.J.; Park, S.; Lim, S.; Kim, J.Y.; Ko, H.; Lee, S. Ultrathin, lightweight and flexible perovskite solar cells with an excellent power-per-weight performance. *J. Mater. Chem. A* **2019**, *7*, 1107–1114. [CrossRef]
206. Kaltenbrunner, M.; Adam, G.; Głowacki, E.D.; Drack, M.; Schwödiauer, R.; Leonat, L.; Apaydin, D.H.; Groiss, H.; Scharber, M.C.; White, M.; et al. Flexible high power-per-weight perovskite solar cells with chromium oxide–metal contacts for improved stability in air. *Nat. Mater.* **2015**, *14*, 1032–1039. [CrossRef]
207. Liu, Z.; Qiu, L.; Juarez-Perez, E.J.; Hawash, Z.; Kim, T.; Jiang, Y.; Wu, Z.; Raga, S.R.; Ono, L.K.; Liu, S. (Frank); et al. Gas-solid reaction based over one-micrometer thick stable perovskite films for efficient solar cells and modules. *Nat. Commun.* **2018**, *9*, 3880. [CrossRef]
208. Li, Y.; Meng, L.; Yang, Y.; Xu, G.; Hong, Z.; Chen, Q.; You, J.; Li, G.; Yang, Y.; Li, Y. High-efficiency robust perovskite solar cells on ultrathin flexible substrates. *Nat. Commun.* **2016**, *7*, 10214. [CrossRef]
209. Zhang, H.; Cheng, J.; Li, D.; Lin, F.; Mao, J.; Liang, C.; Jen, A.K.-Y.; Grätzel, M.; Choy, W.C.H. Toward All Room-Temperature, Solution-Processed, High-Performance Planar Perovskite Solar Cells: A New Scheme of Pyridine-Promoted Perovskite Formation. *Adv. Mater.* **2017**, *29*, 1604695. [CrossRef] [PubMed]

210. Xie, M.; Wang, J.; Kang, J.; Zhang, L.; Sun, X.; Han, K.; Luo, Q.; Lin, J.; Shi, L.; Ma, C.-Q. Super-flexible perovskite solar cells with high power-per-weight on 17 μ m thick PET substrate utilizing printed Ag nanowires bottom and top electrodes. *Flex. Print. Electron.* **2019**, *4*, 034002. [CrossRef]
211. Khan, S.A.; Rahman, A. Efficiency of thin film photovoltaic paint: A brief review. *Int. J. Recent Technol. Eng.* **2019**, *7*, 163–169.
212. Kim, S.-S.; Na, S.-I.; Jo, J.; Tae, G.; Kim, D.-Y. Efficient Polymer Solar Cells Fabricated by Simple Brush Painting. *Adv. Mater.* **2007**, *19*, 4410–4415. [CrossRef]
213. Burkitt, D.; Swartwout, R.; McGettrick, J.; Greenwood, P.; Beynon, D.; Brenes, R.; Bulović, V.; Watson, T. Acetonitrile based single step slot-die compatible perovskite ink for flexible photovoltaics. *RSC Adv.* **2019**, *9*, 37415–37423. [CrossRef]
214. Mohammad, A.A.; AlKhaldi, K.H.; AlTuwaim, M.S.; Al-Jimaz, A.S. Effect of temperature and chain length on the viscosity and surface tension of binary systems of *N*,*N*-dimethylformamide with 1-octanol, 1-nonanol and 1-decanol. *J. Chem. Thermodyn.* **2014**, *74*, 7–15. [CrossRef]
215. Henni, A.; Hromek, J.J.; Tontiwachwuthikul, P.; Chakma, A. Volumetric Properties and Viscosities for AqueousN-Methyl-2-pyrrolidone Solutions from 25 °C to 70 °C. *J. Chem. Eng. Data* **2004**, *49*, 231–234. [CrossRef]
216. Giuri, A.; Masi, S.; Listorti, A.; Gigli, G.; Colella, S.; Corcione, C.E.; Rizzo, A. Polymeric rheology modifier allows single-step coating of perovskite ink for highly efficient and stable solar cells. *Nano Energy* **2018**, *54*, 400–408. [CrossRef]
217. De Bastiani, M.; Saidaminov, M.I.; Dursun, I.; Sinatra, L.; Peng, W.; Buttner, U.; Mohammed, O.F.; Bakr, O.M. Thermochromic Perovskite Inks for Reversible Smart Window Applications. *Chem. Mater.* **2017**, *29*, 3367–3370. [CrossRef]
218. Su, J.; Cai, H.; Yang, J.; Ye, X.; Han, R.; Ni, J.; Li, J.; Zhang, J. Perovskite Ink with an Ultrawide Processing Window for Efficient and Scalable Perovskite Solar Cells in Ambient Air. *ACS Appl. Mater. Interfaces* **2019**, *12*, 3531–3538. [CrossRef]
219. Park, N.-G.; Segawa, H. Research Direction toward Theoretical Efficiency in Perovskite Solar Cells. *ACS Photonics* **2018**, *5*, 2970–2977. [CrossRef]
220. Xiao, Z.; Yan, Y. Progress in Theoretical Study of Metal Halide Perovskite Solar Cell Materials. *Adv. Energy Mater.* **2017**, *7*, 1701136. [CrossRef]
221. Stranks, S.D.; Plochocka, P. The influence of the Rashba effect. *Nat. Mater.* **2018**, *17*, 381–382. [CrossRef]
222. Butler, K.T.; Frost, J.M.; Walsh, A. Ferroelectric materials for solar energy conversion: Photoferroics revisited. *Energy Environ. Sci.* **2015**, *8*, 838–848. [CrossRef]
223. Zhu, C.; Niu, X.; Fu, Y.; Li, N.; Hu, C.; Chen, Y.; He, X.; Na, G.; Liu, P.; Zai, H.; et al. Strain engineering in perovskite solar cells and its impacts on carrier dynamics. *Nat. Commun.* **2019**, *10*, 815. [CrossRef] [PubMed]
224. Busipalli, D.L.; Nachimuthu, S.; Jiang, J.-C. Theoretical study on halide and mixed halide Perovskite solar cells: Effects of halide atoms on the stability and electronic properties. *J. Chin. Chem. Soc.* **2019**, *66*, 575–582. [CrossRef]
225. Marszal-Pomianowska, A.; Heiselberg, P.; Bourrelle, J.S.; Musall, E.; Voss, K.; Sartori, I.; Napolitano, A. Zero Energy Building—A review of definitions and calculation methodologies. *Energy Build.* **2011**, *43*, 971–979. [CrossRef]
226. Walker, L.; Hofer, J.; Schlueter, A. High-resolution, parametric BIPV and electrical systems modeling and design. *Appl. Energy* **2019**, *238*, 164–179. [CrossRef]
227. Assoa, Y.B.; Mongibello, L.; Carr, A.; Kubicek, B.; Machado, M.; Merten, J.; Misara, S.; Roca, F.; Sprenger, W.; Wagner, M.; et al. Thermal analysis of a BIPV system by various modelling approaches. *Sol. Energy* **2017**, *155*, 1289–1299. [CrossRef]
228. Chakkaravarthy, A.N.; Subathra, M.S.P.; Pradeep, P.J.; Kumar, N.M. Solar irradiance forecasting and energy optimization for achieving nearly net zero energy building. *J. Renew. Sustain. Energy* **2018**, *10*, 035103. [CrossRef]
229. Poulek, V.; Matuska, T.; Libra, M.; Kachalouski, E.; Sedláček, J. Influence of increased temperature on energy production of roof integrated PV panels. *Energy Build.* **2018**, *166*, 418–425. [CrossRef]
230. Sprenger, W.; Wilson, H.R.; Kuhn, T.E. Electricity yield simulation for the building-integrated photovoltaic system installed in the main building roof of the Fraunhofer Institute for Solar Energy Systems ISE. *Sol. Energy* **2016**, *135*, 633–643. [CrossRef]

231. Gong, X.; Chen, Y.; Liang, M. Theoretical study of building-integrated photovoltaics based on perovskite single junction and perovskite/silicon tandem solar cells. *Energy Explor. Exploit.* **2019**, *38*, 723–732. [CrossRef]
232. Meng, L.; You, J.; Yang, Y. Addressing the stability issue of perovskite solar cells for commercial applications. *Nat. Commun.* **2018**, *9*, 5265. [CrossRef] [PubMed]
233. Song, Z.; McElvany, C.L.; Phillips, A.B.; Celik, I.; Krantz, P.W.; Watthage, S.C.; Liyanage, G.K.; Apul, D.; Heben, M.J. A technoeconomic analysis of perovskite solar module manufacturing with low-cost materials and techniques. *Energy Environ. Sci.* **2017**, *10*, 1297–1305. [CrossRef]
234. Ghosh, A.; Norton, B.; Duffy, A. Conceptualization of a Photovoltaic Powered Electrochromic Switching of a Multifunctional Glazing. In Proceedings of the ISES Solar Energy Congress, Cancun, Mexico, 3–7 November 2013.
235. Ghosh, A.; Norton, B.; Duffy, A. Multifunctional Glazing System—Solution for Modern Smart Glazing Multifunctional Glazing System: Solution for Modern Smart Glazing. In Proceedings of the Passive Low Energy Architecture, Ahemdabad, India, 16–18 December 2014.
236. Ghosh, A.; Norton, B.; Duffy, A. Daylighting performance and glare calculation of a suspended particle device switchable glazing. *Sol. Energy* **2016**, *132*, 114–128. [CrossRef]
237. Ghosh, A.; Mallick, T. Evaluation of optical properties and protection factors of a PDLC switchable glazing for low energy building integration. *Sol. Energy Mater. Sol. Cells* **2018**, *176*, 391–396. [CrossRef]
238. Xu, X.; Li, S.; Zhang, H.; Shen, Y.; Zakeeruddin, S.M.; Graetzel, M.; Cheng, Y.-B.; Wang, M.-K. A Power Pack Based on Organometallic Perovskite Solar Cell and Supercapacitor. *ACS Nano* **2015**, *9*, 1782–1787. [CrossRef]
239. Li, C.; Islam, M.; Moore, J.; Sleppy, J.; Morrison, C.; Konstantinov, K.; Dou, S.X.; Renduchintala, C.; Thomas, J. Wearable energy-smart ribbons for synchronous energy harvest and storage. *Nat. Commun.* **2016**, *7*, 13319. [CrossRef] [PubMed]
240. Gao, K.; Ti, D.; Zhang, Z. A photocapacitor with high working voltage and energy density. *Sustain. Energy Fuels* **2019**, *3*, 1937–1942. [CrossRef]
241. Zhou, F.; Ren, Z.; Zhao, Y.; Shen, X.; Wang, A.; Li, Y.Y.; Surya, C.; Chai, Y. Perovskite Photovoltachromic Supercapacitor with All-Transparent Electrodes. *ACS Nano* **2016**, *10*, 5900–5908. [CrossRef] [PubMed]
242. Zhang, F.; Li, W.; Xu, Z.; Ye, M.; Xu, H.; Guo, W.; Liu, X. Highly flexible and scalable photo-rechargeable power unit based on symmetrical nanotube arrays. *Nano Energy* **2018**, *46*, 168–175. [CrossRef]
243. Li, C.; Cong, S.; Tian, Z.; Song, Y.; Yu, L.; Lu, C.; Shao, Y.; Li, J.; Zou, G.; Rümmeli, M.H.; et al. Flexible perovskite solar cell-driven photo-rechargeable lithium-ion capacitor for self-powered wearable strain sensors. *Nano Energy* **2019**, *60*, 247–256. [CrossRef]
244. Ahmad, S.; George, C.; Beesley, D.J.; Baumberg, J.J.; De Volder, M. Photo-Rechargeable Organo-Halide Perovskite Batteries. *Nano Lett.* **2018**, *18*, 1856–1862. [CrossRef]
245. Xu, J.; Chen, Y.; Dai, L. Efficiently photo-charging lithium-ion battery by perovskite solar cell. *Nat. Commun.* **2015**, *6*, 8103. [CrossRef]
246. Gurung, A.; Chen, K.; Khan, R.; Abdulkarim, S.S.; Varnekar, G.; Pathak, R.; Naderi, R.; Qiao, Q. Highly Efficient Perovskite Solar Cell Photocharging of Lithium Ion Battery Using DC-DC Booster. *Adv. Energy Mater.* **2017**, *7*, 1602105. [CrossRef]
247. Kin, L.-C.; Liu, Z.; Astakhov, O.; Agbo, S.N.; Tempel, H.; Yu, S.; Kungl, H.; Eichel, R.-A.; Rau, U.; Kirchartz, T.; et al. Efficient Area Matched Converter Aided Solar Charging of Lithium Ion Batteries Using High Voltage Perovskite Solar Cells. *ACS Appl. Energy Mater.* **2019**, *3*, 431–439. [CrossRef]
248. Rong, Y.; Hu, Y.; Mei, A.; Tan, H.; Saidaminov, M.I.; Seok, S.I.; McGehee, M.D.; Sargent, E.H.; Han, H. Challenges for commercializing perovskite solar cells. *Science* **2018**, *361*, eaat8235. [CrossRef]
249. Wojciechowski, K.; Forgacs, D.; Rivera, T. Industrial Opportunities and Challenges for Perovskite Photovoltaic Technology. *Sol. RRL* **2019**, *3*. [CrossRef]
250. Reese, M.O.; Glynn, S.; Kempe, M.; McGott, D.L.; Dabney, M.S.; Barnes, T.M.; Booth, S.; Feldman, D.; Haegel, N.M. Increasing markets and decreasing package weight for high-specific-power photovoltaics. *Nat. Energy* **2018**, *3*, 1002–1012. [CrossRef]
251. Grancini, G.; Roldán-Carmona, C.; Zimmermann, I.; Mosconi, E.; Lee, X.; Martineau, D.; Narbey, S.; Oswald, F.; De Angelis, F.; Graetzel, M.; et al. One-Year stable perovskite solar cells by 2D/3D interface engineering. *Nat. Commun.* **2017**, *8*, 15684. [CrossRef]

252. Saifullah, M.; Gwak, J.; Yun, J.H. Comprehensive review on material requirements, present status, and future prospects for building-integrated semitransparent photovoltaics (BISTPV). *J. Mater. Chem. A* **2016**, *4*, 8512–8540. [CrossRef]
253. Gullbrekken, L.; Kvande, T.; Time, B. Roof-integrated PV in Nordic Climate—Building Physical Challenges. *Energy Procedia* **2015**, *78*, 1962–1967. [CrossRef]
254. Agathokleous, R.A.; Kalogirou, S.A. Status, barriers and perspectives of building integrated photovoltaic systems. *Energy* **2020**, *191*, 116471. [CrossRef]
255. Yang, R. Overcoming technical barriers and risks in the application of building integrated photovoltaics (BIPV): Hardware and software strategies. *Autom. Constr.* **2015**, *51*, 92–102. [CrossRef]
256. Yang, R.; Zou, P. Building integrated photovoltaics (BIPV): Costs, benefits, risks, barriers and improvement strategy. *Int. J. Constr. Manag.* **2015**, *16*, 1–15. [CrossRef]
257. Smestad, G.P.; Germer, T.A.; AlRashidi, H.; Fernández, E.F.; Dey, S.; Brahma, H.; Sarmah, N.; Ghosh, A.; Sellami, N.; Hassan, I.A.I.; et al. Modelling photovoltaic soiling losses through optical characterization. *Sci. Rep.* **2020**, *10*, 1–13. [CrossRef]
258. Chanchangi, Y.N.; Ghosh, A.; Sundaram, S.; Mallick, T.K. Dust and PV Performance in Nigeria: A review. *Renew. Sustain. Energy Rev.* **2020**, *121*, 109704. [CrossRef]
259. Chanchangi, Y.N.; Ghosh, A.; Sundaram, S.; Mallick, T.K. An analytical indoor experimental study on the effect of soiling on PV, focusing on dust properties and PV surface material. *Sol. Energy* **2020**, *203*, 46–68. [CrossRef]
260. Nundy, S.; Ghosh, A.; Mallick, T.K. Hydrophilic and Superhydrophilic Self-Cleaning Coatings by Morphologically Varying ZnO Microstructures for Photovoltaic and Glazing Applications. *ACS Omega* **2020**, *5*, 1033–1039. [CrossRef]
261. Hemaida, A.; Ghosh, A.; Sundaram, S.; Mallick, T.K. Evaluation of thermal performance for a smart switchable adaptive polymer dispersed liquid crystal (PDLC) glazing. *Sol. Energy* **2020**, *195*, 185–193. [CrossRef]
262. Ghosh, A. Soiling Losses: A Barrier for India's Energy Security Dependency from Photovoltaic Power. *Challenges* **2020**, *11*, 9. [CrossRef]
263. Ghosh, A. Possibilities and Challenges for the Inclusion of the Electric Vehicle (EV) to Reduce the Carbon Footprint in the Transport Sector: A Review. *Energies* **2020**, *13*, 2602. [CrossRef]
264. Cancelliere, P.; Manzini, G.; Traina, G.; Cavriani, M.G. PV modules on buildings—Outlines of PV roof samples fire rating assessment. *Fire Saf. J.* **2020**, 103139. [CrossRef]
265. Manzini, G.; Gramazio, P.; Guastella, S.; Liciotti, C.; Baffoni, G.L. The Fire Risk in Photovoltaic Installations—Test Protocols for Fire Behavior of PV Modules. *Energy Procedia* **2015**, *82*, 752–758. [CrossRef]
266. Snaith, H.J. Present status and future prospects of perovskite photovoltaics. *Nat. Mater.* **2018**, *17*, 372–376. [CrossRef]
267. European Council. Restriction of Hazardous Substances in Electrical and Electronic Equipment. European Union. 2003. Available online: https://eur-lex.europa.eu/LexUriServ/LexUriServ.do?uri=CELEX:32002L00 (accessed on 1 June 2020).
268. Cannavale, A.; Martellotta, F. *Smart Perovskite-Based Technologies for Building Integration*; Elsevier BV: Amsterdam, The Netherlands, 2019; pp. 441–466.
269. Cannavale, A.; Ierardi, L.; Hörantner, M.; Eperon, G.E.; Snaith, H.J.; Ayr, U.; Martellotta, F. Improving energy and visual performance in offices using building integrated perovskite-based solar cells: A case study in Southern Italy. *Appl. Energy* **2017**, *205*, 834–846. [CrossRef]
270. Cannavale, A.; Hörantner, M.; Eperon, G.E.; Snaith, H.J.; Fiorito, F.; Ayr, U.; Martellotta, F. Building integration of semitransparent perovskite-based solar cells: Energy performance and visual comfort. *Appl. Energy* **2017**, *194*, 94–107. [CrossRef]

© 2020 by the authors. Licensee MDPI, Basel, Switzerland. This article is an open access article distributed under the terms and conditions of the Creative Commons Attribution (CC BY) license (http://creativecommons.org/licenses/by/4.0/).

MDPI
St. Alban-Anlage 66
4052 Basel
Switzerland
Tel. +41 61 683 77 34
Fax +41 61 302 89 18
www.mdpi.com

Buildings Editorial Office
E-mail: buildings@mdpi.com
www.mdpi.com/journal/buildings

www.ingramcontent.com/pod-product-compliance
Lightning Source LLC
Chambersburg PA
CBHW040816250926
51850CB00007BA/1187
* 9 7 8 3 0 3 6 5 2 3 3 9 2 *